Climate Change in India

The climate over the Indian subcontinent is influenced by complex interactions between the atmosphere, ocean, and land, along with human interventions that are influencing heat extremes, changing monsoon patterns, sea-level rise, and posing serious threats to lives and livelihoods among populations in India. This book, based on recent studies and research, explains how and why the climate is changing across India and how these changes are expected to evolve in the future. It takes a holistic view of the climate from India's perspective and discusses important themes such as groundwater, land use, livestock, and natural disasters. Readers will have an in-depth understanding of impacts and possible adaptations.

Features

- Includes case studies of groundwater and the effect of climate change.
- Identifies land use and land cover changes in different regions in India.
- Discusses the implications of past and present climate change on various livestock types.
- Explains various natural hazards and their relationship to climate change.
- Describes technological intervention in climate change.

Researchers, academics, and graduate students in Earth sciences, climatology, meteorology, or environmental studies working on projects related to climate change will find this an invaluable reference. Governmental organizations, institutions, and global NGOs will also benefit from the insights presented in this book.

Maritime Climate Change: Physical Drivers and Impact
Series Editor: Neloy Khare

As global climate change continues to unfold, the two-way links between the tropical oceans and the poles will play key determining factors in these sensitive regions' climatic evolution. Now is the time to take a detailed look at how the tropical oceans and the poles are coupled climatically. The signatures of environmental and climatic conditions are well preserved in many natural archives available over land and ocean. Many efforts have been made to unravel such mysteries of climate through many natural geological archives from tropics to the polar region. This series makes an effort to cover in pertinent time various depositional regimes, different proxies—planktic, benthic, pollens and spores, invertebrates, geochemistry, sedimentology etc. and emerged teleconnections between the poles and tropics at regional and global scale, besides sea-level changes and neo tectonism. This book series will review theories and methods, analyze case studies, and identify and describe the evolving spatial-temporal variations in climate and provide a better process-level understanding of these patterns. It will discuss significantly, generalizable insights that improve our understanding of climatic evolution across time—including the future. It aims to serve all professionals, students and researchers, scientists alike in academia, industry, government, and beyond.

Climate Change in the Arctic: An Indian Perspective
Neloy Khare

Climate Change and Geodynamics in Polar Regions
Edited by Neloy Khare

Polar Ice and Global Warming in Cryosphere Regions
Edited by Neloy Khare

Climate Change in India: Impacts and Assessments
Edited by Neloy Khare

Climate Change in India

Impacts and Assessments

Edited by
Neloy Khare

CRC Press
Taylor & Francis Group
Boca Raton London New York

CRC Press is an imprint of the
Taylor & Francis Group, an **informa** business

Designed cover image: © Shutterstock, titoOnz

First edition published 2025
by CRC Press
2385 NW Executive Center Drive, Suite 320, Boca Raton FL 33431

and by CRC Press
4 Park Square, Milton Park, Abingdon, Oxon, OX14 4RN

CRC Press is an imprint of Taylor & Francis Group, LLC

Library of Congress Cataloging-in-Publication Data
Names: Khare, Neloy, editor.
Title: Climate change in india : impacts and assessments / Edited by Neloy Khare.
Description: First edition | Boca Raton : CRC Press, Taylor & Francis Group, 2025. |
Series: Maritime climate change |
Includes bibliographical references and index.
Identifiers: LCCN 2024022456 (print) | LCCN 2024022457 (ebook) |
ISBN 9781003485995 (ebook)
Subjects: LCSH: Climatic changes–India.
Classification: LCC QC903.2.I4 C555265 2025 (print) | LCC QC903.2.I4 (ebook) |
DDC 577.270954–dc23/eng20240816|
LC record available at https://lccn.loc.gov/2024022456
LC ebook record available at https://lccn.loc.gov/2024022457

ISBN: 978-1-032-78037-5 (hbk)
ISBN: 978-1-032-78038-2 (pbk)
ISBN: 978-1-003-48599-5 (ebk)

DOI: 10.1201/9781003485995

Typeset in Times
by Newgen Publishing UK

This book is dedicated to the late
Prateek Khare

(25.09.1987–04.11.2023)

Contents

Sarah Sarah, Ira Khan, Rikza Imtiyaz, Atiqur Rahman
and Shakeel Ahmed

Navneet Ranjan, Akhouri Bishwapriya, Gautam Chand Garg,
Shambhu Nath, Mili Singh and Abhishek Kumar Chaurasia

Darve Sabina Iqbal, Gohar Bilal Wani and Asra Imtiyaz Mattoo

S.D. Attri and S. Naresh Kumar

Vandana Prajapati, Shailender Kumar Verma, Anwar Alam
and Neloy Khare

Dilip Kumar Jha, Josephine Anthony, G. Dharani
and R. Kirubagaran

Foreword

The consequences of climate change now include, among others, accentuated extreme weather events, severe fires, warmer and rising sea levels, flooding, melting polar ice, catastrophic storms, and declining biodiversity. The adversity of ongoing climate change becomes more prominent in coastal countries like India, which is surrounded by about 7500-kilometer-long coastline. Indubitably, climate change is considered as a threat to a country such as India.

India is ranked fourth among the list of countries most affected by climate change in 2015. According to an estimate, temperatures in India have risen by 0.7°C (1.3°F) between 1901 and 2018 and there is an increased likelihood of both the severity and number of droughts in India by the end of the present century. A warmer Tibetan Plateau is causing Himalayan glaciers to retreat, threatening the flow rate of the Ganga, Brahmaputra, Sindhu, Yamuna, and other major rivers. Increased frequency and intensity of heat waves are taking a big toll not only on human but entire biota. Severe landslides and floods are reported to become increasingly common.

Climate change is associated with various adverse impacts on agriculture, water resources, forest and biodiversity, health, coastal management, and increase in temperature. Decline in agricultural productivity is threatening the food security of the nation. In order to respond to the impacts of climate change, the Government of India has launched many measures. The National Mission for Green India (GIM) is one of the eight Missions outlined under the National Action Plan on Climate Change (NAPCC). It aims to protect, restore, and enhance India's diminishing forest cover and respond to climate change by a combination of adaptation and mitigation measures.

Many significant data on various climate impacts over the Indian peninsula are available but in a scattered manner and they need to be collated and brought to a focal point to help understand the severe impacts of the ongoing climate change on India in a better holistic manner. In this context, *Climate Change in India: Impacts and Assessments* is very timely. It addresses some of the vital impacts of climate change on fisheries, agriculture, groundwater potential, forest cover, soil evapotranspiration, human health, among others, in a comprehensive manner. It also aims to evaluate climate-induced natural hazards such as landslides, glaciers outburst, and so forth, along with an assessment of climate change and its impact on human inhabitation.

The book is composed of 11 specific chapters devoted to different climate change topic. The book begins with a chapter by Sarah et al. on the groundwater potential in India and discusses the expected challenges and threats of climate change scenario to groundwater resource. This is further augmented by the groundwater vulnerability assessment using GIS-based water quality index (WQI) in parts of Jamui and Nawada districts, Bihar, India, by Ranjan et al. While the impact of climate change on cold water fisheries has been detailed by Iqbal et al., implications of climate change in agriculture and adaptation strategies in India have been discussed by Attri and Kumar. Similarly, Prajapati et al. focused on the impact of climate change in India and the resulting challenges to human health, which is well supported by yet another chapter dealing with the role of biotechnology in assessing climate impact over the Indian subcontinent by Jha et al.

It is well agreed that the ongoing climate change is also attributed to the land use and land cover vis-à-vis a forest cover and may lead to various disasters like landslides, glaciers outburst, and so forth, and therefore, Yunus and Narayana made a logical attempt to understand interannual and seasonal patterns of evapotranspiration in response to vegetation dynamics owing to the ongoing climate change through satellite observations. In addition, Bora et al. identified the hotspots of land use land cover change of Upper Meghna River basin, India, and Walia assessed the rainfall-induced landslide hazards and their forecast in response to the changing climatic patterns. Similarly, mapping of mountain glacier facies has been attempted by using high-resolution optical satellite data by Wankhede et al. and an assessment of climate change and its impact on human inhabitation of Gujarat during the Mid–Late Holocene period was ably made by Das et al.

I compliment the Editor of the book, Dr. Neloy Khare, for bringing out an important thematic publication focusing on the pertinent and vital issues of the prevailing global warming and assessing the dire consequences of climate change on different spheres of life. I am sure this book shall act as a catalyst and a ready reckoner for any researcher and avid reader.

Laxman Singh Rathore

Former Director General of Meteorology & Permanent Representative of India with World Meteorological Organization (WMO)
Chair, Inter-Governmental Board on Climate Services, WMO, United Nations
April, 2024
Jaipur

Preface

"Climate change" refers to the rapid changes in the global climate induced by human activities such as emissions of greenhouse gases (GHGs) by burning fossil fuels, deforestation, and urbanization. Climate change is the significant variation of average weather conditions becoming, for example, warmer, wetter, or drier over several decades or longer. It is the longer-term trend that differentiates climate change from natural weather variability. Climate change refers to significant changes in global temperature, precipitation, wind patterns, and other measures of climate that occur over several decades or longer. The seas are rising. The foods we eat and take for granted are threatened. Ocean acidification is increasing. Millions of poor people associated are adversely affected. Melting ice caps, ice sheets and glaciers, as well as expanding warming waters, lead to a rise in sea levels. This might hurt coastal communities and numerous ecosystems, as well as contaminate fresh water supplies.

Scientists estimate that temperatures in South Asia are around 2 degrees hotter than pre-industrial temperatures thanks to climate change. This is one of the reasons that South Asia (of which India is a part) is considered to be one of the most vulnerable regions when it comes to climate change. According to NOAA's 2021 Annual Climate Report, the combined land and ocean temperature has increased at an average rate of 0.14 °F (0.08 °C) per decade since 1880; however, the average rate of increase since 1981 has been more than twice as fast: 0.32 °F (0.18 °C) per decade.

In pursuit of its development and poverty eradication goals, India is an emerging economy where GHG emissions are set to increase, albeit from a low base. As a matter of fact, India's energy mix is carbon intensive. Coal is its primary source of fuel, accounting for approximately 70 percent of electricity generation, and the country also powers much of its transport through oil. As a result, India is the world's third-largest emitter of CO_2, despite low per capita CO_2 emissions. India has the world's highest social cost of carbon. Some reports that India may lose anywhere around 3–10% of its GDP annually by 2100 and its poverty rate may rise by 3.5% in 2040 due to climate change.

It is important to note that India's historical cumulative emissions from 1850 to 2019 amount to less than 4 percent of cumulative carbon dioxide emissions of the world from the pre-industrial era, despite being home to 17 percent of the world's population. Hence, India's responsibility for global warming thus far has been minimal and even today its annual per capita emissions are only about one-third of the global average. According to some reports, India, at the 26th session of the United Nations Framework Convention on Climate Change (COP 26) in November, 2021, announced its target to achieve net zero by 2070. In recognition of the Para 19 of Article 4 of the Paris Agreement, India's long-term low-carbon development strategy, has been submitted to the United Nations Framework Convention on Climate Change, and it reaffirms the goal of reaching net-zero by 2070.

According to published reports, India's long-term low-carbon development strategy is based on the principles of equity and climate justice and the principle of Common but Differentiated Responsibilities and Respective Capabilities. It also rests on many key transitions to low-carbon development pathways, such as low-carbon development of electricity systems consistent with development, develop an integrated, efficient and inclusive transport system, promote adaptation in urban design, energy and material efficiency in buildings, and sustainable urbanization, promoting economy-wide decoupling of growth from emissions and development of an efficient, innovative low-emission industrial system, development of carbon dioxide removal and related engineering solutions, enhancing forest and vegetation cover consistent with socioeconomic and ecological considerations and economic and financial needs of low-carbon development. With respect to each of these transitions, India's low-carbon development strategy document has elaborated the international and national context as relevant, the current policies and programmes already being implemented as well as the key elements for each transition, potential benefits and challenges.

Indubitably, there are several indices which rank countries on their performance to combat the challenge of climate change. However, there are many differences and disagreements with respect to the methodology, the conceptual framework as well as outcome of these indices both for India and the World.

Government is poised to take several suitable actions to address the rapidly growing environmental problems in the country. Besides others, National Mission for Enhanced Energy Efficiency and the National Action Plan for Climate Change (NAPCC) recommends mandating specific energy consumption decreases in large energy-consuming industries, with a system for companies to trade energy-saving certificates, financing for public–private partnerships to reduce energy consumption through demand-side

It may not be incongruous to submit that the climate change is the most significant problem facing the world. Global warming is increasing day by day. If we cannot prevent it as soon as possible, our world will face undesirable consequences. Climate change is expected to have major health impacts in India, such as increasing malnutrition and related health disorders and child stunting. No doubt that the poor is likely to be affected most severely. Child stunting is projected to increase by 35% by 2050 compared to a scenario without climate change.

Climate change policies in India have primarily focused on supporting synergies between development and outcomes for the climate. India was one of the few countries that passed the Energy Conservation Act in 2001, which underwent an amendment in August 2022. To better understand the risks of climate change to development, the Climate Impact Research and Climate Analytics are required to look at the likely impacts of temperature increases from 2 °C to 4 °C in different regions. The climate scientists should use the best available evidence and supplement it with advanced computer simulations to arrive at likely impacts on agriculture, water resources, cities and coastal ecosystems over Indian peninsula.

It is therefore, most appropriate to review and collate climate impacts over Indian region. With such need the present book titled **"Climate Change in India: Impacts and Assessments"** has been conceived and conceptualized to assimilate scientific insights and data related to multi-faceted climate change over India and their measurable impacts. This book is composed of 11 chapters dedicated and focused to specific issue of ongoing climate change. The book also places emphasis on deciphering the climate records of geological archives

The book begins with a chapter on the groundwater potential in India and identifying the challenges and threats of climate change scenario to this natural resource by Sarah *et al.* As a matter of fact, India comparatively, is not a water scarce country, various water crises either in the form of water resource or water quality are merely due to mismanagement and unscientific decision-making. The management of water and/or groundwater is of prime importance due to the fact that this resource is not evenly distributed. The groundwater occurs in the geological formations that are highly variable and complex. In addition, the annual rainfall in the country exhibits extremely high variation ranging from 200 mm to 4000 mm approximately as well as the population density is an added constraint. Adequate research and development have been made to unearth the groundwater and manage the water resource combining the surface and groundwater resources in a conjunctive manner. Although groundwater is a hidden resource, it is blessed with many other advantages also and India's water resources could be managed following a scientific and judicial approach.

However, the climate change that is evidently observed and the impact of it is visible, has the most bearing on water resources with groundwater the greatest affected. The entire hydrological cycle is impacted by the climate change. In this chapter they have focused on the impact of climate change on the components of the hydrological cycle, in general, and on groundwater, in particular, so that the cause of impact is understood, analyzed, and adopted for management. This chapter is followed by another chapter by Ranjan *et al.* on groundwater vulnerability assessment using GIS-based water quality index (WQI). They rightly pointed out that the groundwater resources of the Earth are vital to drinking needs and their fast depleting quantity and diminishing quality is a matter of global concern. The aquifer layers of Jamui and Nawada districts are located within fissured rocks of Chhotanagpur Gneissic Complex (CGC) and Bihar Mica Belt (BMB) and within Quaternary sediments. But water quality has become a matter of concern owing to anthropogenic activities and marginally due to contaminants of geological origin. They are altering the concentrations of physico-chemical constituents present in the groundwater not complying to drinking water standards established by national and international health agencies. The water quality index (WQI) method chiefly reflects the combined influence of different water quality affecting parameters at individual sample stations, and has been adopted in the present study for the assessment of the groundwater quality and demarcating the zones of potable groundwater available to the local habitants.

In their chapter, 19 quality parameters (pH, EC, TDS, Ca^{2+}, Mg^{2+}, Na^+, K^+, HCO_3^-, Cl^-, NO_3^-, SO_4^{2-}, PO_4^{3-}, F^-, Fe, Cd, Ni, Co, Pb, and Zn) for each 52 sample stations were considered to determine the water quality index calculations by assigned weightage through "decision-making approach" and validating existing knowledge as per their effect on the human health. This has been done for computation of WQI values and ascertaining suitability of groundwater for drinking as per the standards defined by the Bureau of Indian Standards. The spatial variability of "quality parameters" along with their correlation analyses have been determined for understanding their interrelation with allied geological and anthropogenic aspects. The study reveals that in pre-monsoon, the groundwater quality is "Excellent to good" in whole area, however in post-monsoon, only ~38% of the area yields "Excellent to good" quality groundwater for drinking and in the rest ~68% the quality is "Poor to unsuitable for drinking" as per the groundwater quality index probably due to escalation in the concentrations of toxic elements like F^-, NO_3^-, Pb, Ni, Co, and Cd in the groundwater. They suggested that by using WQI, the zones of excellent, good, poor, very poor, and not suitable quality of groundwater in pre- and post-monsoon seasons could be demarcated, which would be helpful for the authorities to identify such localities and introduce mitigation measures for availability of safe drinking water.

On the other hand, Iqbal *et al.* touched upon the climate change, the most important global environmental challenges as it involves changes in the variability or average state of the atmosphere over durations ranging from decades to millions of years. It has both direct and indirect impacts on fisheries. Direct effects act on physiology like behavior, alter growth, reproductive capacity, mortality and distribution. Indirect effects alter the productivity, structure, and composition of the ecosystems on which fish depend for food. The coldwater fisheries resources comprises unique biodiversity with valuable indigenous germplasm and maintaining environmental quality. Temperature has an impact on almost all biological and chemical activities in freshwater habitats. Coldwater fish species require water with low temperature with sufficient *levels of dissolved oxygen (DO), they are especially vulnerable to climate change. Due to temperature and oxygen requirements of coldwater fish, climate change may restrict available habitat.* Any changes in thermal regime occasioned by climate change will have major consequences for fish reproduction, and that these effects will be exercised across all stages of the reproductive process. Global climate change is increasingly and profoundly threatening fishes, resulting in an uncertain future for fish diversity and global fisheries. Climate change, no doubt effects the whole fisheries of the world and its negative effect is visual in coldwater fisheries.

Similarly, Attri and Kumar ably realized the importance of climatic impacts on human kind, as the implications of climate change are being observed in different regions of the world and in different sectors of socio-economic activities especially in agriculture productivity and food security. The situation is also alarming in India on account of poverty, the rainfed nature of farming in more than half of fragmented and smaller land holdings, high cost of inputs, increase of weather, and climate extremes, etc. The situation requires concerted actions to deal with the climate change impacts. Experimental evidences have indicated that elevated CO_2 concentrations significantly improve the yield of C_3 crops, such as rice, wheat, mustard, and soybean while, high temperatures reduced crop growth duration, caused pollen sterility, reduced grain weight and number and consequently affected yield. Since climate change is a multi-factor phenomenon influencing the productivity of crops in combination with genotype, soil and management, crop simulation models are the potential tools for delineating the impacts and developing adaptation strategies. These systems not only integrate the future climate but also numerous management options and genotypes. Accordingly, their chapter focuses on global and national status of climate change, future projections and strategies related to research, operations and policies to address the impacts and sustain crop productivity. Whereas, Prajapati *et al.* stressed on the adverse consequences of climate change that are alarmingly pervasive for the environment, which also affects the components of the entire ecosystem, including human health. These changes apart from increasing the incidence of non-communicable diseases, favor spread of communicable diseases by modifying the vector habitats, ultimately increasing the disability adjusted life years (DALY) burden. Enhanced DALY burden adversely impacts the labor productivity and economic output of a country. India lost 5.4% of its gross domestic product (GDP) due to extreme heat in 2021 and is further estimated to lose upto 6% of the GDP by 2050 and 12.6% of the GDP by 2070. Therefore, it is pertinent that policies made worldwide for economic development must prioritize ways that will have minimum impact on environment and in sustaining the natural resources. Interestingly, Jha *et al.* touched upon the use of biotechnological tools in assessing climate impact on mariculture. It is beyond doubt that the climate change is one of the globally significant challenges to growth and sustainability including the Indian subcontinent. Alarming climatic impact, degradation of natural resources, and the enhanced public demand for fisheries have ignited the need for biotechnological tools to combat its impact on mariculture, protection of natural resources, and augment fish production. As it is directly evident that microbes produce and consume three dominant gases that are responsible for 98% of global warming (carbon dioxide, methane, and nitrous oxide). Further, climate impact could pose a serious threat to the marine ecosystem, especially mariculture and associated activities. Climate change is mainly associated with a temperature rise, change in shoreline due to ice melting, sea surface salinity change, and global change in ocean pH, which largely affect mariculture production. To combat the alarming situation of climate change, there is a need to understand the process and apply the latest genetic engineering methods as transgenic marine organisms are now widely explored to target the genes that regulate the synthesis of natural defensive proteins to overcome microbial diseases, increase fish resilience to temperature fluctuations, and enhance fish growth. Heat shock proteins (Hsps) as the agent of being an immune booster and increasing disease resistance will present a significant advancement in reducing stressful conditions in the aquaculture system. Apart from this, microbes, corals, and microalgae are being genetically modified to combat the changing environmental conditions for providing better productivity. These authors review the climatic impact on mariculture in the Indian subcontinent and the plausible biotechnological role in protecting it from adverse effects due to climatic shifts.

It is well known that the soil evaporation, a major contributor to the evapotranspiration (ET), stands as a vital component within terrestrial ecosystems, and serves as a pivotal driver in assessing different environment stressors. In view of this, Yunus and Narayana conducted a comprehensive analysis of the variations in soil evaporation (E_s) spanning from 2000 to 2020 in Peninsular India (PI) with the aim to quantify the long-term trends and sensitivity of E_s variations to changes in vegetation growth in six major river basins. They utilized the Penman-Monteith-Leuning evapotranspiration model derived E_s, satellite derived vegetation indices and climate variables to evaluate the interannual and seasonal patterns in E_s. Their results show that over the course of the past 20 years, the average annual soil evaporation exhibited a statistically significant decreasing trend ($p < 0.05$) in the coastal stretches of Eastern

Peninsular India. The largest decreases in these areas were attributed to the increased vegetation growth. Other areas of the PI exhibits increasing trend overall, but without any significance level ($p > 0.05$). Seasonal trends of E_s indicates that the temperature also plays a driving factor of E_s in the Peninsular region.

On the contrary, Walia focused on the Meghalaya in particular, because of its structural complexity, rock types, and climatic conditions, is susceptible and highly prone to landslides, debris flow, ground subsidence, and associated damages. The increasing population and rapid urbanization with changing climatic patterns have considerably enhanced the vulnerability of landslide occurrences and have increased risk to human life and property and needs adaptation. The landslide is mostly triggered by events of heavy rainfall (frequent incessant and torrential), unstable slope, anthropogenic interference, and seismicity. The incidences of landslides have been on the increase in the recent times due to rapid pace of infrastructural development, especially of roads in the hilly regions.

According to him, the study area is one of the seismically active regions and displays high seismic activity, tilting and settlement of the river channel. Based on investigations, it is observed that the watershed shows trellis to sub-trellis drainage patterns, indicating high structural control on the drainage development. Low drainage density, low frequency, and coarse drainage texture indicates the presence of permeable geologic material with high infiltration having low groundwater recharging characters and vegetated slopes. The high bifurcation ratio indicates structural control of drainage representing tectonic activity in the study area. The circulatory ratio and form factor depict diversified slope, low discharge of runoff, high permeability of the sub-soil, and a structurally controlled drainage system. The linear aspects, aerial aspects, and the relief aspects reveal that the watershed is in the process of evolution as the basin is in the process of tilting. Further, the asymmetry factor indicates that watershed is under tilting and hence vulnerable to geohazard.

The study indicates that landslides in the region would occur when the daily rainfall crosses 150 mm and 3 days antecedent rainfall (R_{3ad}) exceeding 84 mm. These thresholds can be used as a predictive tool, but due care needs to be taken when used as an early warning system. When the actual rainfall reaches the threshold limit, it is recommended to give an alert for the people in that region to safely move from the risk area. Rainfall threshold correlation will be useful in order to save lives and property loss by issuing advance alerts. To further augment this aspect it has been well known to many that the river basins always play a vital role in human habitation and settlements. Land use land cover changes with the sustainable use of land resources in the river basins. River Meghna is a trans-boundary river shared by India and Bangladesh. However, a small portion of the basin also comes under Myanmar. The total geographical area of the basin is 82,000 sq. km including 29 trans-boundary sub-basins. India shares a significant part of Upper Meghna basin, habited by around 10 million indigenous populations, spread over six North-Eastern states. It has been observed that the topography of the region and demographic structure has an impact on land use pattern of the region, whereas some land areas are hotspots and undergoes periodically repeated land use changes. Boro *et al.* in their chapter demarcate such hotspots of land use land cover change (LULCC) of the upper Meghna River basin based on the frequency of change in different time frames from 2005 to 2019 and to assess the rate of land use land cover change (LUCC) of the region by calculating annual land use change rate (ALUCR) within the study period.

Similarly, glacier facies are the end-of-summer characterizable and mappable distribution of accumulation and ablation snow and ice zones. Satellite remote sensing (RS) of glacier facies presents tremendous opportunities for exploring image acquisition, processing, and mapping techniques for precise monitoring. To address this aspect of the ongoing climate change Wankhede *et al.* first reviewed the available literature that explored different facets of mapping facies. Followed by an experiment where three customized spectral index ratios (SIR 1/2/3) were evaluated with varying thresholds through four different rule sets in an object-based image analysis (OBIA) environment. In their chapter they utilized very high-resolution WorldView-2 satellite imagery to capitalize on the sensor's fine resolution. Different rule sets were utilized to determine the necessity and extent of additional contextual parameters. OBIA rule sets were named as OBIA 1/2/3/4. The OBIA results were then compared against pixel-based image analysis (PBIA) algorithms, such as Mahalanobis distance (MHD), maximum-likelihood (MXL), minimum distance (MD), spectral angler mapper (SAM), and winner takes all (WTA). Every OBIA rule set outperformed all the PBIA methods. OBIA 1 achieved a maximum Kappa score (KS) of 0.94 and SAM yielded the lowest KS of 0.06. The order of algorithm performance is OBIA 1 > OBIA 3 > OBIA 2 = OBIA 4 > MHD = WTA > MXL > MD > SAM. Reliance on contextual parameters appears to increase with a reduction in threshold for a particular index. As OBIA is subjective to the skill of the operator, this increase can vary wildly based on the operator, image, and glacier. Therefore, further testing is needed to determine the consistency of this increasing reliance on contextual parameters. In the future, these findings will be helpful in realizing the spatial-spectral potential of OBIA for precise monitoring of glacier facies. It is an established fact that the climatic changes in association with sea-level changes have long been subjective on human performance in terms of settlement pattern, style, and adaptation methods. To further discuss this aspect Das *et al.* discuss the impact of the climate change and its associated environmental change, that affected one of the most advanced civilizations in the Gujarat region of western India. Based on the regional comparative study, it can be inferred that the palaeoclimatic and palaeoenvironmental shifts, played an important role in the sustenance of these cultural centers. This is very well observed from the Lothal, Dholavira, and Khirsara sites, which witnessed a surge in the site abandonment owing to the widespread aridification witnessed by the region. This widespread aridity is also associated with the drying up of the rivers and sea water retreat ultimately leading to their collapse. However, pertaining to the emergence and submergence of the sea-level stands, the region requires a more systematic and robust approach, with a special focus on the coastal ancient settlers of the region. Furthermore, the continental archives serve the potential parameter to elucidate the response of the monsoonal strength within the region. Although, the fluvial archives too, gives prominent evidence to study the landscape impacts but are not considered significant owing to its preservation in the discrete patches. The landscape of Gujarat henceforth illustrates an excellent palaeoecological domain that has cradled human civilizations, which have been impacted and affected by pertinent climate changes during the Mid-Late Holocene period.

It is hoped that the present book will serve as a ready reference to any one, interested in understanding climate change, its science, assessment and impacts on the society at large. It is of interest to policymakers, economic planners, and scientific institutions.

Neloy Khare
New Delhi
April, 2024

Acknowledgements

Kind support and the help received from all contributing authors are gratefully acknowledged, without their valuable inputs on various facets of the Quaternary climate over Indian sub-continent, the book would not have been possible. Various learned experts who have reviewed different chapters are sincerely acknowledged for their timely, important, and critical reviews.

I express my sincere thanks to the Ministry of Earth Sciences, Government of India, New Delhi (India) for various inputs, support, and encouragements. Dr. M. Ravichandran, Secretary, Ministry of Earth Sciences, and Government of India and Dr. Arbind Mitra, former Secretary, Office of the Principal Scientific Advisor to Prime Minister, Government of India, New Delhi have always been the source of inspiration and are acknowledged for their kind support. Shri. D. Senthilpandian, Joint Secretary to Government of India at Ministry of Earth Sciences, New Delhi (India), Dr. Prem Chan Pandey, former Director, National Centre for Polar and Ocean Research, Goa (India) and Dr. K.J. Ramesh, former Director General, India Meteorological Department (IMD), New Delhi (India) have always been supportive as a true well-wisher and are deeply acknowledged for providing many valuable suggestions at various stages of the preparation of this book. This book has received blessings of eminent Quaternary climate experts of the country namely Prof. Anil Kumar Gupta, from the Indian Institute of Technology, Kharagpur, (India) and Dr. Rajiv Nigam, former Deputy Director, National Institute of Oceanography, Goa (India).

My wife, Dr. Rajni Khare, along with my sons Flight Lieutenant (Dr.) Akshat Khare and Ashmit Khare have unconditionally supported enormously during various stages of this book. Dr. Upasana, S. Banerjee, and Dr. Sidharth Dey of Ministry of Earth Sciences, New Delhi have helped immensely in formatting the text and figures of this book and bring the book to its present form. Publishers (Taylor and Francis) have done a commendable job and are sincerely acknowledged.

Neloy Khare
April, 2024
New Delhi

About the Editor

Neloy Khare, presently Adviser and Scientist G to the Government of India at MoES has a very distinctive acumen not only of administration but also of quality science and research in his areas of expertise covering a large spectrum of geographically distinct locations like the Antarctic, Arctic, Southern Ocean, Bay of Bengal, Arabian Sea, Indian Ocean, etc. Dr. Khare has almost 30 years of experience in the field of paleoclimate research using paleobiology (palaeontology)/teaching/science management /administration/coordination for scientific programs (including the Indian Polar Programme) etc. Having completed his doctorate (Ph.D) on tropical marine regions and Doctor of Science (D.Sc) on Southern High latitude marine regions towards environmental/climatic implications using various proxies, including foraminifera (micro-fossil), he has made significant contributions in the field of paleoclimatology of Southern high latitude regions (Antarctic and Southern Ocean) using micropaleontology as a tool. These studies coupled with his paleoclimatic reconstructions from tropical regions helped understand causal linkages and tele-connections between the processes taking place in Southern high latitudes with that of climate variability occurring in tropical regions. Dr. Khare has been conferred Honorary Professor and Adjunct Professor by many Indian Universities. He has a very impressive list of publications to his credit Government of India and many professional bodies have bestowed him with many prestigious Awards for his humble scientific contributions to past climate changes/oceanography/polar science and Southern oceanography. The most coveted award is Rajiv Gandhi National Award-2013 conferred by the Honourable President of India. Others include ISCA Young Scientist Award, Boyscast Fellowship, CIES French Fellowship, Krishnan Gold Medal, Best Scientist Award, Eminent Scientist Award, ISCA Platinum Jubilee Lecture, IGU Fellowship and Rajbhasha Gaurav Puraskar-2023. Dr. Khare has made tremendous efforts to popularize ocean science and polar science across the country by way of delivering many invited lectures, radio talks, and publishing popular science articles. Many books authored/edited on thematic topics and published by reputed International Publishers are testimony to his commitment to popularizing science among the masses.

Dr. Khare sailed in the Arctic Ocean as a part of "Science PUB" in 2008 during the International Polar Year campaign for scientific exploration and gained the prestigious distinction of being the first Indian to sail in the Arctic Ocean.

Contributors

Shakeel Ahmed
School of Sciences
Maulana Azad National Urdu University
Hyderabad, India

Anwar Alam
Indian Institute of Technology
Delhi, India

Josephine Anthony
Central Research Laboratory
Meenakshi Academy of Higher
    Education and Research (MAHER)
Chennai, India

S.D. Attri
India Meteorological Department
Ministry of Earth Sciences
New Delhi, India

Keshava Balakrishna
Dept. of Civil Engineering
Manipal Institute of Technology
Manipal Academy of Higher Education
Manipal, Karnataka, India

Akhouri Bishwapriya
Geological Survey of India
SU: Bihar, Patna, Bihar

Amritee Bora
Department of Geography
North-Eastern Hill University
Shillong

Abhishek Kumar Chaurasia
Assistant Geologist
Geological Survey of India
SU: Bihar, Patna, Bihar

Sumer Chopra
Institute of Seismological Research
Gandhinagar, India

Archana Das
Institute of Seismological Research
Gandhinagar, India

G. Dharani
Ocean Science and Technology for
    Islands
National Institute of Ocean Technology
Ministry of Earth Sciences, Government
    of India
Chennai, India

Gautam Chand Garg
Senior Geologist
Geological Survey of India
SU: Jaipur, Rajasthan

Rikza Imtiyaz
Department of Geography
Jamia Millia Islamia
New Delhi, India

Darve Sabina Iqbal
Faculty of Fisheries
Sher-e-Kashmir University of
    Agricultural Sciences and Technology
    of Kashmir
Shalimar, Srinagar, Jammu and
    Kashmir, India

Shridhar D. Jawak
The Norwegian Climate and
    Environmental Research Institute
    (NILU)
Kjeller, Norway

Dilip Kumar Jha
Ocean Science and Technology for
    Islands
National Institute of Ocean Technology
Ministry of Earth Sciences, Government
    of India
Chennai, India

Ira Khan
School of Sciences
Maulana Azad National Urdu University
Hyderabad, India

Neloy Khare
Ministry of Earth Sciences
Government of India

R. Kirubagaran
Ocean Science and Technology for
    Islands
National Institute of Ocean Technology,
    Ministry of Earth Sciences
Government of India
Chennai, India

S. Naresh Kumar
ICAR-Indian Agricultural Research
    Institute
New Delhi, India

Alvarinho J. Luis
National Centre for Polar and Ocean
    Research (NCPOR)
Ministry of Earth Sciences, Government
    of India
Headland Sada, Vasco-da-Gama,
    Goa, India

Asra Imtiyaz Mattoo
Faculty of Fisheries
Sher-e-Kashmir University of
    Agricultural Sciences and Technology
    of Kashmir
Shalimar, Srinagar, Jammu and Kashmir
    (India)

B.S. Mipun
Department of Geography
North-Eastern Hill University
Shillong

A.C. Narayana
Centre for Earth, Ocean, and
    Atmospheric Sciences
University of Hyderabad
Hyderabad, India

Shambhu Nath
Director
Geological Survey of India
SU: Uttar Pradesh, Northern Region

Vandana Prajapati
Department of Biotechnology
Ministry of Science and
    Technology
Government of India

S.P. Prizomwala
Institute of Seismological Research
Gandhinagar, India

Atiqur Rahman
Department of Geography
Jamia Millia Islamia
New Delhi, India

Navneet Ranjan
Assistant Geologist
Geological Survey of India
SU: Bihar, Patna, Bihar

Sarah Sarah
Department of Earth Sciences
University of Kashmir
Srinagar, India

Mili Singh
Director
Geological Survey of India
SU: Assam, Guwahati

Vishwa Ranjan Sinha
Program Officer
Water and Wetlands
South Asia Natural Resources
    Group, IUCN

Aashima Sodhi
Institute of Seismological Research
Gandhinagar, India

Shailender Kumar Verma
Department of Environmental
    Studies
University of Delhi

Devesh Walia
North-Eastern Hill University,
Shillong

Gohar Bilal Wani
Faculty of Fisheries
Sher-e-Kashmir University of
    Agricultural Sciences and Technology
    of Kashmir
Shalimar, Srinagar, Jammu and Kashmir
    (India)

Sagar F. Wankhede
Dept. of Civil Engineering
Manipal Institute of Technology
Manipal Academy of Higher Education
Manipal, Karnataka, India

Ali P. Yunus
Indian Institute of Science Education
    and Research
Mohali, India

1 Groundwater Potential in India

Challenges and Threats of Climate Change Scenario

Sarah Sarah, Ira Khan, Rikza Imtiyaz, Atiqur Rahman and Shakeel Ahmed

1.1 INTRODUCTION

Water that provides life to all living things circulates in the combined earth-atmospheric system, which is called the hydrological cycle. Its circulation is in a very large system but it is finite. Also since its cycle of circulation is large, it is found in all states with active transformation. Water resource is a natural resource, potentially useful for agricultural, industrial, household, recreational and environmental activities. All living things require water to survive, grow and reproduce. However, it is very important to note the following:

> The concept of virtual groundwater loss suggests if a part of groundwater in the system, remains contaminated beyond acceptable limits for a significantly longer period it cannot be considered a usable resource. Thus such groundwater becomes conditionally renewable.
>
> Sarah *et al.* (2014); Sarah *et al.* (2021)

World water resources are observed in Figure 1.1; although there are approximate and diverging estimates, the fact is that we have very limited freshwater resource globally. India's water resources too have similar conditions.

1.2 WATER AND GROUNDWATER RESOURCES OF INDIA

For the utilization of this resource, water in atmosphere as well as in the form of ice is not practically or directly useable, however, water on the land surface as well as in the sub-surface can be practically utilized. The utilizable water on the surface is found in rivers, ponds, lakes, wetlands, etc., including dams and canals. The sub-surface water is in the form of soil moisture and groundwater with a very straight relation where groundwater is directly utilized. Various components of the water resources in India are estimated and given in Table 1.1, while Table 1.2 provides the approximate

DOI: 10.1201/9781003485995-1

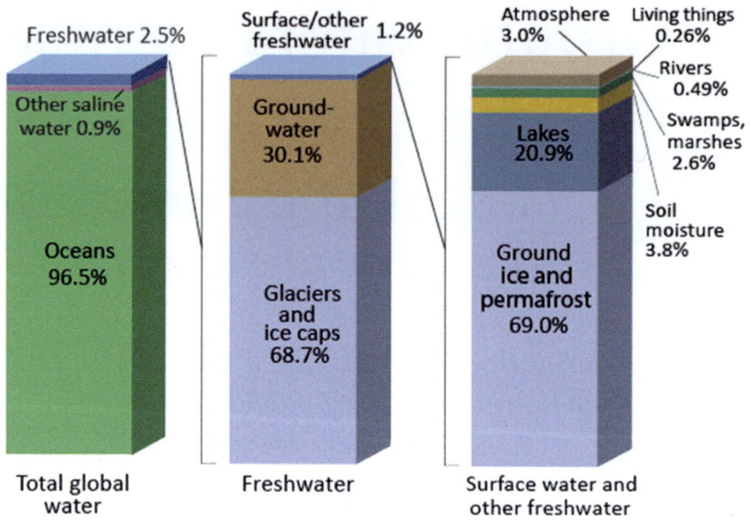

FIGURE 1.1 The earth is full of water but very little freshwater can be used. A 3D graph shows where different types of water lie and we are managing with very small amounts. It also shows the importance of the availability of freshwater.

A 3D graph illustrating the distribution of water on Earth. It highlights the scarcity of freshwater in the liquid form compared to the abundant but unusable saltwater and ice. Freshwater sources, crucial for human use, are depicted as limited.

TABLE 1.1

Depicts the Sketchy Estimates of Various Water Components for India in a Nutshell

1	Average annual precipitation	4000 BCM
2	Average precipitation during monsoon	3000 BCM
3	Surface runoff	1986 BCM
4	Estimated utilizable surface water resource	690 BCM
5	Estimated utilizable groundwater resource	433 BCM
6	Total annual utilizable resources	1132 BCM

amount of utilizable surface as well as groundwater in different river basins. Since groundwater is found in the sub-surface in different types of aquifers, depending on the geological formations, dividing the country into distinct aquifer systems is not feasible.

Although Figure 1.2 depicts various aquifer systems based on their potential as well as the geological formations, the estimation of water resources is mostly based on the river basins or watershed approach.

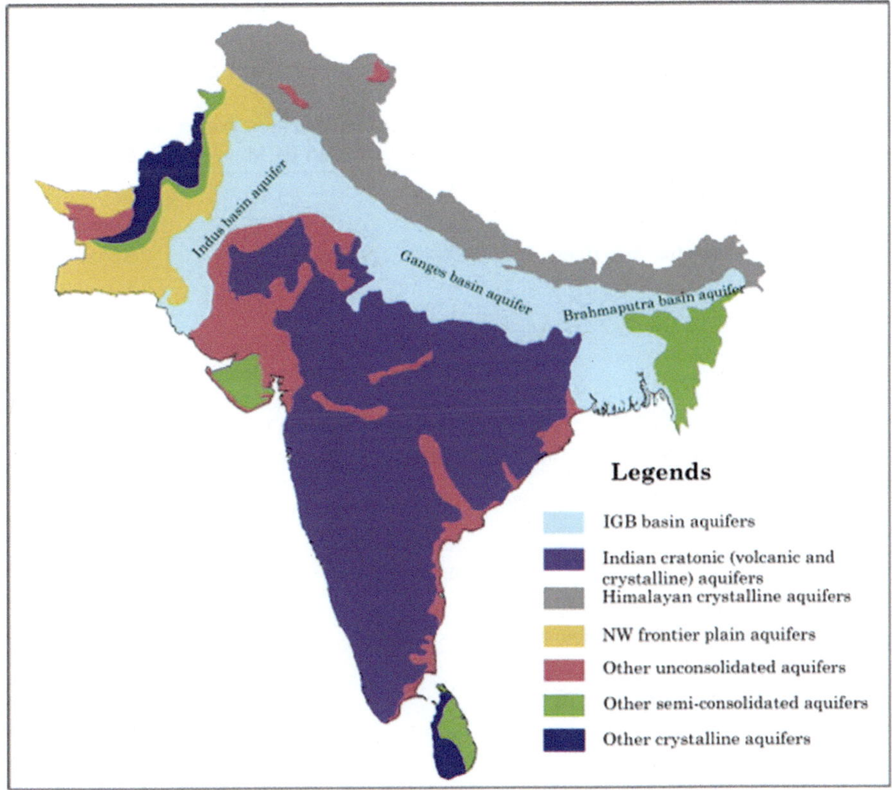

FIGURE 1.2 Groundwater resides in geological formations called aquifers, which are different at different places with different characteristics. Various aquifers distributed in India are shown depicting the variation in lithology and hence water availability also.

A hydrogeological map of India displaying various aquifers in different geological formations. The map shows lithological variations among aquifers across regions. These lithological variations are responsible for differences in groundwater availability from these aquifers as geological characteristics impact groundwater availability.

India has been and still continues to be an intensely agriculture-dominated country and obviously an immense water user as seen in Figure 1.3. It is also a fact that scientific research and practical management of this resource have not gone hand-in-hand, except the civil engineering and planning at administrative or political levels. But it has now become essential to use the scientific acumen for water management due to obvious reasons. The research and development of water resources strongly relates to the hydrological cycle and all its components. However, for practical purposes, as well as for applied research surface water, soil moisture and groundwater have been the major components and most of the research has focused on developing their physio-chemical properties and their dynamics. Later the exploration, exploitation and management of these components will become pivotal.

TABLE 1.2
The Approximate Assessment of the Water Resources That Can Be Utilized

	Basin	Catchment in km²	Surface Water in BCM	Groundwater in BCM
1	Indus Basin (within India)	321289	73.31	26.49
2	Ganga Basin	861452	525.02	170.99+7.19
3	Brahamputra & other NE rivers	194413	585.60	26.55
4	Meghna	41723		8.52
5	Godavari	312812	110.54	40.65
6	Krishna	258948	78.12	26.41
7	Cauvery	81155	21.36	12.30
8	Pennar	55213	6.32	4.93
9	Mahanadi	141589	66.88	16.46
10	Brahmani & Baitami	39033	28.48	4.05
11	Subernarekha	29196	12.37	1.82
12	Sabarmati	21674	3.81	
13	Mahi	34842	11.02	
14	Narmada	98796	45.64	10.83
15	Tapi	65145	14.88	8.27
16	East flowing rivers b/w Mahanadi and Pennar		22.52	
17	East flowing rivers b/w Pennar and Kanyakumari		16.46	18.22
18	West flowing river from Tapi to Tadri		87.41	
19	West flowing rivers between Tadri to Kanyakumari		113.53	17.69
20	West flowing rivers of Kutch, Sabarmati including Luni		15.10	11.23
21	Minor rivers draining into Bangladesh and Burma		31.00	18.84
	Total	2527280	1869.37	431.44

Source: http://iwrs.org.in/indias-water-resources/

To adapt a top-down approach, advanced research starts from GRACE, which provides information on a very large scale. In addition, a number of satellites, for example, the applications of remote sensing, provide considerable information and additional satellites with sophisticated features are devising new and improved applications. Although remote-sensing techniques are widely used and are extremely helpful in providing physical properties related to water and land resources, unfortunately, they do not provide sub-surface information even to the shallower depth.

Geophysical and hydrogeological investigations compliment this research, particularly for groundwater and soil moisture. Investigations related to surface water are confined to estimating the rainfall, its spatio-temporal variability, runoff, formation of streams and their characteristics. Rainfall–runoff models have contributed a

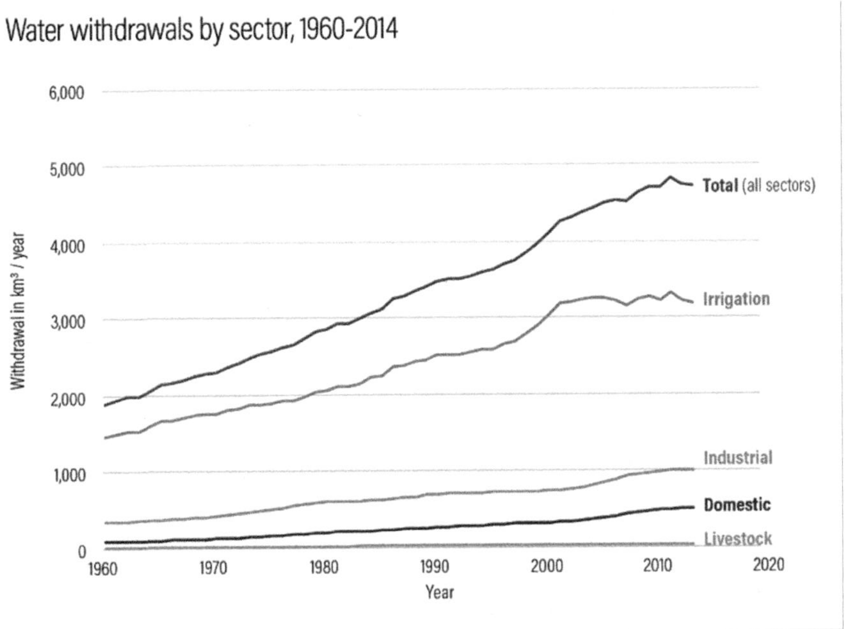

FIGURE 1.3 Freshwater is utilized for various purposes so we compare their utilization and need. The graph shows the major part of water used for irrigation (India being an agricultural country).

A graph illustrating the utilization of freshwater in various sectors. India being an agrarian economy, the largest consumer of groundwater is agriculture sector.

lot in determining the magnitude of the surface water its spatio-temporal variation. Evaporation and evapotranspiration are important components, particularly in the semi-arid regions of the country. Geological investigations are the initial ones to understand hydrogeology and the aquifer dynamics, but for quantification, geophysical approaches are required.

The major issue related to resources, particularly water resources of India, is the unequal and uneven distribution of its availability and demand. Here demand can be represented by the population and the availability by the water and/or land resource.

Population (India): ~17.7% of the world population
(population density: ~464 person/km²)
Water resources (India): 4% of the total in the world

Figure 1.4 shows the mismatch in space also.

We may presume, for instance, that the mismatch of the two parameters in space or the geographical location could be managed by the **virtual water** transfer but it is

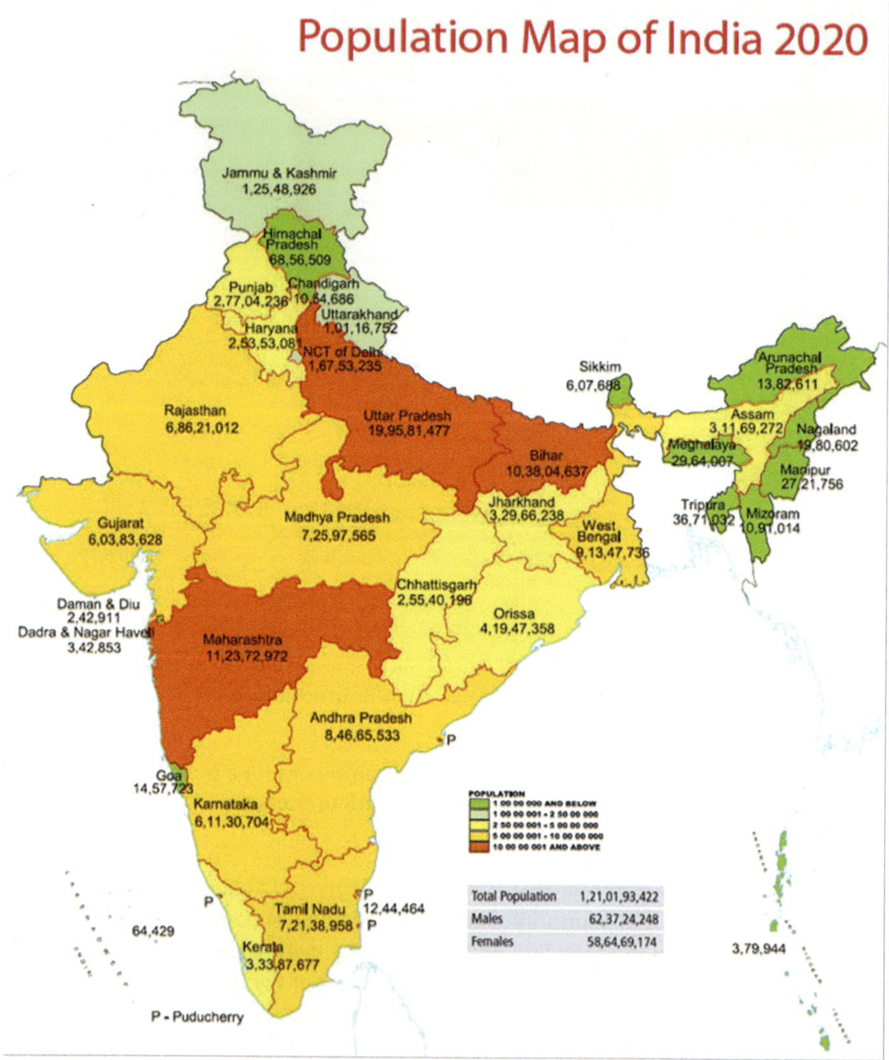

FIGURE 1.4 India has variations in all aspects of water resource management but in a different way. Population density and rainfall as the source of water are related but they do not vary with the same variability. So different management strategies are needed in different places.

A map of India showing variations in population density and rainfall distribution across different regions. The map highlights that while rainfall and population density are related, their patterns of variability differ, indicating the need for diverse water resource management

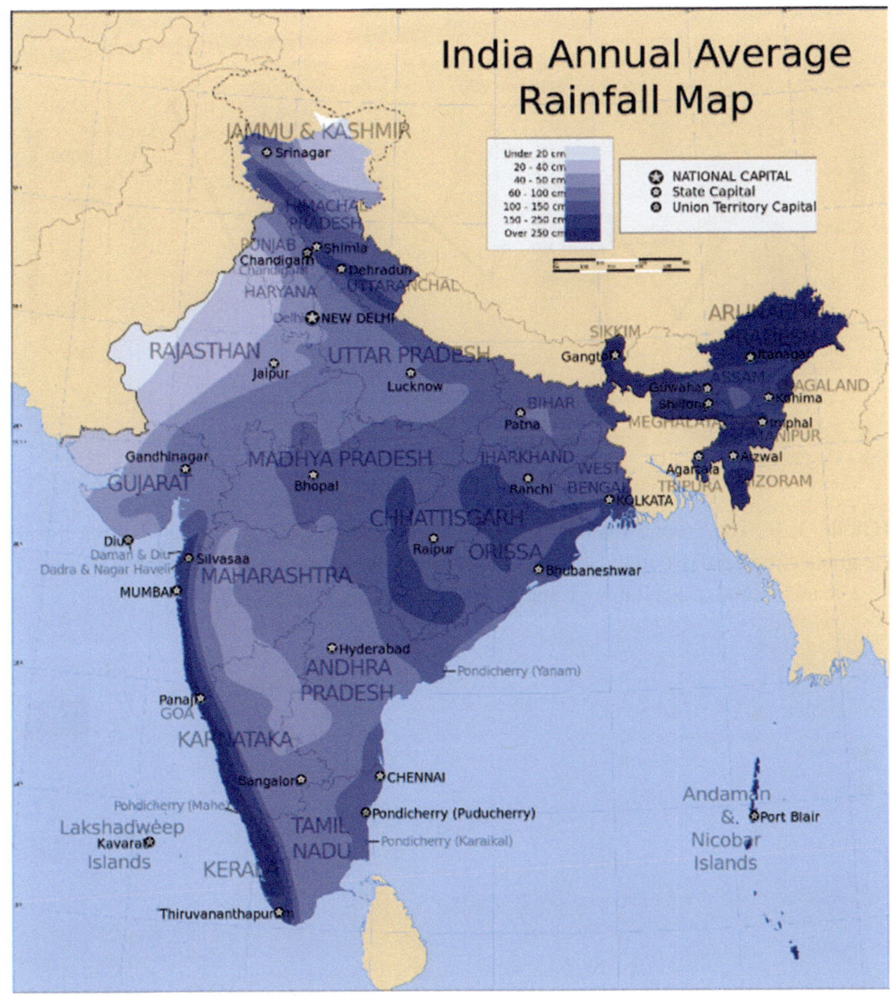

FIGURE 1.4   Continued

extremely crucial to intervene and manage the mismatch between demand and availability scientifically.

The above situation is incredibly alarming simply because the trend for availability is the opposite (Figure 1.5). On the other hand, the rainfall pattern and magnitude both are visibly affected by climate change as well as human intervention.

In spite of the best efforts from several NGOs as well as government, the population of India is increasing and has become the most populated country in the world (Figure 1.6).

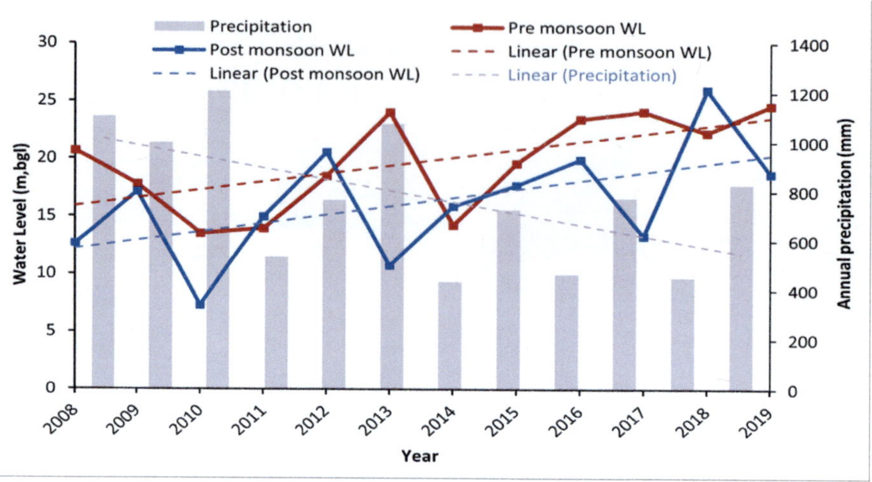

FIGURE 1.5 The groundwater level changes due to infiltration of rainfall but not linearly. The graph shows the rise in the groundwater levels following the rainfall and then the decline. In 12 years, a trend of declining groundwater levels has been observed.

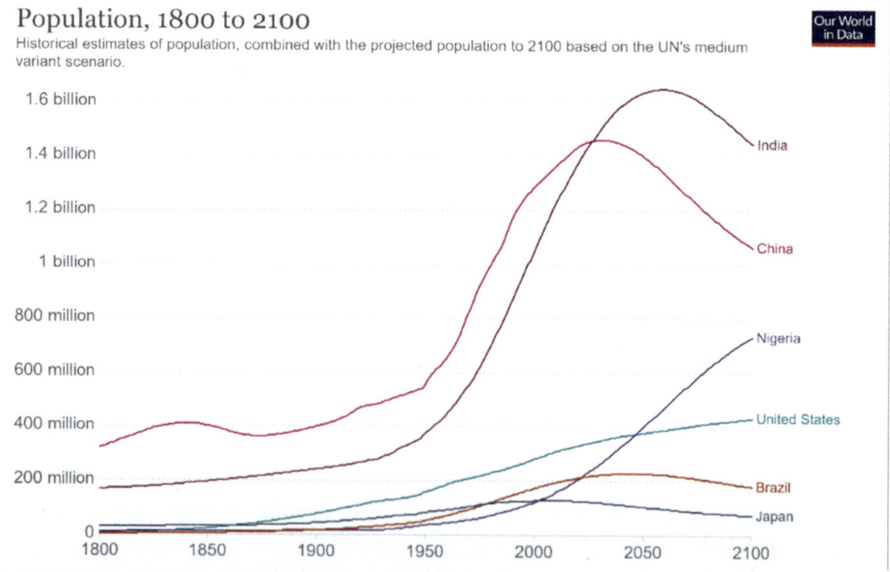

FIGURE 1.6 Population in most of the countries increases but the rates of growth are different. India and China are much faster. However, India has already crossed the limit with China becoming the most populated country globally but is expected to decline now.

1.3 THE COMPONENTS OF THE HYDROLOGICAL CYCLE

Water can occur in three physical phases: solid, liquid and gas and is found in nature in all of these phases in large quantities. Depending upon the environment of the place of occurrence, water can quickly change its phase. The hydrological cycle, also known as the water cycle, is one such cycle which forms the fundamental concept in hydrology. The hydrological cycle is defined as "the pathway of water as it moves in its various phases to the atmosphere, to the earth, over and through the land, to the ocean and back to the atmosphere". This cycle has no beginning or end and water is present in all three states (solid, liquid and gas). A pictorial view of the hydrological cycle is given in Figure 1.7. The science of hydrology primarily deals with the land portion of the hydrologic cycle; interactions with the oceans and atmosphere are also studied. NRC (1991) called the hydrologic cycle the integrating process for the fluxes of water, energy and the chemical elements.

The hydrologic cycle can be visualized as a series of storages and a set of activities that move water among these storages. Among these, oceans are the largest reservoirs, holding about 97% of the earth's water. Of the remaining 3% freshwater, about 78% is stored in ice in Antarctica and Greenland. About 21% of freshwater on the earth is groundwater, stored in sediments and rocks below the surface of the earth. Rivers, streams and lakes together contain less than 1% of the freshwater on

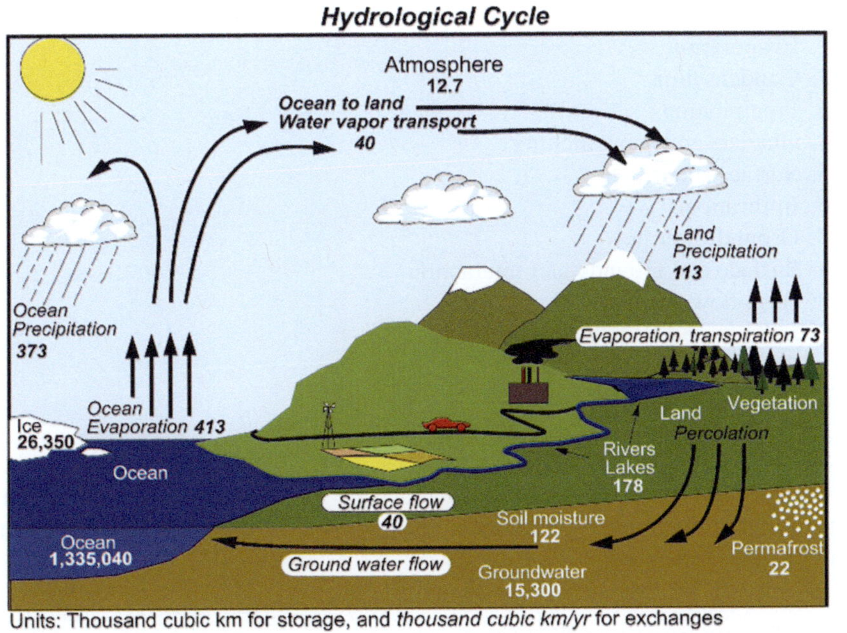

FIGURE 1.7 Water is found in all three states on the earth and atmosphere. However, since it is dynamic, it circulates in different forms called the state in cycle. Various components of the water cycle are shown.

the earth and less than 0.1% of all the water on the earth (Figure 1.1). Since water is a very good solvent, chemistry is an integral part of the hydrological cycle. Rain and snow usually are supposed to be the purest form of water although these may also be contaminated with pollutants present in the atmosphere. During the journey on earth, many chemical compounds are mixed with water and subsequently the quality of water gets changed.

The study of the complete hydrological cycle is very special and interdisciplinary; the atmospheric part is studied by the meteorologists, the pedosphere part by soil scientist, the lithological part by geologists/hydrogeologists and the final part falling in the ocean by the oceanographers. Other professionals who study the water cycle are agriculture scientists, energy managers, ecologists, environmentalists, public health officers, industrialists, chemists and inland navigators. There are three major systems: oceans are the major reservoir and source of water, the atmosphere functions as the carrier as well as deliverer of water and land uses water. On a large and global scale, the hydrological cycle is a closed system but on a local scale it is an open system.

1.4 IMPACT OF CLIMATE CHANGE ON THE COMPONENTS OF THE HYDROLOGICAL CYCLE

The components of the hydrological cycle consist of a number of physical processes, as listed below:

1. Evaporation
2. Condensation
3. Precipitation
4. Glaciers and snow melting
5. Surface runoff
6. Infiltration
7. Groundwater flow
8. Surface and groundwater interaction
9. Base-flow
10. Submarine groundwater discharge
11. Groundwater withdrawal
12. Irrigation return flow
13. Transpiration and evapotranspiration

Although all the components mentioned above are impacted and affected by climate change, a few important and prominent components are discussed below with their behaviors under changing climate.

1.4.1 EVAPORATION AND EVAPOTRANSPIRATION

Evaporation is a type of vaporization that occurs on the surface of a liquid as it changes into the gas phase. The surrounding gas must not be saturated with the evaporating substance. When the molecules of the liquid collide, they transfer energy to each other based on how they collide with each other. Evaporation is caused by the sun. The sun warms up the water and the water turns into gas. The rate of evaporation depends on the surface area, temperature and humidity. A few associated and closely related components are as follows:

- Evaporation is an essential part of the water cycle because it allows large amounts of water to move from bodies of water on Earth up into the atmosphere. Water vapor in the atmosphere condenses into clouds, which move around the globe and release precipitation.
- Evaporation is the process by which water changes from a liquid to a gas or vapor. Evaporation is the primary pathway that water moves from the liquid state back into the water cycle as atmospheric water vapor.
- Evaporation causes cooling but at the same time transfer of energy takes place. The surface of the substance at which evaporation is happening, gets cooled. This increase in the surrounding energy, increases the humidity content in air and also the contained heat.

The rate of evaporation depends on various factors, which are given below.

- Temperature: when the temperature of the liquid is increased, it results in the increase of the kinetic energy of the molecules present in the liquid. By increasing the energy, the liquid molecules overcome the intermolecular force of attraction and escape in the form of gas in the atmosphere. Thus, the greater the temperature of the liquid, the faster the rate of evaporation.
- Surface area: evaporation takes place on the surface of the liquid. During the evaporation process, the molecules present on the surface overcome the intermolecular force of attraction and break away from the liquid particles and skip to the environment. Thus, the greater the surface area occupied, the quicker the evaporation.
- Humidity: the greater the humidity of the atmosphere, which is surrounding the water molecule, the slower the rate of evaporation.
- Transpiration: transpiration is the evaporation occurring through plant leaves (stomatal openings). Transpiration is affected by plant physiology and environmental factors, such as, type of vegetation, stage and growth of plants, soil conditions (type and moisture) and climate and weather.
- Evapotranspiration: combined "loss" of water vapor from within the leaves of plants ("transpiration") and evaporation of liquid water from water surfaces, bare soil and vegetative surfaces. Globally, about 62% of the precipitation that falls on the continent is evapotranspired (~72,000 km³/yr); 92% of which from land surfaces evapotranspiration and 3% from open water evaporation (Dingman, 2015, "*Physical Hydrology*"). In practice, the terms E and ET are often used to mean the same thing – the evaporation from the land surface. Therefore, one must use the context to determine what the term evaporation

means in a specific case (i.e., is it just from an open water surface or the entire land surface?).

- Potential evaporation (PE): PE is the climate-controlled evaporation from an open water surface with unlimited supply (and no thermal capacity).

It is estimated that open water reservoirs lose more than 40% of their total water storage capacity per year due to evaporation. While this loss is of significant concern, the threat of a changing climate has been directing greater focus to how much water will be lost from reservoirs in the future. Obviously the rise in temperature due to climate change is a known fact and this is one of the most important parameters causing evaporation and transpiration.

Warmer temperatures associated with climate change and increased carbon dioxide levels may speed plant growth in regions with ample moisture and nutrients. This could lead to increased transpiration: the release of water vapor into the air by plants as a result of photosynthesis.

1.4.2 PRECIPITATION

After condensation, liquid or solid water falls to Earth, which is known as precipitation. Rain and snow are precipitation and are important in the water cycle. It is how water moves from the atmosphere back to the Earth. Vapor molecules collide and join to form droplets that fall as rain, snow sleet or hail, etc.

Climate change is likely causing parts of the water cycle to speed up as warming global temperatures increase the rate of evaporation worldwide. More evaporation is causing more precipitation, on average. We are already seeing impacts of higher evaporation and precipitation rates, and the impacts are expected to increase over this century as climate warms.

Higher evaporation and precipitation rates are not evenly distributed around the world. Some areas may experience heavier than normal precipitation, and other areas may become prone to droughts, as the traditional locations of rain belts and deserts shift in response to a changing climate. Some climate models predict that coastal regions will become wetter and the middle of continents will become drier. Also, some models forecast more evaporation and rainfall over oceans, but not necessarily over land.

1.4.3 RUNOFF AND INFILTRATION AS WELL AS GROUNDWATER RECHARGE

Soon after the precipitation, particularly in the form of rainfall, three major processes take place simultaneously and all are extremely important for our utilization.

1. Infiltration
2. Runoff
3. Evaporation

When the rain falls on the uncovered land, water enters the soil and further down due to gravity and continues until it meets the groundwater. Of course, it is a slow process and is governed by a few forces in the medium. This infiltration process is slower and often much slower than the speed of rainfall. This is why the excess water that remains on the land surface starts flowing towards the lower elevation. This flow and the process is called surface runoff. Thus, the two processes take place quite simultaneously. Another process that is evaporation, as discussed inearlier, takes place anyway. The process of evaporation takes place slowly and a substantial part of the rainfall returns to the atmosphere again in the form of evaporation. The surface runoff becomes the surface water and continues flowing in the lower elevations. The in-filtered water that initially remains as soil moisture in the upper sub-surface, called unsaturated zones, and then shifts down to the saturated zone, called groundwater. Groundwater of course flows from the higher heads towards lower heads but flows very slowly as it is further governed by another geological parameter called permeability.

1.4.4 BASE-FLOW AND AQUIFER-RIVER INTERACTION

Base flow is a portion of the stream flow that is not runoff; it is water from the aquifer, flowing into the channel over a long time and with a certain delay. The surface water and groundwater are always linked and depend upon each other at all time and everywhere. This is very well known as aquifer river interaction (Figure 1.8). Depending on the situation the river water contributes to the groundwater and then the stream is known as influent streams and otherwise they are known as effluent streams. This process of exchange of water is rarely visible and very difficult to estimate.

The estimation of base flow can be made using Darcy's law but it is very difficult to estimate the intervening transmissivity due to access and different geological condition. This quantity can be estimated using the aquifer model but then all other parameters that are usually unknowns should be separately estimated and taken as known, which is practically very difficult. However, through geophysical methods, e.g., water-borne or heli-borne, a geophysical survey can be deployed to determine the inventing transmissivity. In India there are two types of river system, viz., Himalayan and Peninsular depending on their origin. During the period of monsoon both the river system gets water due to runoffs. However, the Himalayan rivers receive water from the glacial melts during the non-rainy and summer season. However, the Peninsular rivers receive water only from the base-flow and any reduction or stoppage of base-flow to the streams makes the stream ephemeral from perennial. Such a situation leads to sever disasters.

1.4.5 IRRIGATION AND RETURN FLOW

Irrigation return flow is defined as the excess of irrigation water that is not evapotranspirated or evacuated by direct surface drainage, and which returns to the aquifer or surface water body such as lake or a stream. Irrigation return flow coefficient, i.e., the ratio of the irrigation return flow and the abstracted flow, varies from more than 50% for rice cultivation to about 15% for sugarcane cultivation as shown in Figure 1.9 (Jalota and Arora, 2002 and Dewandel *et al.*, 2008), and are usually nil in the case of drip irrigation technique. Thus, irrigation return flows can be a significant component of the water or groundwater balance particularly in the case of intensively

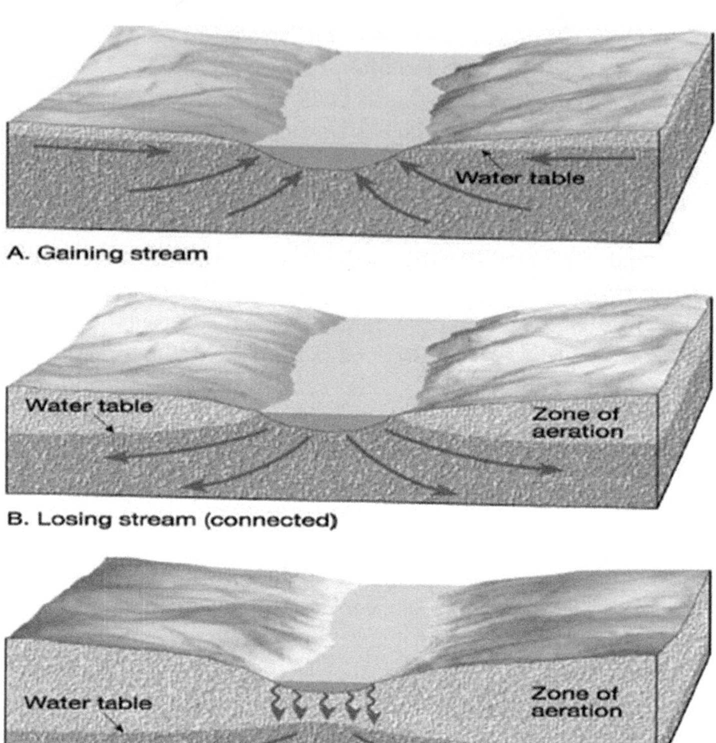

FIGURE 1.8 Water is found on the surface as well as in the sub-surface that are mostly connected. However, the interaction of the two waters depends on a number of parameters that are depicted.

irrigated areas. To avoid any inaccurate aquifer budgeting and/or modeling, and consequently any erroneous water resource management, this irrigation return flow component therefore needs to be accurately assessed at an adapted time-step.

Estimation of such a component is usually difficult and requires complex flow modeling of the unsaturated zone at the scale of one or a few experimental fields (e.g., Chen and Liu, 2002) or at the watershed scale (Gosain *et al.*, 2005). Such experiments require numerous field data, such as pressure head profiles, water content profiles, soil texture analyses and hydraulic conductivities, etc., in addition to crop irrigation characteristics and climatic data (e.g., Singh *et al.*, 2001; Jalota and Arora, 2002; Chen *et al.*, 2002). Moreover, such studies generally require heavy field investments and may only be conducted in experimental farms, which may not be available where water or groundwater balance is needed. Moreover, even if the irrigation return flow estimated by such techniques is accurately assessed at the scale of the experiment, it may not be representative at the watershed scale where irrigated crops, soil characteristics and farming practices (e.g., irrigation techniques) may widely

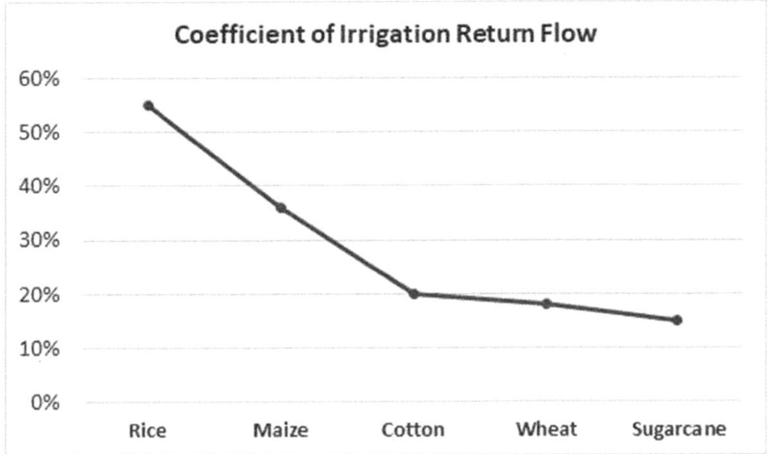

FIGURE 1.9 When water is applied to the crops as irrigation, part of it percolates down to meet the groundwater. The amount of the returned water is different for different crops and varies significantly.

vary. To assess the groundwater balance of a pilot and agricultural watershed in south India, Marèchal *et al.* (2006) worked out a methodology to assess the seasonal irrigation return flow coefficients of the various irrigated crops at the watershed scale with corresponding uncertainties. The method has been used for estimating the seasonal irrigation return flows of rice (paddy field), vegetables and flower fields at watershed and seasonal scale for a rural watershed.

1.5 SCALES OF HYDROLOGICAL CYCLE

1.5.1 GLOBAL SCALE

The hydrological cycle can be considered to be comprised of three major systems, from a global perspective, viz., precipitation, runoff and evaporation. These are the principal processes that transmit water from one system to the other and shows the interaction between the earth, ocean and atmosphere. The result of these studies form important inputs to water resources planning for a national, regional water resource assessment, weather forecasting and study of climate changes.

1.5.2 CATCHMENT SCALE

The spatial coverage can range from a few square km to thousands of square km while studying the hydrological cycle on a catchment scale. The time scale could be a storm lasting for a few hours to a study spanning many years. When the attention is focused on the hydrological cycle of the land system, the dominant process is precipitation, evapotranspiration, infiltration and surface runoff. The land system itself comprises three subsystems: vegetation, structural and soil subsystems. The water in this subsystem is either lost to the atmospheric system or enters into the subsurface system.

The time required for the movement of water through various components of the hydrological cycle varies considerably. The velocity of streamflow is much higher compared to the velocity of groundwater. The time-steps depend upon the purpose of study, the availability of data and how detailed the study is. The estimated period of renewal of water resources in water bodies on the earth is given in Table 1.3. The time step should be sufficiently small so that variations in the processes can be captured in detail but at the same time it should not put undue burden on data collection and computational efforts. Figures 1.10, 1.11 and 1.12 are depicting various cycles in the system.

Shiklomanov (1999) has provided an informative table on the approximate period of renewal of various water bodies. However, this will also be impacted by climate change and further assessment will be needed.

1.6 CLIMATE CHANGE AND ITS IMPACT ON WATER RESOURCES

Recent study has exposed that climate change leads to the altered precipitation and evapotranspiration rate globally, so it may be considered as an additional major threat

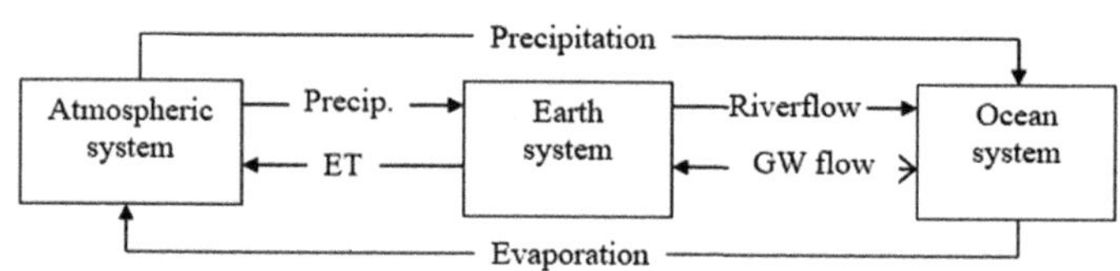

FIGURE 1.10 On global scale, some smaller components are ignored and only major components are taken at this scale.

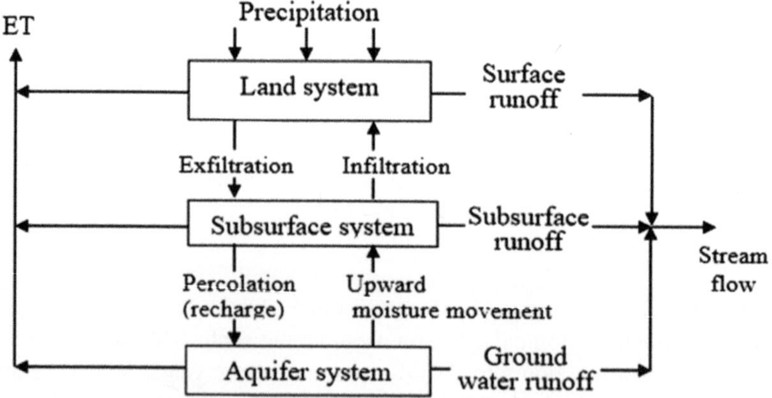

FIGURE 1.11 Various other components of the hydrological cycle at the Earth and near-Earth scale.

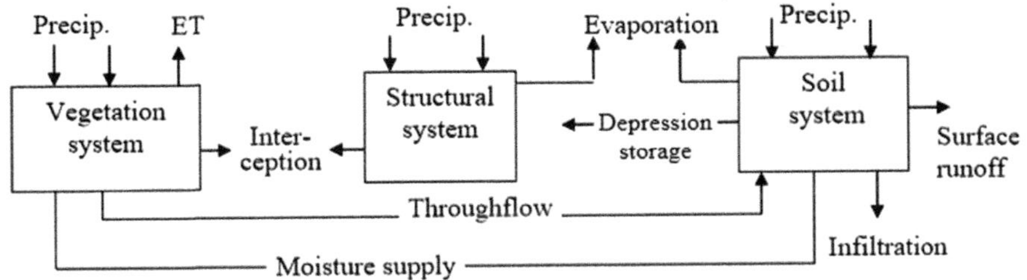

FIGURE 1.12 The components of the hydrological cycle at the land scale.

TABLE 1.3
Period of Water Renewal in Different Water Bodies

#	Water of Hydrosphere	Period of Renewal
1	World ocean	2500 years
2	Groundwater	1400 years
3	Polar ice	9700 years
4	Mountain glaciers	1600 years
5	Ground ice of the permafrost zone	10000 years
6	Lakes	17 years
7	Bogs	5 years
8	Soil moisture	1 year
9	Channel network	16 days
10	Atmospheric moisture	8 days
11	Biological water	Several hours

Source: Shiklomanov (1999).

to this resource. Hence, understanding the behavior and dimensions of the groundwater regime under future climatic and other changes is significant in adopting a necessary water management strategy (Tiwari and Ahmed, 2022).

Precipitation is commonly downscaled in climate change impact studies; however, the reliability of the downscaled result is often poor or unreliable, as there is often little correlation between the predictors and the predictands. A poor correlation is often attributed to mesoscale processes occurring at the site-scale that are not represented in regional models due to their representative spatial and temporal sizes in comparison to larger-scale regional precipitation. Mesoscale precipitation processes generally occur in the summer season in the form of convective clouds, which are a result of local-scale evapotranspiration from elevated temperatures and solar radiation magnitudes. As a result, global-scale models may underestimate the summer precipitation measured at a site.

Although climate change has been widely recognized, research on the impacts of climate change on the groundwater system is relatively limited. The reasons may be that long historical data are required to analyze the characteristics of climate change.

These data are not always available. Also, the driving forces that cause such changes are yet unclear. The climatic abnormality may occur frequently and last for a period of time. Even if the required data exist, uncertainty is embedded in model parameters, structure and driving force of the hydrological cycle. Predicting the long-term effect of a dynamic system is very difficult because of limitations inherent in the models, and the unpredictability of the forces that drive the earth. A physically based model of a groundwater system under possible climate change based on available data is very important to prevent the deterioration of regional water-resource problems in the future. Although uncertainties are inevitable, new response strategies in water-resource management based on the model may be useful.

The investigation of the relationship between climate change and loss of fresh groundwater resources is important for understanding the characteristics of the different regions. The impact of future climatic change may be felt more severely in developing countries such as India, whose economy is largely dependent on agriculture and is already under stress due to current population increase and associated demands for energy, freshwater and food. In spite of the uncertainties about the precise magnitude of climate change and its possible impacts, particularly on regional scales, measures must be taken to anticipate, prevent or minimize the causes of climate change and mitigate its adverse effects.

If the likely consequences of future changes of groundwater recharge, resulting from both climate and socio-economic change, are to be assessed, hydrogeologists must increasingly work with researchers from other disciplines, such as socio-economists, agricultural modelers and soil scientists.

Climate change poses uncertainties to the supply and management of water resources. The Intergovernmental Panel on Climate Change (IPCC) estimates that the global mean surface temperature has increased 0.6 ± 0.2 °C since 1861, and predicts an increase of 2 to 4 °C over the next 100 years. Temperature increases also affect the hydrologic cycle by directly increasing evaporation of available surface water and vegetation transpiration. Consequently, these changes can influence precipitation amounts, timings and intensity, and indirectly impact the flux and storage of water in surface and sub-surface reservoirs (i.e., lakes, soil moisture, groundwater). In addition, there may be other associated impacts, such as glacier melting, sea water intrusion, water quality deterioration, etc.

While climate change affects surface water resources directly through changes in the major long-term climate variables, such as air temperature, precipitation and evapotranspiration, the relationship between the changing climate variables and groundwater is more complicated and poorly understood. The greater variability in rainfall could mean more frequent and prolonged periods of high or low groundwater levels, and saline intrusion in coastal aquifers due to sea level rise and resource reduction. Groundwater resources are related to climate change through the direct interaction with surface water resources, such as lakes and rivers, and indirectly through the recharge process. The direct effect of climate change on groundwater resources depends upon the change in the volume and distribution of groundwater recharge. Therefore, quantifying the impact of climate change on groundwater resources requires not only reliable forecasting of changes in the major climatic variables, but also accurate estimation of groundwater recharge.

A number of global climate models (GCMs) are available for understanding climate and projecting climate change. There is a need to downscale GCM on a basin scale and couple them with relevant hydrological models considering all components of the hydrological cycle. Output of these coupled models, such as quantification of the groundwater recharge will help in taking appropriate adaptation strategies due to the impact of climate change. This chapter presents the likely impact of climate change on groundwater resources, climate change scenario for groundwater in India, status of research studies carried out at national and international level, and methodology to assess the impact of climate change on groundwater resources.

1.7 IMPACT OF CLIMATE CHANGE ON GROUNDWATER QUANTITY

The groundwater regime is most affected due to various changes under climate change. Various parts of the aquifer system are affected, as described below.

Soil water zone: this zone is important as it supports vegetation and all biogeochemical reactions. Climate change has an adverse effect on this zone. Higher temperature leads to higher evapotranspiration rates, resulting in less moisture content in this zone. Little or no moisture in the soil leads the penetration of solar radiation into the deeper soils and increased dryness in soils, resulting in severe droughts. The high precipitation in wet climate change scenario will increase surface runoff and in promoting rapid soil erosion. Less infiltration, high evapotranspiration and high runoff will have a great impact on the water availability in this zone, which will affect the entire plant and animal kingdom. Because of change in evapotranspiration patterns in this zone, the rainfall pattern will also be affected. The transpiration process, which holds 80–90% of overall evapotranspiration on Earth, will show various changes depending on the regional vegetation.

Vadose zone: changes in vadose zone due to climate change can be computed by studying the variations in major cations, anions, trace elements and isotopes from the pore water. Due to increase in surface temperature, groundwater temperature will increase. The change in temperature will affect pore water chemistry, residence time and volume of water in matrix and fractures, and thus the composition of the water. The increase in recharge rate will help in mobilizing the contaminants into greater depths. As an example, in semiarid and arid regions, increased infiltration can mobilize large, pore-water chloride and nitrate reservoirs affecting the quality of water. The diurnal temperature fluctuations may be detectable at depths of less than 1 m in the unsaturated zone and seasonal fluctuations at depths of 10 m or more, indicating that the climate change effects depend on depth and are slow in the deep vadose zone.

Groundwater in the saturated zone is important as it is less polluted and has no effects of evapotranspiration. The sensitivity of this zone depends on the depth of the water table; shallow aquifers are more vulnerable to climate change than deeper aquifers. This zone responds to climate change by showing changes in its amount, quality and flow of water depending on the trends of precipitation, evapotranspiration, recharge and discharge. It is generally observed that climate change has less effect on this zone in comparison to human activities on groundwater exploitation, such as excessive pumping, reduction in recharge rate and contamination.

Confined and unconfined aquifers will show alterations in some of their properties during climate change. It has been noticed that severe dry periods can alter the properties of aquifers such as transmissivity and storativity. During dry periods, the conductive channels such as fractures and fissures may become desaturated and the pressure pulse of water within the aquifer will be transmitted slowly, whereas during the wet season, fractures get fully saturated and transmit pressure pulse rapidly. Thus, the changes in recharge pattern will affect the specific storage of an aquifer. In the case of extreme aridity, their vulnerability becomes higher if the potentiometric surface falls below the upper confining beds and results in converting confined aquifers to unconfined aquifers.

Groundwater recharge: the changes in recharge patterns will affect discharge patterns, which will have a direct impact on groundwater supplies and on surface water availability. In a cold or wet and cold–wet scenario, the relationship is directly proportional, i.e., high temperature results in high precipitation and high recharge. In the case of dry scenario, the temperature and precipitation are inversely related to each other, as high temperature will result in less rainfall. Hence, the effects in the case of dry scenario will be severe, implying that the aquifers of semiarid and arid regions are more vulnerable to climate change. Changes in recharge patterns will also alter the quality of water by affecting geochemical reactions and movement of water in the vadose zone. Various types of aquifers will be recharged differently. An unconfined aquifer is recharged directly by local rainfall, rivers, and lakes, and the rate of recharge will be influenced by the permeability of overlying rocks and soils. Unconfined aquifers are sensitive to local climate change, abstraction and seawater intrusion. However, quantification of recharge is complicated by the characteristics of the aquifers themselves as well as overlying rocks and soils. A confined aquifer, on the other hand, is characterized by an overlying bed that is impermeable, and local rainfall does not influence the aquifer. It is normally recharged from lakes, rivers and rainfall that may occur at distances ranging from a few kilometers to thousands of kilometers. Determining the potential impact of climate change on groundwater resources, in particular, is difficult due to the complexity of the recharge process, and the variation of recharge within and between different climatic zones.

Groundwater discharge: under wet climate scenarios, runoff is considered as the most sensitive component and the combined effect of increased precipitation and high discharge will increase the risk of flooding. Under dry climate scenarios, recharge will be the most sensitive component as evapotranspiration will increase while both recharge and discharge will decrease in all seasons, resulting in decline in groundwater level. Increased discharge from melting of glaciers in the Himalayas will increase the risk of flooding in the catchment areas affecting major parts of North India, Pakistan and Bangladesh. Due to changes in discharge, the quality of groundwater will be adversely affected, since during high discharge all the pollutants will be mobilized and may reach groundwater level. In the case of a dry climate scenario, generally the water level will fall and this will affect the needs of the people and may result in increased use of energy to extract water. The increase in groundwater pumping and loss of groundwater storage from aquifers resulted in land subsidence in many Asian cities, such as Osaka and Bangkok. In the future, the increase in discharge and decrease in recharge will make land subsidence a much bigger problem.

1.8 IMPACT OF CLIMATE CHANGE ON GROUNDWATER QUALITY

The three major reasons for the sea-level rise are as follows: expansion of oceans on warming, increase in discharge due to melting of glaciers and excessive pumping due to human settlements along the coasts. The rise in sea level is said to have a great impact on the mangrove forests of the world and aquatic life, affecting the fish stocks and planktons. It is estimated that 30% of coral reefs could be lost in the next 10 years, which will affect the food web of the aquatic environment.

The effect of climate change on Indian groundwater resources means the levels are continuously falling. The groundwater level in Gujarat, Rajasthan, Punjab, Haryana and Tamil Nadu has shown a critical decline. Groundwater decline has been registered in 289 districts of India. The water table in Ahmedabad is falling at a rate of 4–5 m every year; in some parts of Delhi a lowering of 10 m has been noticed. Due to melting of the Himalayan glaciers, the Indo-Gangetic Plains will experience increased water discharge until the 2030s but will face gradual reductions thereafter. Sea-water intrusion has been observed in several coastal states of India, such as Tamil Nadu, Puducherry and Gujarat (Saurashtra), which is not only engulfing the land but also the groundwater reservoirs of India are highly sensitive to climate change in terms of its effect on water supply for irrigation needs.

1.9 THE IMPORTANCE AND REQUIREMENTS FOR SCIENTIFIC RESEARCH

A number of advanced researches are being carried out on many aspects of hydrology, hydrogeology and earth sciences but a few of the following ones could be prioritized.

a. Assessing the existing groundwater system and exploring new sources of groundwater.

b. Adequate groundwater monitoring using an optimal monitoring network.

c. Adopting the climate change model relevant to the country and downscaling to the smaller areas.

d. Delineation of zones and areas of artificial recharge and surface water availability.

e. Appropriate technique of artificial recharge to maximize the benefit. Predictive study to be taken before the artificial recharge is planned and carried out.

f. Reduction in demands using new methods of irrigation as well as public awareness.

g. Tackling coastal aquifers in detail as well as advanced research on recycling waste water.

h. Frequent review of policies and effective governance with practical implementations.

1.9.1 Water Conservation using Scientific Approach

The natural system itself takes care of water conservation in the sub-surface but the additional conservation has been practised since a long time ago, at least in the arid and semi-arid regions. Farmers were capable of drilling several bore wells and connecting them with a channel in their bottom (Figure 1.13) very commonly known as Qanat in Iran.

The time series of the rainfall for a sufficiently long time illustrates that the amount of annual rainfall has not departed much from the long-term mean. Thus, the water received by the rainfall is invariable but the process is altered. So, to cope with the situation, conservation of water through artificial means is important and necessary. Since most regions in the country are in semi-arid type, the conservation needs to be practiced in the sub-surface thus *scientific knowledge of the highly variable sub-surface is a must.*

The system of Qanat or a similar one depicted in the figure serves two purpose, viz., (1) transporting conserved water underground through gravity and (2) recharging the groundwater as well.

Fervent examples of traditional water management systems, several hundred years old, such as Qanat, Karaj, etc. for example (Figure 1.14), in Burhanpur, Bidar and Bijapur (Wahurwagh and Dongre, 2015) exist, and with modern technology they could very well be revived. Secondly, the conservation cannot be one sided, a benchmark has to be set to maintain the minimum natural and ecological conditions.

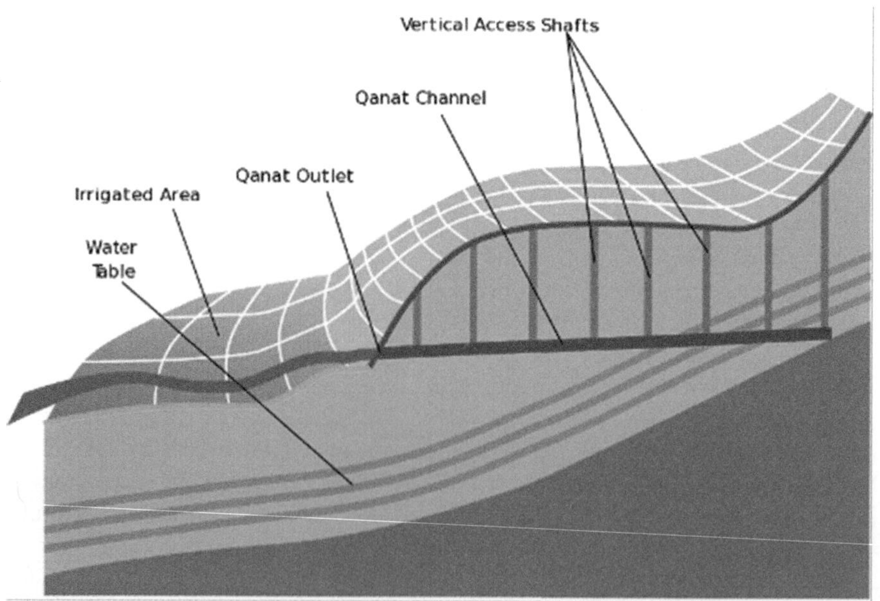

FIGURE 1.13 The figure shows a traditional way of conserving rainwater underground to save it from evaporation losses as well as from contamination. This system is needed to reduce the impact of extreme events.

1.9.2 Groundwater Regulation and Strategic Planning

In the past, due to a number of factors, the availability of water resources was much more than the demand and the balance was mostly positive. In that case, regulating its use was redundant. However, with a number of parameters, the demand has increased much more and has surpassed the availability, and hence its management is necessary. For an effective management, certain rules and regulations have to be formed. However, the country has endorsed failures and non-implementation of a large number of regulations and water laws. Deep thinking and damage control approaches are required before outlining and executing a regulation. Water is the basic need for life and a right also. Thus, the regulations must be flexible, practical, based on scientific reasoning and frequently evaluated. Groundwater should no longer be a private property and all drilling should be controlled and monitored. Water education and hygiene education must be made mandatory in schools. A combination of traditional or natural cum advanced scientific approaches by involving stakeholders should be adopted.

1.10 CONCLUSIONS

The impact of climate change that is visible and felt clearly has impacted the groundwater resources the most. Although various scientific research is being carried out, certain measures need to be taken urgently. The modified pattern of rainfall with reduced rainy days as well as the enhanced gap between two rainy days has affected the groundwater recharge. The increased temperature has impacted the evaporation and evapotranspiration. However, the changed rainfall pattern has certainly increased the surface water but in the form of excess runoff causing the unmanaged floods every year.

India has been suffering with water management issues. The most dominant one is the mismatch between the availability represented by the occurrence of rainfall and the demand represented by the population. Through critical analysis, Figure 1.4 shows the mismatch clearly. Climate change is certainly going to deteriorate this gap.

FIGURE 1.14 Real pictures of the existing underground rainwater harvesting system, like Qanat in the Burhanpur region.

Although it is not very simple, this issue to some extent can be tackled by a virtual water concept. Since agriculture is the main user of water resources, agriculture practice needs to be modified. A revised assessment of water resources and their future availability should be made with error bars so that the cropping pattern change can be decided. Of course, this is an integrated task for hydrogeologists, climatologists, agriculture scientists as well as socio-economists, etc. Transport plays an important role as crops grown somewhere else need to be consumed elsewhere.

The above proposal has been part of the solution and has not been an easy task. The most important point is the rainwater conservation through artificial recharge to groundwater. Rainwater harvesting and conservation into surface water bodies are not beneficial due to a high rate of evaporation and want of the space. Thus, storing the excess rainfall into the sub-surface is a viable solution. This requires a detailed geoscientific knowledge of the sub-surface. Fortunately, advanced studies, e.g., geophysical aquifer mapping, which is being carried out, will be of greater help in deciding the location for the sub-surface storage of the rainfall or the runoff water. As discussed in the above sections, the years' long acumen on Qanat as well as the recent advanced technology together could be useful and effective. In addition, a few points to be taken care of in the current research and development can improve the situations.

- Reducing uncertainties in the numerical models incorporating a more process-based approach.
- Capturing the variability of the system parameters to have a reliable spatial prediction.
- Extensive monitoring and refinement of climate models, etc.
- Exhaustive application of remote-sensing techniques.
- Integration of disciplines, data and organizations in discussing the issues, finding solutions and involvement of stakeholders in the implementation of policies.

Another new research area is to be opened to monitor and study the viruses likely to flow in the river water due to glacial melt as they are inactive in the glacier at present.

REFERENCES

Abdullahi MG, Toriman ME, Gasim, MB (2014) The application of Vertical Electrical Sounding (VES) for groundwater exploration in Tudun Wada Kano State, Nigeria. *Int J Eng Res Rev* 2(4):51–55.

Adiat KAN, Nawawi MNM, Abdullah K (2012) Assessing the accuracy of GIS-based elementary multi criteria decision analysis as a spatial prediction tool – A case of predicting potential zones of sustainable groundwater resources. *J Hydrol* 440:75–89. https://doi.org/10.1016/j.jhydrol.2012.03.028

Aguilera H, Murillo J (2009) The effect of possible climate change on natural groundwater recharge based on a simple model: A study of four karstic aquifers in SE Spain. *Environ Geol* 57(5):963–974.

Ahmed S (2014) A new chapter in groundwater geophysics in India: 3D aquifer mapping through Heliborne Transient Electromagnetic Investigations. *Geol Soc India* 84(4):501–503.

Ahmed S (2015) 3D imaging and national aquifer mapping program. *Geography and You* July-August 2015:10–13. www.geographyandyou.com/3d-imaging-and-national-aquifer-mapping-programme/

Aizebeokhai AP, Oyeyemi KD, Adeniran A (2017) An overview of the potential impacts of climate change on groundwater resources. *J Inform Math Sci* 9(2):437–453.

Aizebeokhai AP, Oyeyemi KD, Joel ES (2016) Electrical resistivity and induced polarization imaging for groundwater exploration. *Proceedings of SEG International Exposition and 87th Annual Meeting*, pp. 2487–2491. https://doi.org/10.1190/segam2016-13857737.1

Amadi J (2016) *Application of the Electrical Resistivity Method in Groundwater Exploration.* GRIN Verlag, Munich. www.grin.com/document/501569

Anupam K, Shinjiro K (2013) Climate change and groundwater: Vulnerability, adaptation and mitigation opportunities in India. *Int J Environ Sci Dev* 4(3):272–276. https://doi.org/10.7763/IJESD.2013.V4.352Google Scholar

Aribisala JO, Awopetu MS, Ademilua OI, Okunadel EA, Adebayo WO (2015) Effect of climate change on groundwater resources in Southwest Nigeria. *J Environ Earth Sci* 5(12):2224–3216.

Arkoprovo B, Adarsa J, Prakash SS (2012) Delineation of groundwater potential zones using satellite remote sensing and geographic information system techniques: A case study from Ganjam district, Orissa, India. *Res J Recent Sci* 1(9):59–66.

Arsène M, BidichaelWahile WE, Gouet D, Ndougsa-Mbarga T, Kuiate K, Ngoh JD (2018) Hydrogeophysical investigation for groundwater resources from electrical resistivity tomography and self-potential data in the Méiganga area, Adamawa, Cameroon. *Int J Geophys*, 2018(1):1–14, DOI:10.1155/2018/2697585

Attri SD, Tyagi A (2010) Climate profile of India. Met Monograph No. Environment Meteorology-01/2010, Contribution to the Indian Network of Climate Change Assessment (NATIONAL COMMUNICATION-II), Ministry of Environment and Forests, India Meteorological Department.

Baltassat JM, Krishnamurthy NS, Girard JF, Dutta S, Dewandel B, Chandra S, Descloitres M, Legchenko A, Robain H, Ananda Rao V, Shakeel A (2006) Proton magnetic resonance technique in weathered-fractured aquifers. Report of IFCPAR Project 2700-W1.

Barthel R, Sonneveld BGJS, Goetzinger J, Keyzer MA, Pande S, Printz A, Gaiser T (2009) Integrated assessment of groundwater resources in the Oueme basin, Benin, West Africa. *Phys Chem Earth* 34(4–5):236–250.

Bello R, Ballogun AO, Nwosu, UM (2019) Evaluation of Dar Zarrouk parameters of parts of Federal University of Petroleum Resources, Effurun, Nigeria. *J Appl Sci Environ Manage* 23(9):1709–1715. https://dx.doi.org/10.4314/jasem.v23i9.16

Bhaskar Rao YJ, Ahmed S, Gahalaut V, Kumar A, Ravi Kumar M (2015) CSIR-National Geophysical Research Institute during the year 2014. Curr Sci 108(11):2010–2013.

Brooks GA, Olyphant GA, Harper D (1991) Application of electromagnetic techniques in survey of contaminated groundwater at an abandoned mine complex in south-western Indiana, U.S.A. *Environ Geol Water Sci* 18:39–47. https://doi.org/10.1007/BF01704576

Burhanpur – A Medieval Water Oasis (2017) VIRASAT – E – HIND FOUNDATION. https://blogvirasatehind.wordpress.com/2017/07/10/burhanpur-a-medieval-water-oasis/

Buselli G, Davis GB, Barber C, Height MI, Howard SHD (2018) The application of electromagnetic and electrical methods to groundwater problems in urban environment. *Exploration Geophys* 23(4):543–555. https://doi.org/10.1071/EG992543

CGWB (2011) Dynamic ground water resources of India. Central Ground Water Board, Ministry of Water Resources, Government of India, New Delhi.

CGWB (2017) Report of the Ground Water Resource Estimation Committee (GEC-2015). Central Ground Water Board, Ministry of Water Resources, RD & GR, Government of India, New Delhi.

CGWB (2018) Ground water quality in shallow aquifers of India. Central Ground Water Board, Ministry of Jal Shakti, Department of Water Resources, RD & GR, Government of India, Faridabad.

CGWB (2019a) National compilation on dynamic ground water resources of India, 2017. Central Ground Water Board, Ministry of Jal Shakti, Department of Water Resources, RD & GR, Government of India, Faridabad.

CGWB (2019b) Ground water year book India 2018-19. Central Ground Water Board, Ministry of Jal Shakti, Department of Water Resources, RD & GR, Government of India, Faridabad.

Chandra S, Ahmed S, Auken E, Pedersen JB, Singh A, Verma SK (2016a) 3D aquifer mapping employing airborne geophysics to meet India's water future. *Lead Edge* 35(9):770–774. http://dx.doi.org/10.1190/tle35090770.1

Chandra S, Choudhury J, Maurya PK, Ahmed S, Auken S, Verma SK (2019) Geological significance of delineating paleochannels with AEM. *J Explor Geophys* 51:74–83. https://doi.org/10.1080/08123985.2019.1646098

Chandra S, Esben A, Maurya PK, Ahmed S, Verma, SK (2019) Large scale mapping of fractures and groundwater pathways in crystalline hardrock by AEM. *Sci Rep* 9:398. https://doi.org/10.1038/s41598-018-36153-1

Chandra S, Nagaiah E, Veerababu N, Mondal NC, Somvanshi VK, Ahmed S (2016b) Advanced geophysical investigation including Heliborne TEM in high-resolution aquifer mapping with special emphasis to crystalline hard rocks. *J Geol Soc India* 5:87–96. https://doi.org/10.17491/cgsi/2016/95954

Chaterjee R, Jain AK, Chandra S, Tomar V, Parchure PK, Ahmed S (2018) Mapping and management of aquifers suffering from over-exploitation of groundwater resources in Baswa-Bandikui watershed, Rajasthan, India. *J Environ Earth Sci.* https://doi.org/10.1007/s12665-018-7257-1

Chatterjee K (1999) Water Resources of India – Climate Change Centre, Development Alternatives. Water Resources Initiating and Sustaining Water Sector Reform: A Synthesis; World Bank in Collaboration with the Govt. of India, Ministry of Water Resources. Allied Publishers. www.climatechangecentre.net › pdf ›

Chegbeleh LP, Akudago JA, Nishigaki M, Edusei SNK (2009) Electromagnetic geophysical survey for groundwater exploration in the Voltaian of Northern Ghana. *J Environ Hydrol* 17:1–16.

Chen L, Feng Q (2013) Geostatistical analysis of temporal and spatial variations in groundwater levels and quality in the Minqin Oasis, Northwest China. *Environ Earth Sci* 70:1367. https://doi.org/10.1007/s12665-013-2220-7Google Scholar

Chen SK, Liu CW (2002) Analysis of water movement in paddy rice fields (I) simulation studies. *J Hydrol* 260:206–215.

Chen SK, Liu CW, Huang HC (2002) Analysis of water movement in paddy rice fields (II) simulation studies. *J Hydrol* 268:259–271.

Chen Z, Grasby SE, Osadetz KG (2004) Relation between climate variability and groundwater levels in the upper carbonate aquifer, Southern Manitoba, Canada. *J Hydrol* 290(1–2):43–62.

Chinnaiah Y (2015) Analysis of Ground Water Levels and Water Quality in Musi River Basin, Telangana State (Unpublished Doctoral Thesis). Sri Krishnadevaraya University, Anantapuram, AP, India.

Christensen JH, Krishna Kumar K, Aldrian E, An S-I, Cavalcanti IFA, de Castro M, Dong W, Goswami P, Hall A, Kanyanga JK, Kitoh A, Kossin J, Lau N-C, Renwick J, Stephenson DB, Xie S-P, Zhou T (2013) Climate phenomena and their relevance for future regional climate change. In: Stocker TF, Qin D, Plattner G-K, Tignor M, Allen SK, Boschung J, Nauels A, Xia Y, Bex V, Midgley PM (eds). *Climate Change 2013: The Physical Science Basis, Contribution of Working Group I to the Fifth Assessment Report of the Intergovernmental Panel on Climate Change.* Cambridge University Press, Cambridge, UK and New York.

CPCB (2008) Status of groundwater quality in India, part II, Groundwater quality series: GWQS/ 10/2007–2008. CPCB, Ministry of Environment and Forests, New Delhi.

Dains SR, Pawar JR (1987) Economic returns to irrigation in India. Report prepared by SDR Research Groups Inc. for the US Agency for International Development, New Delhi.

Das S (2011) *Groundwater Resources of India.* National Book Trust, New Delhi.

Dettinger M, Earman S (2007) Western groundwater and climate change – Pivotal to supply sustainability or vulnerable in its own right? *Ground Water News* 4:4–5.

Dewandel B, Gandolfi JM, de Condappa D, Ahmed S (2008) An efficient methodology for estimating irrigation return flow coefficients of irrigated crops at watershed and seasonal scales. *Hydrol Process* 22(11):1700–1712.

Dingman, S. L. (2014) *Physical Hydrology,* 3rd ed. Waveland Press, Long Grove, IL. ISBN: 978-1478611189.

Eckhardt E, Ulbrich U (2003) Potential impacts of climate change on groundwater recharge and streamflow in a central European low mountain range. *J Hydrol* 284(1–4):244–252.

Fauzia, Surinaidu L, Rahman A et al. (2021) Distributed groundwater recharge potentials assessment based on GIS model and its dynamics in the crystalline rocks of South India. *Sci Rep* 11:11772. https://doi.org/10.1038/s41598-021-90898-w

Gandhi VP, Bhamoriya V (2011) Groundwater irrigation in India: Growth, challenges, and risks. *India Infrastruct Rep* 90:91–117.

Glassley WE, Nitao JJ, Grant CW, Johnson JW, Steefel CI, Kercher JR (2003) The impact of climate change on vadose zone pore waters and its implication for long-term monitoring. *Comput Geosci* 29:399–411.

Glover PWJ, Hole MJ, Pous J (2000) A modified Archie's law for two conducting phases, *EPSL* 180:369–383.

Goldman M, Neubauer FM (1994) Groundwater exploration using integrated geophysical techniques. *Surv Geophys* 15:331–361. https://doi.org/10.1007/BF00665814

Goldman M, Rabinovich B, Rabinovich M, Gilad D, Gev I, Schirov M (1994) Application of integrated NMR-TDEM method in groundwater exploration in Israel. *J Appl Geophys* 31:27–52.

Gómez LS, Fuentes OE (2010) The application of electrical methods in exploration for ground water resources in the River Malacatoya sub-basin, Nicaragua. *Geofísica Internacional* 49(1):27–41.

Gosain AK, Rao S, Srinivasan R, Gopal Reddy N (2005) Return-flow assessment for irrigation command in the Palleru river basin using SWAT model. *Hydrol Process* 19:673–682. https://doi.org/10Ð1002/hyp.5622

Green TR (2016) Linking climate change and groundwater. In: Jakeman AJ et al. (eds). *Integrated Groundwater Management.* Springer, Cham, pp. 97–141. https://doi.org/ 10.1007/978-3-319-23576-9_5

Gun JV, Lipponen A (2010) Reconciling groundwater storage depletion due to pumping with sustainability. *Sustainability* 2(11):3418–3435. https://doi.org/10.3390/su2113418

Gurdak JJ, Hanson RT, McMahon PB, Bruce BW, McCray JE, Thyne GD, Reedy RC (2007) Climate variability controls on unsaturated water and chemical movement, high plains aquifer, USA. *Vadose Zone J* 6:533–547.

Gürer A, Bayraka M, Gürerb, ÖF (2009) A VLF survey using current gathering phenomena for tracing buried faults of Fethiye–Burdur Fault Zone, Turkey. *J Appl Geophys* 68:437–447. doi:10.1016/j.jappgeo.2009.03.011

Hanson RT, Dettinger MD, Newhouse V (2006) Relations between climatic variability and hydrologic time series from four alluvial basins across the southwestern United States. *Hydrogeol J* 14(7):1122–1146.

Holman IP (2006) Climate change impacts on groundwater recharge – Uncertainty, shortcomings, and the way forward? *Hydrogeol J* 14(5):637–647.

Hoque MA, Khan AA, Shamsudduha M, Hossain MS, Islam T, Chowdhury SH (2009) Near surface lithology and spatial variation of arsenic in the shallow groundwater: Southeastern Bangladesh. *Environ Geol* 56:1687–1695. https://doi.org/10.1007/s00254-008-1267-3Google Scholar http://hydrology-project.gov.in/SurfaceWater.html#Q1 http://iwrs.org.in/indias-water-resources/ https://svs.gsfc.nasa.gov/vis/a010000/a011700/a011784/IMERG_Final_Cut_v2_ipod_lg.m4v www.indiawaterportal.org/sites/default/files/iwp2/aquifer_systems_of_india_cgwb_2012.pdf

Hubbard SS, Linde N (2011) Hydrogeophysics. In: Wilderer P (ed). *Treatise on Water Science,* Vol. 3. Elsevier, Amsterdam, Netherlands, pp. 401–434. ISBN: 978-0444531995.

IPCC (2007) Understanding and attributing climate change. In: Solomon S et al. (eds). *Climate Change 2007: The Physical Science Basis, Contribution of Working Group I to the Fourth Assessment Report of the Intergovernmental Panel on Climate Change.* Cambridge University Press, Cambridge/New York.

IPCC (2013) Climate phenomena and their relevance for future regional climate change. In: Stocker TF, Qin D, Plattner G-K, Tignor M, Allen SK, Boschung J, Nauels A, Xia Y, Bex V, Midgley PM (eds). *Climate Change 2013: The Physical Science Basis, Contribution of Working Group I to the Fifth Assessment Report of the Intergovernmental Panel on Climate Change.* Cambridge University Press, Cambridge, UK and New York, pp. 1217–1308.

IPCC (2014) In: Barros VR, Field CB, Dokken DJ, Mastrandrea MD, Mach KJ, Bilir TE, Chatterjee M, Ebi KL, Estrada YO, Genova RC, Girma B, Kissel ES, Levy AN, MacCracken S, Mastrandrea PR, White LL (eds). *Climate Change 2014: Impacts, Adaptation, and Vulnerability. Part B: Regional Aspects, Contribution of Working Group II to the Fifth Assessment Report of the Intergovernmental Panel on Climate Change.* Cambridge University Press, Cambridge, UK and New York.

Jalota SK, Arora VK (2002) Model-based assessment of water balance components under different cropping systems in North-West India. *Agric Water Manage* 57:75–87.

Jeet I (2005) *Groundwater Resources of India: Occurrence, Utilization and Management.* Mittal Publications, New Delhi.

Jha BM (2007) Groundwater development and management strategies in India. *Bhu-Jal News* 22:1–15.

Jha MK, Chowdhury A, Chowdary VM, Peiffer S (2007) Groundwater management and development by integrated remote sensing and geographic information systems: Prospects and constraints. *Water Resour Manage* 21(2):427–467.

Johnson DG (1992) Use of Ground-Penetrating Radar for Water-Table Mapping, Brewster and Harwich, Massachusetts, U.S. Geological Survey Water-Resources Investigations Report 90–4086, p. 32.

Karl TR, Melillo JM, Peterson TC (2009) *Global Climate Change Impacts in the United States.* Cambridge University Press, New York.

Khan HH, Khan A (2019) Groundwater and surface water interaction (Chapter 14). In: Senapathi Venkatramanan, Viswanathan Prasanna M, Yong Chung S (eds). *GIS and Geostatistical Techniques for Groundwater Science.* Elsevier, Amsterdam, Netherlands, pp. 197–207.

Khan GD, Waheedullah, Bhatti AS (2013) Groundwater investigation by using resistivity survey in Peshawar, Pakistan. *J Resour Develop Manage Open Access Int J* 2:9–20.

Khullar DR (2005) *India: A Comprehensive Geography*. Kalyani, New Delhi.

Konikow LF (2011) Contribution of global groundwater depletion since 1900 to sea-level rise. *Geophys Res Lett* 38:L17401.

Kothawale DR, Rupa Kumar K (2005) On the recent changes in surface temperature trends over India. *Geophys Res Lett* 32:L18714. https://doi.org/10.1029/2005GL023528

Kumar CP (2014) *Impact of Climate Change on Groundwater Resources*. GRIN Verlag, Munich. www.grin.com/document/281606

Kumar KR, Sahai AK, Kumar KK, Patwardhan SK, Mishra PK, Revadekar JV, Kamala K, Pant GB (2006) High-resolution climate change scenarios for India for the 21st century. *Curr Sci* 90(3):334–345.

Kumar V, Jain SK (2010) Trends in rainfall amount and number of rainy days in river basins of India (1951–2004). *Hydrol Res* 42(4):290–306.

Kumar CP (2014) Impact of climate change on groundwater resources. GRIN Verlag, Munich. www.grin.com/document/281606

Kundzewicz ZW, Mata LJ, Arnell NW, Doll P, Kabat P, Jimenez B, Miller KA, Oki T, Sen Z, Shiklomanov AI (2007) Freshwater resources and their management. In: Parry ML, Canziani OF, Palutikof JP, van der Linden PJ, Hanson CE (eds). *Climate Change 2007: Impacts, Adaptation and Vulnerability*. Cambridge University Press, Cambridge.

Lal M (2001) Climatic change – Implications for India's water resources. *J Soc Econ Dev* III(1):57–87.

Machiwal D, Singh PK (2015) Understanding factors influencing groundwater levels in hardrock aquifer systems by using multivariate statistical techniques. *Environ Earth Sci* 74(7):5639–5652. https://doi.org/10.1007/s12665-015-4578-11

Maheswari K, Senthil Kumar P, Mysaiah D, Ratnamala K, Sri Hari Rao M, Seshunarayana T (2013) Ground penetrating radar for groundwater exploration in granitic terrains: A case study from Hyderabad. *J Geol Soc India* 81:781–790.

Majumdar P, Tiwari VM (eds.) *Water futures in India: Status of Science and Technology*. INSA, IISc Press, India.

Mall RK, Gupta A, Singh R, Singh RS, Rathore LS (2006) Water resources and climate change: An Indian perspective. *Curr Sci* 90(12):1610–1624.

Manap MA, Sulaiman WNA, Ramli MF, Pradhan B, Surip N (2013) A knowledge-driven GIS modeling technique for groundwater potential mapping at the Upper Langat Basin, Malaysia. *Arab J Geosci* 6(5):1621–1637.

Maréchal JC, Dewandel B, Ahmed S, Galeazzi L, Zaidi FK (2006) Combined estimation of specific yield and natural recharge in semi-arid groundwater basin with irrigated agriculture. *J Hydrol* 329: 281–293.

McNeill JD (1990) Use of electromagnetic methods for groundwater studies, geotechnical and environmental geophysics (Chapter 7). In: Ward SH (ed). *Geotechnical and Environmental Geophysics: Volume I – Review and Tutorial*. Society of Exploration Geophysicists, pp. 191–218. https://doi.org/10.1190/1.9781560802785.ch7

Meijerink, AMJ (1996) Remote sensing applications to hydrology: Groundwater. *Hydrol Sci J* 41(4):549–561. https://doi.org/10.1080/02626669609491525

Meinzen-Dick R (1996) *Groundwater Markets in Pakistan: Participation and Productivity*. International Food Policy Research Institute, Washington.

Milan Beres Jr, Haeni FP (1991) Application of ground-penetrating-radar methods in hydrogeologie studies. *Ground Water* 29(3):357–386. https://doi.org/10.1111/j.1745-6584.1991.tb00528.x

Mohamaden MII, Ehab D (2019) Application of electrical resistivity for groundwater exploration in Wadi Rahaba, Shalateen, Egypt. *NRIAG J Astron Geophys* 6(1):201–209.

Mohamaden MII, Hamouda AZ, Mansour S (2016) Application of electrical resistivity method for groundwater exploration at the Moghra area, Western Desert, Egypt. *Egypt J Aquatic Res* 42:261–268.

Moore GK (1982) Ground-water applications of remote sensing. U.S. Geological Survey Open-File Report 82–240, p. 61.

Mukherjee A, Bhanja SA (2019) An untold story of groundwater replenishment in India: Impact of longterm policy interventions. In: Singh A et al. (eds). *Water Governance: Challenges and Prospects*. Springer, Singapore. https://doi.org/10.1007/978-981-13-2700-1_11

Mukherjee A, Bhanja SN, Wada Y (2018) Groundwater depletion causing reduction of baseflow triggering Ganges river summer drying. *Sci Rep* 8:12049. https://doi.org/10.1038/s41598-018-30246-7

Mukherjee P, Singh CK, Mukherjee S (2012) Delineation of groundwater potential zones in arid region of India—a remote sensing and GIS approach. *Water Resour Manage* 26:2643–2672.

Mukhopadhyay S, Mandal AK (2021) Impact of climate change on groundwater resource of India: A geographical appraisal. In Islam MN, van Amstel A (eds). *India: Climate Change Impacts, Mitigation and Adaptation in Developing Countries. Springer Climate.* Springer, Cham. https://doi.org/10.1007/978-3-030-67865-4_6.

NITI Aayog (2019 August) Composite water management index (Report). Ministry of Jal Shakti and Ministry of Rural Development, p. 239.

NRC (1991) *Opportunities in the Hydrologic Sciences, in Committee on 'Opportunities in the Hydrologic Sciences' of National Research Council.* National Academy Press, Washington, DC.

Nonner JC (2014) Introduction to hydrogeology, the Qanat: An ancient technology still delivering water today. WaterHistory.org – Qanat.

Novicky O, Kasparek L, Uhlik J (2010) Vulnerability of groundwater resources in different hydrogeology conditions to climate change. In: Taniguchi M, Holman IP (eds). *Groundwater response to changing climate.* International Association of Hydrogeologists, CRC Press/Taylor & Francis, London.

Ohwoghere-Asuma O, Felix Chinyem I, Ejiro Aweto K, Iserhien-Emekeme R (2020) The use of very low-frequency electromagnetic survey in the mapping of groundwater condition in oporoza-gbamaratu area of the Niger Delta. *Appl Water Sci* 10:164. https://doi.org/10.1007/s13201-020-01244-w

Panwar S, Chakrapani GJ (2013) Climate change and its influence on groundwater resources. *Curr Sci* 105(1):37–46.

Pellerin L (2002) Applications of electrical and electromagnetic methods for environmental and geotechnical investigations. *Surveys Geophys* 23:101–132.

Rajaram K (2006) *Geography of India.* Spectrum Books, New Delhi.

Rana K, Shrestha V, Chimoriya R (2020). The effect of housing on health and challenges of demographic changes. *Global J Sci Frontier Res: Interdisciplinary*, 20(5):Version 1.0.

Ranjan SP, Kazama S, Sawamoto M (2006) Effects of climate and land use changes on groundwater resources in coastal aquifers. *J Environ Manag* 80(1):25–35.

Rathore LS, Attri SD, Jaswal AK (2013) State level climate change trends in India. Meteorological Monograph No. ESSO/IMD/EMRC/02/2013, India Meteorological Department.

Ravi Shankar K (1994) Groundwater exploration. *Proceedings of the 20th WEDC Conference on Affordable Water Supply and Sanitation*, Colombo, Sri Lanka, pp. 225–228.

Report of the Expert Group (2007) Ground water management and ownership. Planning Commission, Government of India, New Delhi.

Rodell M, Velicogna I, Famiglietti JS (2009) Satellite-based estimates of groundwater depletion in India. *Nature* 460:999–1002. https://doi.org/10.1038/nature08238

Salvi K, Kannan S, Ghosh S (2013) High-resolution multisite daily rainfall projections in India with statistical downscaling for climate change impacts assessment. *J Geophys Res Atmos* 118:3557–3578. https://doi.org/10.1002/jgrd.50280

Sarah S, Ahmed S, Boisson A, Violette S (2014) Projected groundwater balance as a state indicator for addressing sustainability and management challenges of overexploited crystalline aquifers. *J Hydrol* 519, Part B:1405–1419, ISSN 0022-1694, https://doi.org/10.1016/j.jhydrol.2014.09.016

Sarah S, Ahmed S, Violette S, de Marsily G (2021) Groundwater sustainability challenges revealed by quantification of contaminated groundwater volume and aquifer depletion in hard rock aquifer systems. *J Hydrol* 597(1): 126286, ISSN 0022-1694.

Schirov M., Legchenko A, Creer G. (1991) A new direct non-invasive groundwater detection technology for Australia. *Explor Geophys* 22(2):333–338. https://doi.org/10.1071/EG991333

Seidu J, Ewusi A, Kuma JSY (2019) Combined electrical resistivity imaging and electromagnetic survey for groundwater studies in the Tarkwa Mining Area, Ghana. *Ghana Mining J* 19(1): 29–41.

Semenov AG (1987) NMR Hydroscope for water prospecting. *Proceedings of the Seminar on Geotomography*, Indian Geophysical Union, Hyderabad, pp. 66–67.

Shah T (1993) *Groundwater Markets and Irrigation Development: Political Economy and Practical Policy.* Oxford University Press, Mumbai.

Shah T (2009) Climate change and groundwater: India's opportunities for mitigation and adaptation. *Environ Res Lett* 4(13):175–195.

Sharif MM, Singh VP (1999) Effect of climate change on sea water intrusion in coastal aquifers. *Hydrol Process* 13(8):1277–1287.

Sharma AK (2017) Applications of remote sensing in ground water exploration, geo-science division (GHCAG/EPSA), Space Applications Centre (ISRO), Ahmedabad, Personal Communication.

Shiklomanov IA (1999) World water resources and water use: Modern assessment and outlook for future. *Adv Water Sci* 10(3): 219–234.

Singh EJ (2015) Trace element analysis of groundwater in Manipur Valley with special reference to arsenic contamination (unpublished doctoral thesis). Assam University, Silchar, Assam, India.

Singh EJ, Singh NR, Gupta A (2017) Hydrochemistry of groundwater and quality assessment of Manipur Valley, Manipur, India. *Int Res J Environ Sci* 6(9):8–18.

Singh KB, Gajri PR, Arora VK (2001) Modelling the effects of soil and water management practices on the water balance and performance of rice. *Agric Water Manage* 49:77–95.

Singh RD, Kumar CP (2010) Impact of climate change on groundwater resources. Retrieved from www.researchgate.net/publication/215973855

Solomon S (2003) Remote Sensing and GIS: Applications for Groundwater Potential Assessment in Eritrea (Doctoral Thesis). Royal Institute of Technology, SE-100 44 Stockholm, Sweden, ISBN 91-7283-457-9, 147p.

Sukhija BS, Reddy DV, Nagabhushanam P (1998) Isotopic fingerprints of paleoclimates during the last 30,000 years in deep confined groundwaters of Southern India. *Quat Res* 50(3):252–260.

Taniguchi M, Burnett WC, Ness GD (2008) Integrated research on subsurface environments in Asian urban areas. *Sci Total Environ* 404:377–392.

Taylor CA, Stefan HG (2009) Shallow groundwater temperature response to climate change and urbanization. *J Hydrol* 375:601–612.

Tiwari VM, Ahmed S (2022) India's groundwater and its sustainability. In: Nayak S (guest ed). Special Volume on "Earth Science for Sustainable Development Goals". *J Ind Geophys Union* 26(4):315–335.

Todd DK (1980) *Groundwater Hydrology*, 2nd ed. Wiley India, New Delhi.

Todd DK, Mays LW (2005) *Groundwater Hydrology*, 3rd edn. Wiley, Hoboken.

Trushkin DV, Shushakov OA, Legchenko AV (1994) The potential of a noise reducing antenna for surface NMR groundwater surveys in the earth's magnetic field. *Geophys Prosp* 42:855–862.

UNESCO (1994) Groundwater, IHP humid tropics programme serial no. 8. UNESCO, Paris, pp. 29–30.

UNFCC (1992) United Nations framework convention on climate change. Article 1. https://unf ccc.int/files/essential_background/background_publications_htmlpdf/application/pdf/ conveng.pdf

Vita DP, Allocca V, Manna F, Fabbrocino S (2012) Coupled decadal variability of the North Atlantic Oscillation, regional rainfall and karst spring discharges in the Campania region (Southern Italy). *Hydrol Earth Syst Sci* 16:1389–1399.

Wahurwagh A, Dongre A (2015) Burhanpur cultural landscape conservation: Inspiring quality for sustainable regeneration. *Sustainability* 7(1):932–946. https://doi.org/10.3390/ su7010932

Wang X, Farooq SH, Keramat M, Lubis RF, Akhter MG (2013) Climate change vulnerability and adaptation in ground water dependent irrigation system in Asia-Pacific Region. Asia-Pacific Network for Global Change Research, Project Reference Number: ARCP2013-25NSY-Shahid.

Wani SP, Sudi R, Pathak P (2009) Sustainable groundwater development through integrated watershed management for food security. International Crops Research institute for the Semi Arid Tropics (ICRISAT), Patancheru.

Woldeamlak ST, Batelaan O, Smedt FD (2007) The effects of climate change on groundwater system in the Grote-Nete catchment, Belgium. *Hydrogeol J* 15(5):891–901.

2 Groundwater Vulnerability Assessment Using GIS-Based Water Quality Index (WQI) in Parts of Jamui and Nawada Districts, Bihar, India

*Navneet Ranjan, Akhouri Bishwapriya,
Gautam Chand Garg, Shambhu Nath,
Mili Singh and Abhishek Kumar Chaurasia*

2.1 INTRODUCTION

Water is a life sustaining and saving component present on Mother Earth. Available water on Earth includes 97% seawater and 3% freshwater. Within 3% freshwater, 30.1% is groundwater and is considered to be the main source of drinking water. However, with time, its scarcity and degrading quality have emerged as issues of concern globally. Geological contaminants and anthropogenic interference largely contribute to this problem.

Research in the field indicates that the elements (in ionic form) are responsible for contamination and pollution of groundwater. Some elements in traces are boon to human health, but the same in excess concentration are the curse for living beings. The economic boom that India has experienced in the past two decades has resulted in increased industrial and urban waste disposal problems. Urban growth, territorial migration of people from rural areas to urban centers, growth in industrial centers and likewise the generation of a copious volume of industrial effluents has accentuated the problem (Verma and Saxena, 2010).

In the developing world ~80% of all diseases in human beings are related to contaminants in water (United Nations, 2013). About 30% of urban and 90% of rural households still depend on untreated surface or groundwater (Palanisamy *et al.*, 2007). The current study attempts to assess the vulnerability of drinking water using

DOI: 10.1201/9781003485995-2

the water quality indexing system for groundwater in parts of Jamui and Nawada Districts, Bihar, India.

The water quality index (WQI) reflects the composite impact of various water quality parameters. Horton first used the concept of WQI (Horton, 1965) followed by Brown *et al.* (1972) and improved by Scottish Development Department (1975). WQI is a mathematical equation used to convert a large amount of water quality data into a single number. Various updations in WQI have also been incorporated in recent years (Nives *et al.*, 1999; Swamee *et al.*, 2000; Harrison *et al.*, 2000; Faisal *et al.*, 2003; Ahmad *et al.*, 2004; Shiow-Mey *et al.*, 2004, Ranjan *et. al.*, 2022).

The current study is to find the groundwater suitability for drinking purposes by determining the classes of all representative 52 samples using the weighted arithmetic index method as per Brown *et al.* (1972). Nineteen parameters such as pH, electrical conductivity (EC), total dissolved solid (TDS), calcium (Ca^{2+}), magnesium (Mg^{2+}), sodium (Na^+), potassium (K^+), bicarbonate (HCO_3^-), chlorides (Cl^-), nitrates (NO_3^-), sulfates (SO_4^{2-}), phosphate (PO_4^{3-}), fluoride (F^-), iron (Fe), cadmium (Cd), nickel (Ni), cobalt (Co), lead (Pb) and zinc (Zn) were analyzed and considered for water quality indexing.

2.2 STUDY AREA

The study area is bounded by latitude 24°50′ to 25°00′ N and longitude 85°45′ to 86°00′ E and dominantly covers the parts of Jamui and Nawada districts of Bihar. The Nawada district forms the upper north western part and the remaining part lies within the jurisdiction of the Jamui district. The northern part, shares the boundaries of the Sheikhpura and Lakhisarai district of Bihar, whilst the southern part falls in the territorial limits of Bihar-Jharkhand states (Figure 2.1).

The area comprises rocks of Chhotanagpur Gneissic Complex (CGC), Bihar Mica Belt (BMB) and the Quaternary sediments (Figure 2.2). Granite Gneiss, Biotite Gneiss, Coarse Grained Porphyritic Granite Gneiss represents CGC, whilst Quartzite, Gritty quartzite, Mica-Schist, Quartz-mica Schist belong to BMB, later intruded by younger granites, dolerites, amphibolites, pegmatite and quartz veins. The Quaternary sediments of Nawada and Diara Formations comprising Caliche rich clay and sand, silt, gravel occur in the northern plain tract of the area (Ghosh, 1980–81). The area exhibits undulating topography with low to moderately high ridges and isolated hillock bounded by the Ganga river towards north. In southern parts, hillocks of Bihar Mica Belt and CGC are exposed having altitude 200–500 m.

The pediplain and pediments follow the rocky upland from south to north along the gradient. The unconsolidated alluvial sediments and pediplain/peneplain possess good water retaining and yielding properties. The groundwater within the rocky upland is controlled by size, depth and degree of joints, bedding and fractures. The natural precipitation is the primary source of groundwater recharge within the ephemeral drainage system. Seepage from the canal network also contributes in replenishing groundwater.

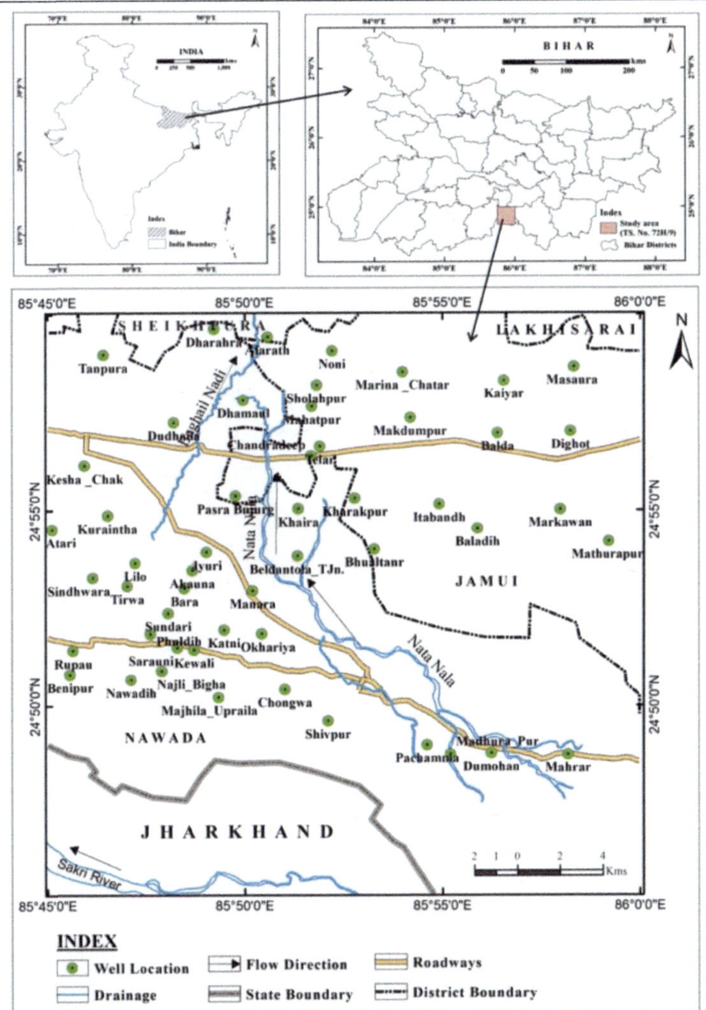

FIGURE 2.1   Map of the study area.

2.3   MATERIAL AND METHOD

A total of 52 nos. of groundwater samples were collected during pre-monsoon and post-monsoon, in the year 2018 from various representative locations of the study area representing a single common aquifer system (with spatial depth control) being used by local population as per the standard protocol prescribed by APHA (1995). The groundwater sample location points were marked by using the Global Positioning System (GPS). Groundwater sample location points are shown in Figure 2.1. In this study, the above mentioned 19 groundwater quality parameters were analyzed using

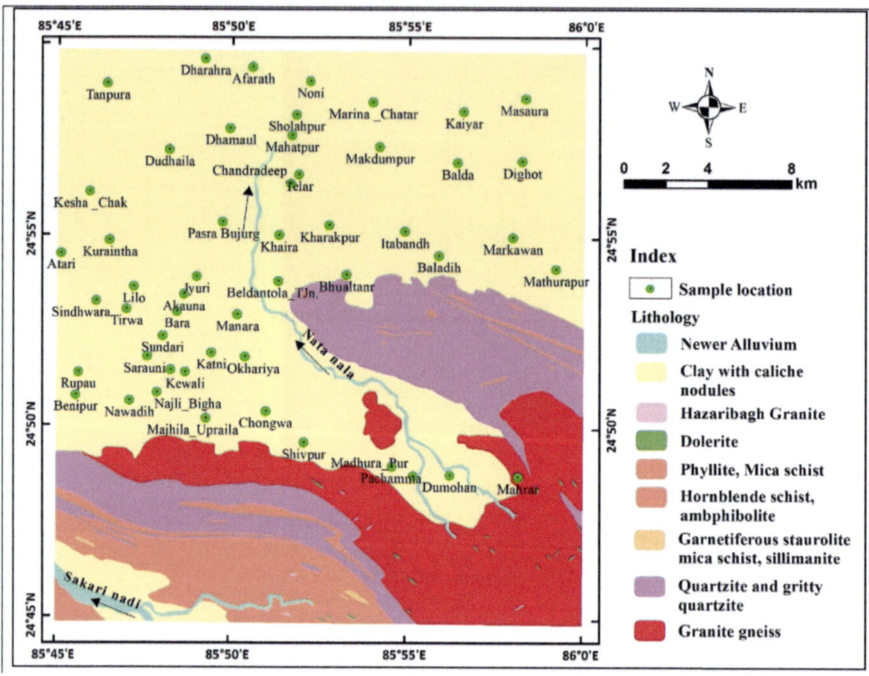

FIGURE 2.2   Lithological map of the study area showing sample locations.

the standard method prescribed by APHA (1995). The statistical analysis of the analyzed groundwater quality is given in Table 2.1.

2.4   WATER QUALITY INDEX (WQI)

In the present study, 19 important parameters were chosen to calculate the water quality index. The WQI has been calculated using the standards of drinking water quality recommended by WHO (2012 and BIS (2012). The Weighted Arithmetic Index method (APHA, 1995) has been used for the calculation of the WQI of the water.

Calculation of quality rating/su-index (q_n): quality rating is calculated by the following equation:

$$q_n = (V_n - V_{io}) / (S_n - V_{io}) \times 100 \tag{2.1}$$

(Let there be *n* water quality parameters and quality rating or sub-index (q_n) corresponding to the *n*th parameter is a number in the polluted water with respect to its standard permissible value.)

q_n = quality rating for the *n*th water quality parameter.

V_n = estimated value of the *n*th parameter at a given sampling station.

S_n = standard permissible value of the *n*th parameter.

TABLE 2.1
Statistical Analysis of Analyzed Physico-Chemical Groundwater Quality Parameter

Parameter	Minimum*	Maximum*	Mean*	Standard Deviation
pH	7.10	8.40	7.65	0.31
Electrical conductivity (µS/cm)	248	3070	955.10	577.40
Total dissolve solids (ppm)	145	1700	512.92	315.44
Calcium (ppm)	13	250	71.47	45.38
Magnesium (ppm)	00	162	39.08	29.17
Sodium (ppm)	16	458	87.28	98.92
Potassium (ppm)	00.10	115.65	3.86	15.87
Bicarbonate (ppm)	171	1820	520.77	300.01
Chloride (ppm)	07	240.80	52.52	54.92
Nitrate (ppm)	0.01	70.95	8.48	15.09
Sulfate (ppm)	1.80	65.80	11.24	15.59
Phosphate (ppm)	0.01	12	0.52	1.68
Fluoride (ppm)	0.32	3.91	1.26	0.73
Iron (ppb)	2.50	6220	615.32	1199.42
Cadmium (ppb)	2.50	60.00	20.24	20.69
Nickel (ppb)	2.50	320.00	82.81	92.60
Cobalt (ppb)	3.50	290.00	93.82	93.96
Lead (ppb)	2.50	670.00	164.10	183.72
Zinc (ppb)	2.50	996.00	176.81	235.39

* All the parameters expressed in ppm except pH and EC (micro S/cm at 25°C)

V_{io} = ideal value of the nth parameter in pure water (i.e., 0 for all other parameters, except the parameter pH {7}).

Calculation of unit weight (W_n): unit weight was calculated by a value inversely proportional to the recommended standard value S_n of the corresponding parameter:

$$W_n = K/S_n$$

W_n = unit weight for the nth parameter.
S_n = standard value for nth parameter.
K = constant for proportionality and it is calculated by using the expression given in equation:

$$K = [1 / (\Sigma 1/ S_n = 1, 2....n)] \tag{2.2}$$

The standard values of water quality parameters and their corresponding standard values and unit weights are given in Tables 2.2 and 2.3.

TABLE 2.2
Weight (w_i) and Relative Weight (W_i) of Each Parameter

Parameter	BIS Standard	Weightage (w_i)	Relative Weight (W_i)
pH	6.5–8.5	1	0.029411765
Electrical conductivity (µS/cm)	300	1	0.029411765
Total dissolve solids (ppm)	200–600	1	0.029411765
Calcium (ppm)	75–200	1	0.029411765
Magnesium (ppm)	30–100	1	0.029411765
Sodium (ppm)	200	1	0.029411765
Potassium (ppm)	10–12	1	0.029411765
Bicarbonate (ppm)	600	1	0.029411765
Chloride (ppm)	100	1	0.029411765
Nitrate (ppm)	45	2	0.058823529
Sulfate (ppm)	400	1	0.029411765
Phosphate (ppm)	50	1	0.029411765
Fluoride (ppm)	1.5	2	0.058823529
Iron (ppb)	3000	2	0.058823529
Cadmium (ppb)	5	4	0.117647059
Nickel (ppb)	20	4	0.117647059
Cobalt (ppb)	50	4	0.117647059
Lead (ppb)	10	4	0.117647059
Zinc (ppb)	5000	1	0.029411765
		$\Sigma w_i = 34$	$\Sigma W_i = 1$

The overall water quality index was calculated by aggregating the quality rating with the unit weight linearly:

$$WQI = \Sigma q_n \times W_n / \Sigma W_n$$

2.5 RESULTS AND DISCUSSION

Nineteen groundwater quality parameters, viz., pH, electrical conductivity (EC), total dissolved solid (TDS), calcium (Ca^{2+}), magnesium (Mg^{2+}), sodium (Na^+), potassium (K^+), bicarbonate (HCO_3^-), chlorides (Cl^-), nitrates (NO_3^-), sulfates (SO_4^{2-}), phosphate (PO_4^{3-}), fluorides (F^-), iron (Fe), cadmium (Cd), nickel (Ni), cobalt (Co), lead (Pb) and zinc (Zn) were analyzed to evaluate the WQI of groundwater for 52 representative locations representing the study area and assess the suitability of groundwater for drinking purpose under different class (Table 2.6).

    pH: water in a pure state has a neutral pH, which shows a concentration of hydrogen ion in water. The desired pH for drinking purpose is 6.5–8.5. In the current study, the pH ranges between 6.90 (minimum) to 7.91 (maximum) in pre-monsoon (Figure 2.3a) and 7.10 to 8.40 in the post-monsoon period, which indicate that it is within the permissible limit (6.5 to 8.5; BIS, 2012) during the pre- and

TABLE 2.3
Standard Values of Water Quality Parameters and their Corresponding Ideal Values and Unit Weights

Parameter	BIS (2012) Standard	K value	Unit Weight (W_n)
pH	6.5-8.5	0.087522	0.010296672
Electrical conductivity (µS/cm)	300	0.087522	0.00029174
Total dissolve solids (ppm)	200–600	0.087522	4.37609E-05
Calcium (ppm)	75–200	0.087522	0.00043761
Magnesium (ppm)	30–100	0.087522	0.00087522
Sodium (ppm)	200	0.087522	0.00043761
Potassium (ppm)	10–12	0.087522	0.007293
Bicarbonate (ppm)	600	0.087522	0.00014587
Chloride (ppm)	100	0.087522	8.75217E-05
Nitrate (ppm)	45	0.087522	0.00194493
Sulfate (ppm)	400	0.087522	0.0002188
Phosphate (ppm)	50	0.087522	0.001750434
Fluoride (ppm)	1.5	0.087522	0.05834781
Iron (ppb)	3000	0.087522	0.000292
Cadmium (ppb)	5	0.087522	0.029174
Nickel (ppb)	20	0.087522	0.004376
Cobalt (ppb)	50	0.087522	0.875217
Lead (ppb)	10	0.087522	0.008752
Zinc (ppb)	5000	0.087522	1.75043E–05
			$\Sigma W_n = 1$

post-monsoon (Figure 2.3a') observation period. The higher values of pH in the down-streamed western flank of Nata nala in the area are primarily influenced by the alkaline rocks exposed in the south and caliche enriched soil.

EC: specific conductance of groundwater varies widely, in chemically inert rocks in the area of abundant precipitation, to over 100,000 µS/cm in desert brines. Desirable limit of EC for drinking purpose is 300 µS/cm (BIS, 2012). In the current study, electrical conductivity ranges from 299 to 2170 µS/cm in pre-monsoon (Figure 2.3b), whereas, in the post-monsoon season it ranges from 240 to 3070 µS/cm (Figure 2.3b').

The higher values of EC may be due to regular discharge of charged particles in the groundwater form prolonged clay–water interactions (clay of Nawada Formation form a ubiquitous sequence), some anthropogenic activities (fertilizers, human excreta, coal ash from brick kiln) in the area, and excessive use of groundwater resources meeting the domestic and irrigational needs of the inhabitants.

TDS: The suitability of groundwater for any purpose is ascertained by classifying the groundwater depending upon its hydrochemical properties based on TDS values (Catroll, 1962; Freeze *et al.*, 1979). The principal constituents are usually calcium, magnesium, sodium, potassium, carbonate, bicarbonate, chloride and sulfate. According to the BIS (2012), the ideal TDS for drinking water is below 500 mg/l and the max permissible limit is 2000 mg/l. In the current study area, it varies from 180 to 1192 mg/l in the pre- (Figure 2.3c) and 145 to 1700 mg/l in the post-monsoon (Figure 2.3c') season. The TDS and EC are directly correlative entities. Thus, higher EC connotes higher TDS in groundwater. Besides, the spatial distributions of their values are comparatively similar on the map (Figure 2.3b-b' and 2.3c-c').

Calcium: the range of calcium content in the groundwater is largely dependent on the solubility of the calcium carbonates and varies widely with the partial pressure of carbon dioxide in the air in contact with the water. Due to their higher concentration, the water causes abdominal ailment and encrustation (Kumar *et al.*, 2014). In the present study, the calcium concentration ranges from 12 to 209 mg/l in pre- (Figure 2.3d) and 13 to 250 mg/l in post-monsoon (Figure 2.3d') seasons. The localities namely Sarauni (201 mg/l), Benipur (202 mg/l) and Kewali (209 mg/l) in pre-monsoon and Mariana Chatar (215 mg/l) and Jyuri (250 mg/l) in post-monsoon are found beyond the permissible limit (200 mg/l). The predominant succession of caliche nodule rich clay may be controlling the concentration of Ca^{2+} in groundwater.

Magnesium: water chemistry of magnesium is similar to that of calcium. The solubility of pure magnesium carbonate is much greater than that of calcium carbonate so that magnesium carbonate is not precipitated under normal temperatures (Davis *et al.*, 1966). Higher concentration of magnesium causes hardness of water. In the study area, the concentration ranges from 10.69 to 142 mg/l in the pre- (Figure 2.3e) and the below detection limit (BDL) to 160 mg/l in the post-monsoon (Figure 2.3e') season. The localities namely Phuldih (108 mg/l), Katni (139 mg/l) and Lilo (140 mg/l) in pre-monsoon and Jyuri (162 mg/l) in post-monsoon are found beyond the permissible limit (100 mg/l).

Sodium: all groundwater contains some sodium because most rocks and soils contain sodium compounds from which sodium is easily leached out. The permissible limit of sodium is 200 mg/l and it varies from BDL to 241.20 mg/l in the pre- (Figure 2.3f) and 16 to 458 mg/l in the post-monsoon (Figure 2.3f') season in the area of interest. The localities namely Katni (241 mg/l) from pre-monsoon and Benipur (273 mg/l), Phuldih (284.4 mg/l), Makdumpur (321 mg/l), Katni (444 mg/l) and Lilo (458 mg/l) are found beyond the permissible limit. Increased intake of sodium in drinking water may be problematic for people with hypertension, heart disease or kidney problems that require them to follow a low sodium diet. The instances of high Na^+ are observed in the southern part of the study area in contact with weathered granitic exposures of Chhotanagpur Gneissic Complex.

Potassium: potassium is common in many rocks. Many of these rocks are relatively soluble and potassium concentrations in groundwater increase with time. In the present study area it varies from 0.07 to 4.80 mg/l in pre- (Figure 2.3g) and 0.10 to 115.7 mg/l in post-monsoon (Figure 2.3g'). At Phuldih, its concentration (115.7 mg/l) is found beyond the permissible limit (12 mg/l).

Bicarbonate: action of carbon dioxide in water on carbonate rocks such as limestone and dolomite, results in formation of bicarbonate (HCO_3^-), producing an alkaline environment. Its permissible limit is 600 mg/l. It ranges in the study area from 216 to 1049 mg/l in the pre-monsoon (Figure 2.3h) and 171 to 1820 mg/l in the post-monsoon season (Figure 2.3h'). Its higher concentrations observed in the area may be a result of dissociation of carbonic acid and further reactions with silicates and carbonates present in the rocks of CGC, BMB, leached out in Quaternaries as well.

Chloride: high concentration of chloride shows an indication of pollution due to high organic waste of animal origin (Singh, 1995). The higher concentration of chlorine in water made it hazardous to human health, which is subjected to laxative effects (Anitha *et al.*, 2011; Sadat-Noori *et al.*, 2014). In the study area it varies from 11 to 213.47 mg/l in the pre- (Figure 2.3i) and 7 to 240.80 mg/l in the post-monsoon season (Figure 2.3i'), which are within the permissible limit (1000 mg/l).

Nitrate: higher concentration of nitrate causes Blue Baby Syndrome, gastric cancer, thyroid disease and diabetes, beyond the permissible limit (45 mg/l) as per BIS, 2012 (Krishna Kumar *et al.*, 2011; Comly, 1945; Gilly *et al.*, 1984). Nitrate concentration in the area ranges from 0.09 to 47.69 mg/l in the pre- (Figure 2.3j) and 0.01 to 68.83 mg/l in the post-monsoon season (Figure 2.3j'). In the area the causative source of nitrate is likely anthropogenic (use of fertilizer and stagnant untreated sewage water).

Sulfate: sulfates (SO_4^{2-}) come from rocks containing gypsum, iron sulfides and other sulfur compounds. Mainly present in mine water and in industrial wastes. In the study area it ranges from 2.50 to 347 mg/l in the pre-monsoon (Figure 2.3k) and 1.80 to 65.80 mg/l in the post-monsoon season (Figure 2.3k'), which is under the permissible limit (400 mg/l).

Phosphate: phosphorus and its oxides are vital for the growth of the biological system. Phosphate enters into the natural water system from human and animal waste, phosphorus-rich bedrock, cleaning waste, industrial effluent and fertilizer runoff. Phosphate is not regulated by the WHO or BIS or U.S. Environmental Protection Agency (EPA) because it is not toxic to humans or animals and it is not considered a problematic chemical in drinking water (National Science Foundation, 2004). Its concentration in groundwater ranges from 0.01 to 2.13 mg/l in the pre-monsoon (Figure 2.3l) and 0.01 to 12 mg/l in the post-monsoon season (Figure 2.3l').

Fluoride: fluoride is one of the main trace elements in groundwater, which generally occurs as a natural constituent. Main source of fluoride in water is geogenic. Very low doses of fluoride (<1 mg/l) in water enhances tooth decay. When consumed in higher doses (>1.5 mg/l), it leads to dental fluorosis or mottled enamel and excessively high concentration (>3.0 mg/l) of F^- lead to skeletal fluorosis (Janardhana Raju *et al.*, 2009). In the study area the F^- concentration ranges from 0.58 to 6.07 mg/l in the pre-monsoon (Figure 2.3m) and 0.32 to 3.91 mg/l in the post-monsoon season (Figure 2.3m'). The highest values of F^-, i.e., 5.20 mg/l from Tanpura and 6.07 mg/l from Chongwa are situated in the north-west and southern parts of the study area where the alkaline granitic rocks are exposed, which in general contains the principal fluoride like apatite, fluorite, amphiboles, micas, etc. (GSI, 1968).

Iron: it is an essential element in the human body (Moore, 1973) and is found in groundwater all over the world; higher concentrations of iron cause a bad taste, discoloration, staining, turbidity, aesthetic and operational problem in water supply systems (Dart, 1974; Vigneswaran *et al.*, 1995). In the current area the iron ranges from 0.075 to 5.590 mg/l in pre-monsoon (Figure 2.3n) and 0.0025 to 6.220 mg/l in post-monsoon (Figure 2.3n').

Cadmium: it found in low concentrations in rocks, coal and petroleum and enters the groundwater and surface water when dissolved with acidic waters. In the study area, it is below the detection limit in pre-monsoon (Figure 2.3o) groundwater samples, while it ranges from 0.0025 to 0.060 mg/l in post-monsoon (Figure 2.3o') groundwater samples, which cross the permissible limit (0.003 mg/l) in villages (Nawadih, Rupau, Benipur etc.), which lie along the granitoid exposures of Chhotanagpur Gneisses Complex. Due to increased uptake of cadmium, it damages the liver and kidney and increases the blood pressure (Sawyer and McCartly, 1967).

Nickel: the primary source of nickel in drinking water is leaching from metals in contact with drinking water, such as pipes and fittings. In the present study area, it ranges from 0.005 to 0.180 mg/l in pre-monsoon (Figure 2.3p) to 0.02 to 0.32 mg/l in post-monsoon (Figure 2.3p'). The localities where it extends beyond the permissible limit (0.02 mg/l) include Benpur, Nawadih, Bhualtanr, etc., which are located in proximity to rocks of BMB and CGC. These areas have extensive cultivable land where use of fertilizers for crop production cannot be ruled out. Increased nickel concentrations in groundwater and municipal tap water (0.1–2.5 mg/l) in polluted and natural nickel mobilized areas have been reported (McNeely *et al.*, 1972; Hopfer *et al.*, 1989).

Cobalt: it occurs in rocks, soil and waters in very low concentrations and is an essential element of Vitamin B12. In the present context, it ranges from <0.007 mg/l to 0.040 mg/l in pre-monsoon (Figure 2.3q) and <0.007 mg/l to 0.280 mg/l in post-monsoon (Figure 2.3q'). Its higher concentration was recorded in the villages, Benipur, Nawadih, Rupau, Katani, which lie along the granitoid exposures of Chhotanagpur Gneisses Complex (southern part), within cultivable lands and close to brick kilns.

Lead: Pb poisoning is accompanied by symptoms of intestinal cramps, peripheral nerve paralysis, anemia and fatigue (Umar *et al.*, 2001). In the area of interest, it ranges from 0.0025 to 0.21 mg/l in pre-monsoon (Figure 2.3r) to 0.0025 to 0.620 mg/l in post-monsoon groundwater samples (Figure 2.3r'), which are above the permissible limit (0.01 mg/l). These areas lie to the north-east where the rocks of Bihar Mica Belt are exposed and there is a predominance of agricultural activity and several brick kilns.

Zinc: zinc is a transition metal, which imparts an undesirable astringent taste to water at concentrations exceeding 3 mg/l (as $ZnSO_4$). However, drinking water seldom contains zinc above 0.1 mg/l. In the present study area it ranges from 0.0 to 1.73 mg/l in pre-monsoon (Figure 2.3s) and 0.002 to 0.996 mg/l in post-monsoon (Figure 2.3s') groundwater samples under the permissible limit (15 mg/l).

The correlation matrices for the different 19 water quality parameters have been analyzed to assess their inter-elemental relationship (Tables 2.4 and 2.5). This

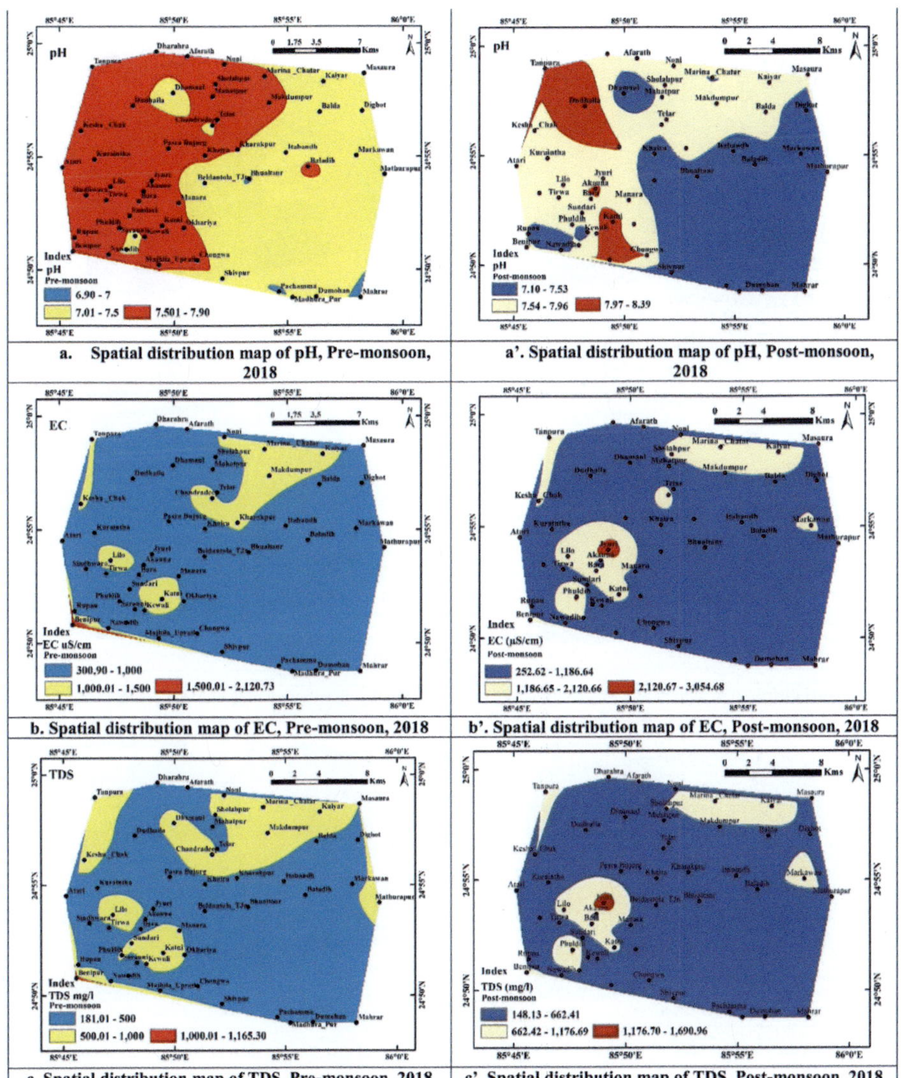

FIGURE 2.3 Spatial distribution map of pH, EC, TDS, Ca, Mg, Na, K,, HCO_3, Cl, NO_3, SO_4, PO_4, F, Fe, Cd, Ni, Co, Pb and Zn for pre- and post-monsoon, 2018. (a) Spatial distribution map of pH, pre-monsoon, 2018. (a') Spatial distribution map of pH, post-monsoon, 2018. (b) Spatial distribution map of EC, pre-monsoon, 2018. (b') Spatial distribution map of EC, post-monsoon, 2018. (c) Spatial distribution map of TDS, pre-monsoon, 2018. (c') Spatial distribution map of TDS, post-vulnsoon, 2018. (d) Spatial distribution map of Ca, pre-monsoon, 2018. (d') Spatial distribution map of Ca, post-monsoon, 2018. (e) Spatial distribution map of Mg, pre-monsoon, 2018. (e') Spatial distribution map of Mg, post-monsoon, 2018. (f) Spatial distribution map of Na, pre-monsoon, 2018. (f') Spatial distribution map of Na, post-monsoon, 2018. (g) Spatial distribution map of K, pre-monsoon, 2018. (g') Spatial distribution map of K, post-monsoon, 2018. (h) Spatial distribution map of HCO_3, pre-monsoon, 2018. (h') Spatial distribution map of HCO_3, post-monsoon, 2018. (i) Spatial distribution map of Cl, pre-monsoon, 2018. (i') Spatial distribution map of Cl, post-monsoon, 2018. (j) Spatial distribution map of NO_3, pre-monsoon, 2018. (j') Spatial

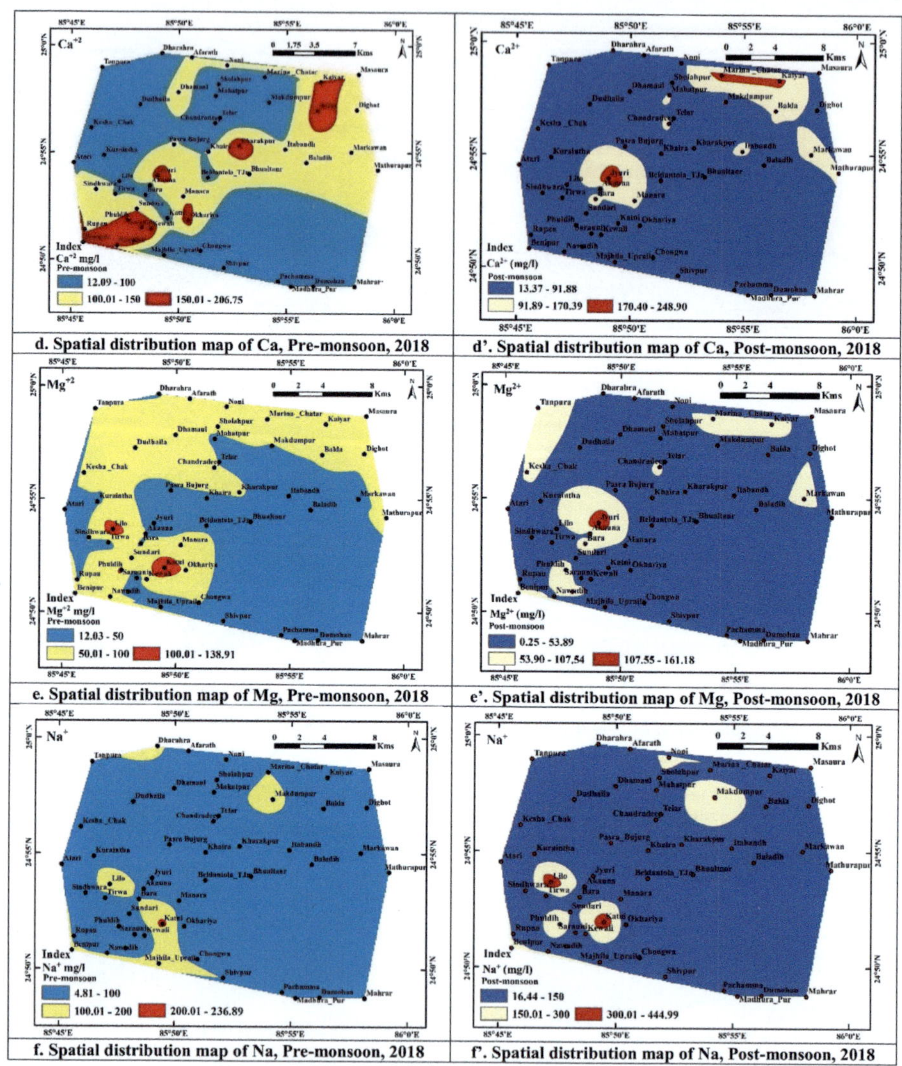

FIGURE 2.3 Continued

Caption for Figure 2.3 Continued

distribution map of NO_3, post-monsoon, 2018. (k) Spatial distribution map of SO_4, pre-monsoon, 2018. (k') Spatial distribution map of SO_4, post-monsoon, 2018. (l) Spatial distribution map of PO_4, pre-monsoon, 2018. (l') Spatial distribution map of PO_4, post-monsoon, 2018. (m) Spatial distribution map of F, pre-monsoon, 2018. (m') Spatial distribution map of F, post-monsoon, 2018. (n) Spatial distribution map of Fe, pre-monsoon, 2018. (n') Spatial distribution map of Fe, post-monsoon, 2018. (o) Spatial distribution map of Cd, pre-monsoon, 2018. (o') Spatial distribution map of Cd, post-monsoon, 2018. (p) Spatial distribution map of Ni, pre-monsoon, 2018. (p') Spatial distribution map of Ni, post-monsoon, 2018. (q) Spatial distribution map of Co, pre-monsoon, 2018. (q') Spatial distribution map of Co, post-monsoon, 2018. (r) Spatial distribution map of Pb, pre-monsoon, 2018. (r') Spatial distribution map of Pb, post-monsoon, 2018. (s) Spatial distribution map of Zn, pre-monsoon, 2018. (s') Spatial distribution map of Zn, post-monsoon, 2018.

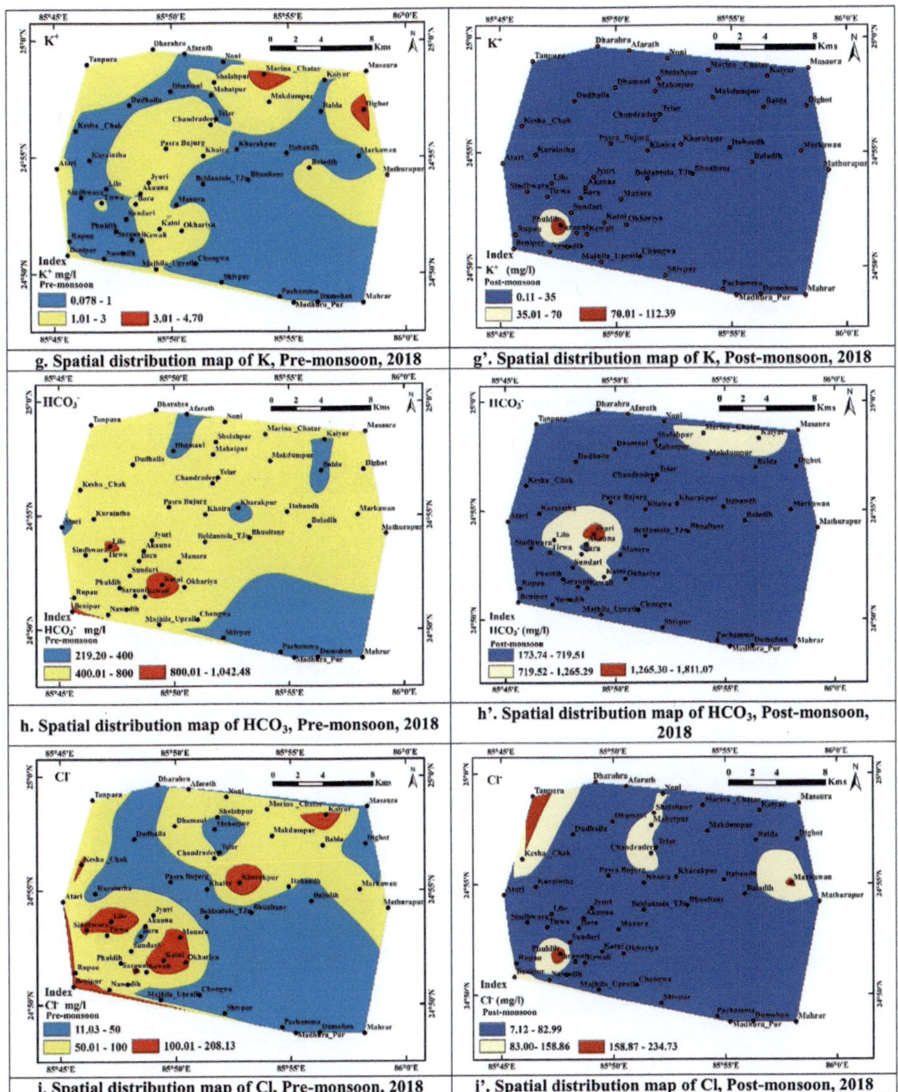

g. Spatial distribution map of K, Pre-monsoon, 2018

g'. Spatial distribution map of K, Post-monsoon, 2018

h. Spatial distribution map of HCO$_3$, Pre-monsoon, 2018

h'. Spatial distribution map of HCO$_3$, Post-monsoon, 2018

i. Spatial distribution map of Cl, Pre-monsoon, 2018

i'. Spatial distribution map of Cl, Post-monsoon, 2018

FIGURE 2.3 Continued

correlation study shows that in the pre-monsoon period nickel-calcium (0.53) and lead-cobalt (0.58) exhibit intermediate positive correlation, whereas, in the post-monsoon period cadmium exhibits positive correlation with nickel (0.93), cobalt (0.96) and lead (0.95). Besides, nickel possesses positive correlation with cobalt (0.91). Also among major ions, F$^-$ exhibits moderate positive correlation with pH (0.34) and Na$^+$ (0.39) and negative correlation with Ca^{+2} (–0.33) in the pre-monsoon season, whereas in both the seasons, NO$_3^-$ possess good to moderate correlations with Cl$^-$ (0.75) and SO$_4^{2-}$ (0.41), however, NO$_3^-$ and SO$_4^{2-}$ both exhibit weak positive correlation (0.11 to 0.13) with heavy metals like Ni, Co and Pb.

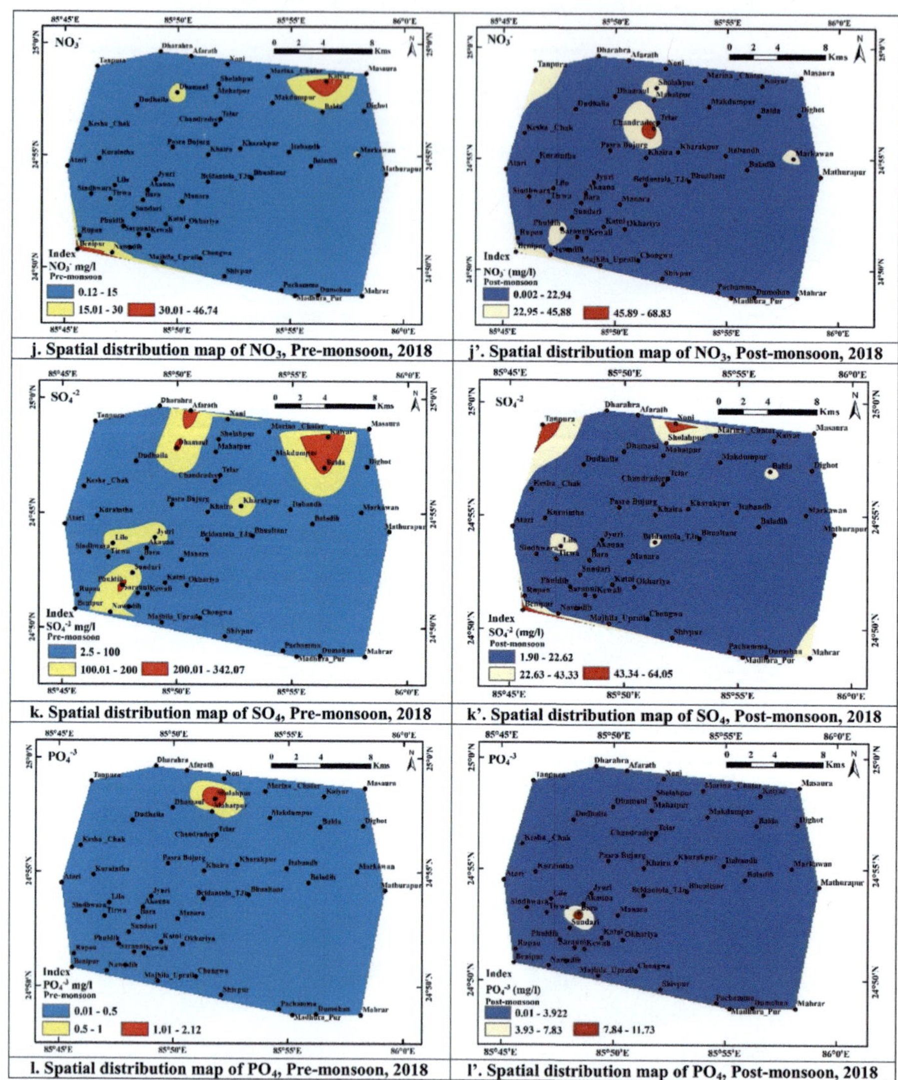

j. Spatial distribution map of NO₃, Pre-monsoon, 2018

j'. Spatial distribution map of NO₃, Post-monsoon, 2018

k. Spatial distribution map of SO₄, Pre-monsoon, 2018

k'. Spatial distribution map of SO₄, Post-monsoon, 2018

l. Spatial distribution map of PO₄, Pre-monsoon, 2018

l'. Spatial distribution map of PO₄, Post-monsoon, 2018

FIGURE 2.3　Continued

The spatial zonation maps have been prepared for each of the 19 quality parameters for understanding their spatial distribution pattern on Euclidian surface, as shown in Figure 2.2. These maps show that the SW part of the study area is affected by the higher concentration of Ca, Cl, F, Ni and Mg. The SE part of the study area is controlled by high concentrations of Cd, Ni and F value. The NW part of the study area is influenced by the high concentration of F, Co and pH. The central part of the study area is affected by the high concentrations of Ca and Pb in groundwater, while the NE portion is influenced by Co and K.

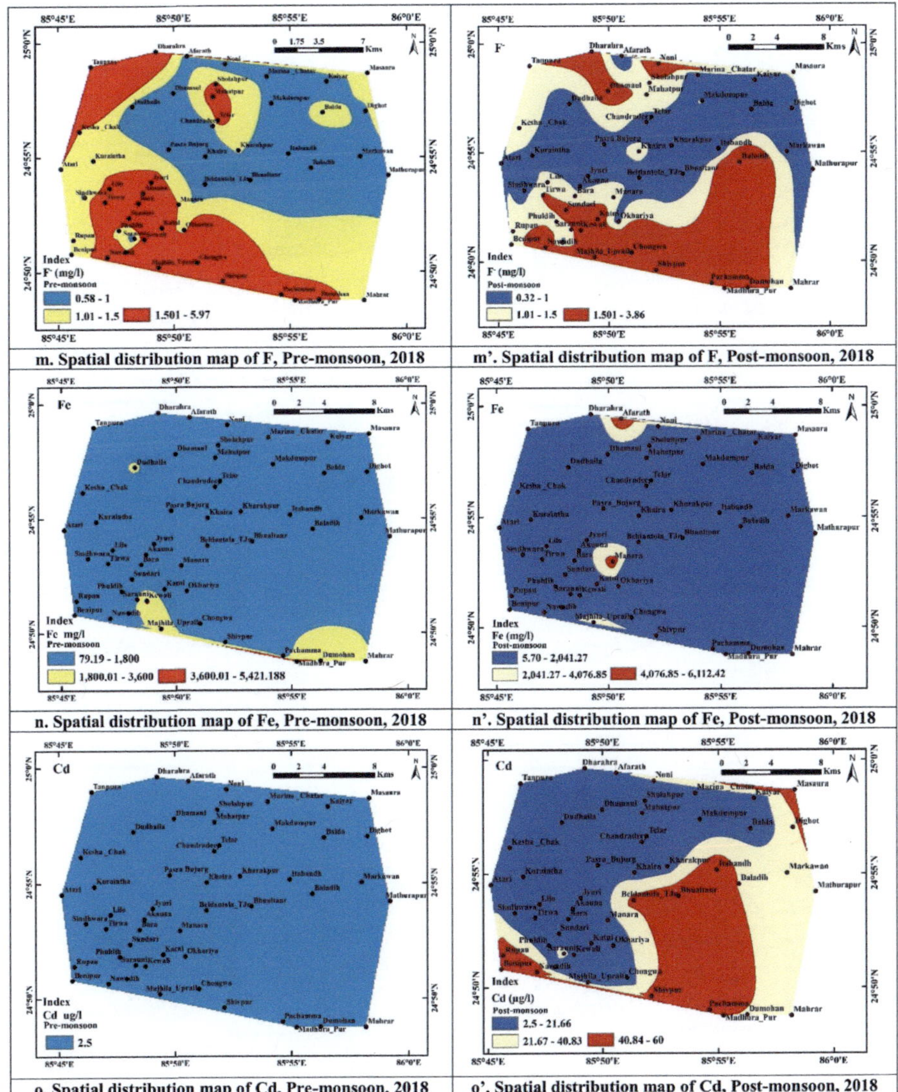

m. Spatial distribution map of F, Pre-monsoon, 2018

m'. Spatial distribution map of F, Post-monsoon, 2018

n. Spatial distribution map of Fe, Pre-monsoon, 2018

n'. Spatial distribution map of Fe, Post-monsoon, 2018

o. Spatial distribution map of Cd, Pre-monsoon, 2018

o'. Spatial distribution map of Cd, Post-monsoon, 2018

FIGURE 2.3 Continued

The WQI zonation maps have been prepared by using the WQI values as the outcome of their calculation for all 19 water quality parameters (Figure 2.3), which clearly depict the suitability pattern of the drinking water in pre- as well as in post-monsoon seasons (Figure 2.4). The type of groundwater vis-a-vis WQI has been tabulated as Table 2.6.

The WQI index has been calculated for the following 52 representative locations from unconfined aquifer range of 10–60 m, which is extensively exploited for the purpose of drinking and domestic purposes. The same has been tabulated in Table 2.7.

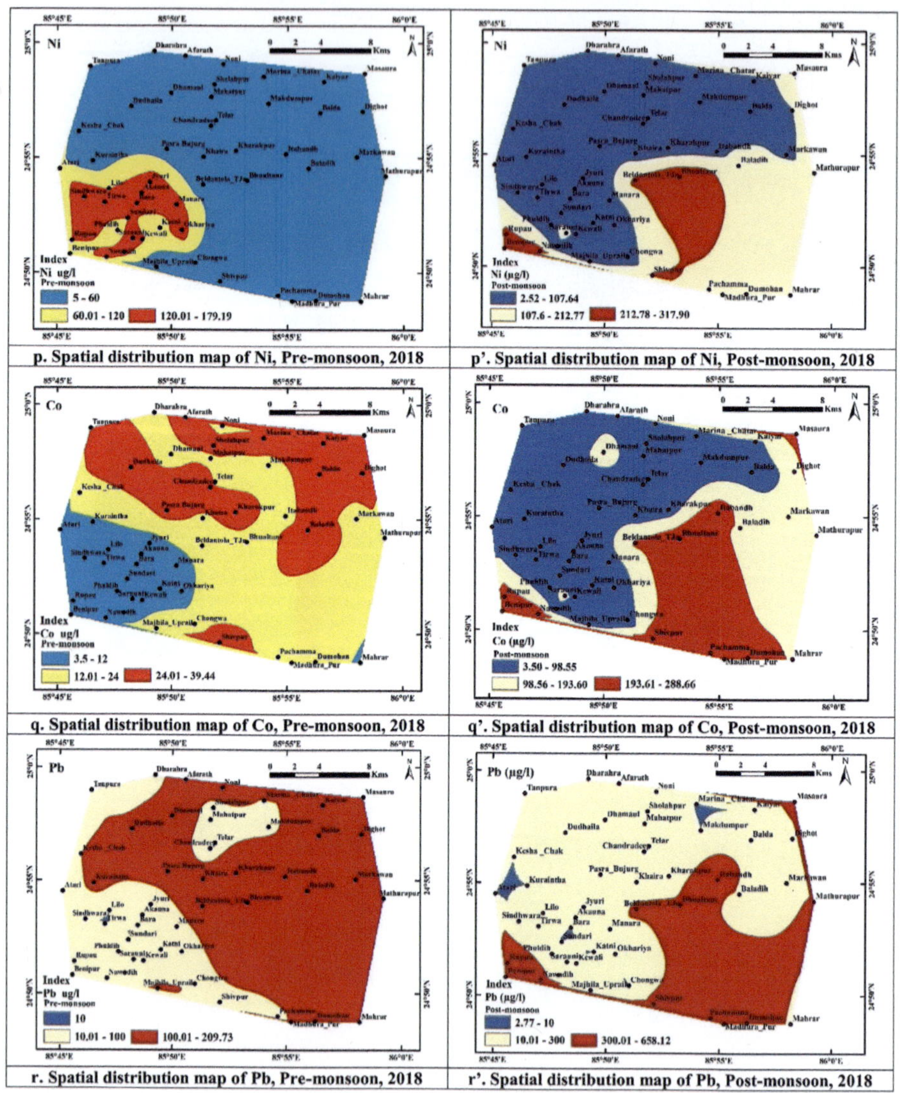

FIGURE 2.3 Continued

WQI is generally used for the detection and evaluation of water pollution and it may be defined as the reflection of composite influence of different quality parameters on the overall quality of water for drinking purposes. This study reveals that >50 % of the study area falls under the category of "not suitable for drinking" to "very poor" with regard to the drinking purposes specially during the post-monsoon season assessment.

The water quality index (WQI) maps (Figure 2.5a,b) have been prepared for the study area for the period of 2018. The spatial distribution map of the WQI clearly

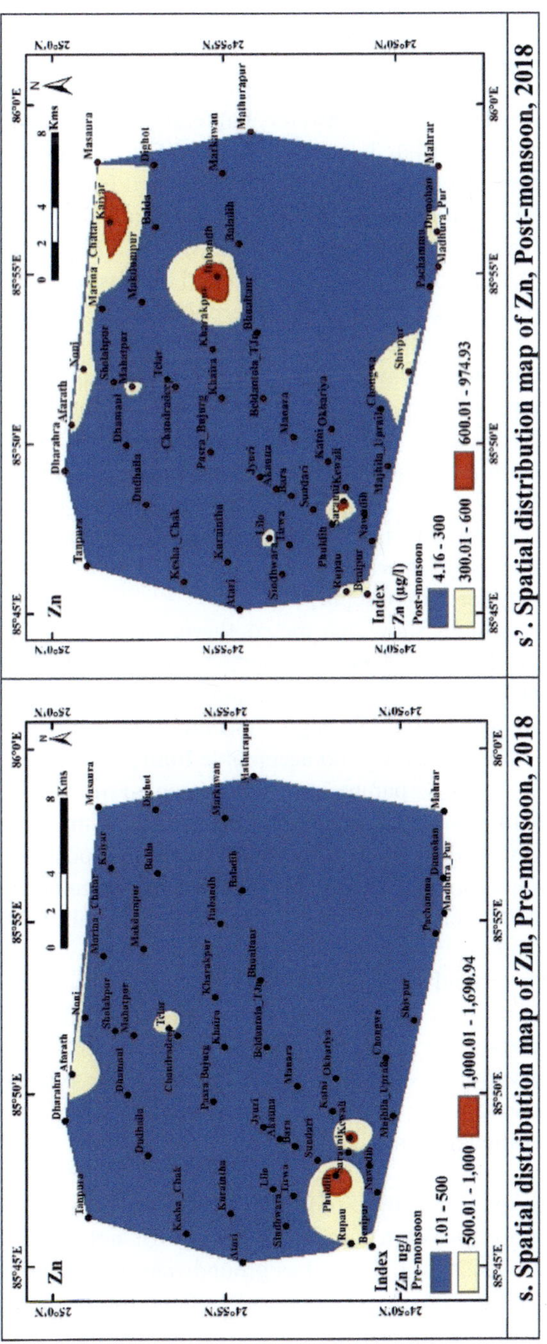

FIGURE 2.3 Continued

shows that the south-eastern part of the study area shows critical contamination with heavy metals, viz., Ni, Co, Cd and Pb.

2.6 CONCLUSIONS

For assessment of qualitative groundwater vulnerability, the analysis and GIS-framework-based GQI model has been implemented for 52 representative groundwater sample locations picked from parts of the Jamui and Nawada districts of Bihar in order to match their qualities as per the BIS standards (2012).

Initially, 19 quality parameters (pH, EC, TDS, Ca^{2+}, Mg^{2+}, Na^+, K^+, HCO_3^-, Cl^-, NO_3^-, SO_4^{2-}, PO_4^{3-}, F^-, Fe, Cd, Ni, Co, Pb and Zn) were analyzed to develop "spatial distribution patterns" using IDW interpolation (geo-statistical analyses). The hydro-geochemistry based on these parameters showed that the concentration of bicarbonate ions in groundwater is far more than other anions (Figure 2.5a,b) that makes the groundwater quality dominantly alkaline, resulting in enhanced concentration of alkali metal (Na^+) over alkali earth metal (Ca^{2+}) in pre- to post-monsoon season (Figure 2.5c,d) owing to ion-exchange reaction. The weak positive correlation among $NO_3^- + SO_4^{2-}$ and Pb, Ni and Co analyses and the sympathetic escalation in their concentrations in pre- to post-monsoon seasons (Figure 2.5a,b and e,f) attest the adverse effect of prevailing anthropogenic activities adversely affecting the groundwater quality for drinking purposes.

The spatial zonation maps show that the southern and north-western parts of the areas are vulnerable to F^- contamination, whereas, south-eastern part is affected by high concentration of Pb, Cd and Ni; north-western part by Co; and the central and north-eastern parts are affected by the high concentrations of Pb and Co in groundwater, which are very hazardous beyond acceptable limits.

Thus, using such "quality parameters" implemented using the GQI tool by assigning each of them a weightage through "decision-making" and validating our knowledge as per their affectivities on the human health appears to be a reasoned approach towards validating such problems. Later, such a mathematical algorithm was used to calculate the "relative weight", "unit weight", "quality rating" and finally compute the "GQI" values.

Accordingly, each groundwater sample was recognized as "excellent," "good", "poor", "very poor" or "not suitable" for drinking as per its "GQI" value. On deciphering the GQI zonation maps, it is indicative that in the pre-monsoon season the groundwater falls within "excellent to good" quality. In the post-monsoon season, due to the role of flowing water through enhanced surficial runoff, precipitation and replenishment of existing aquifers accentuates concentrations of F^-, NO_3^-, Pb, Ni, Co and Cd.

Analytical results show that the ~68% part of the study area (southern, central and south-eastern to north-eastern) are found "poor" to "not suitable" for drinking and domestic purposes. 32% of the area has groundwater representing aquifer depth (10.67 to 61 m) lies in the category of good to excellent, for drinking and daily need of the habitants.

There is a conspicuous shift in water quality in the pre-monsoon observation period from excellent and good category to poor, very poor and unsuitable for

TABLE 2.4
Correlation Matrix Between Various Groundwater Quality Parameters (Pre-Monsoon, 2018)

Parameter	pH	EC	TDS	Ca	Mg	Na	K	HCO_3	Cl	NO_3	SO_4	PO_4	F	Fe	Cd	*Ni*	*Co*	*Pb*	*Zn*
pH	1.00																		
EC	0.36	1.00																	
TDS	0.33	0.99	1.00																
Ca	−0.02	0.30	0.24	1.00															
Mg	0.40	**0.75**	**0.75**	0.21	1.00														
Na	0.44	0.43	0.45	−0.39	0.37	1.00													
K	0.12	0.22	0.25	−0.09	0.18	0.19	1.00												
HCO_3	0.49	**0.70**	**0.71**	0.20	**0.70**	0.55	0.36	1.00											
Cl	0.30	**0.77**	**0.73**	0.50	**0.62**	0.35	0.05	0.48	1.00										
NO_3	−0.10	**0.53**	**0.52**	0.53	0.34	0.00	−0.05	0.11	0.55	1.00									
SO_4	0.06	0.30	0.28	0.45	0.43	−0.08	−0.08	−0.05	0.44	0.55	1.00								
PO_4	0.14	−0.03	−0.01	−0.31	0.01	0.17	0.05	0.00	−0.06	−0.16	−0.10	1.00							
F	0.34	−0.12	−0.10	−0.33	−0.04	0.39	−0.04	0.03	−0.12	−0.18	−0.12	0.17	1.00						
Fe	−0.25	−0.31	−0.30	−0.08	−0.23	−0.08	0.00	−0.13	−0.22	−0.16	−0.18	−0.02	0.10	1.00					
Cd	0.00	0.00	0.00	0.00	0.00	0.00	0.00	0.00	0.00	0.00	0.00	0.00	0.00	0.00	1.00				
Ni	0.41	0.16	0.13	0.52	0.19	0.14	0.01	0.38	0.44	0.11	0.14	-0.02	0.07	0.03	0.00	1.00			
Co	−0.37	−0.10	−0.07	−0.38	−0.11	−0.26	0.23	−0.33	−0.36	−0.06	−0.01	0.11	−0.12	−0.10	0.00	**−0.83**	1.00		
Pb	−0.52	−0.15	−0.17	0.05	−0.20	−0.51	−0.18	−0.41	−0.29	0.11	0.02	−0.30	−0.35	−0.03	0.00	**−0.74**	0.58	1.00	
Zn	0.23	0.06	0.05	0.28	0.21	−0.11	−0.10	0.18	0.08	−0.04	0.16	−0.06	−0.04	0.36	0.00	0.21	−0.25	−0.07	1.00

TABLE 2.5
Correlation Matrix Between Various Groundwater Quality Parameters (Post-Monsoon, 2018)

Parameter	pH	EC	TDS	Ca	Mg	Na	K	HCO_3	Cl	NO_3	SO_4	PO_4	F	Fe	Cd	*Ni*	*Co*	*Pb*	*Zn*
pH	1.00																		
EC	0.17	1.00																	
TDS	0.11	**0.98**	1.00																
Ca	−0.10	**0.58**	**0.60**	1.00															
Mg	0.15	**0.81**	**0.84**	**0.60**	1.00														
Na	0.27	0.46	0.41	−0.30	0.09	1.00													
K	−0.03	0.36	0.37	−0.03	0.23	0.29	1.00												
HCO_3	0.15	**0.90**	**0.88**	**0.67**	**0.74**	0.36	0.10	1.00											
Cl	0.14	0.42	0.41	−0.07	0.35	0.43	0.51	0.03	1.00										
NO_3	−0.02	0.31	0.31	−0.06	0.27	0.26	0.33	−0.03	0.75	1.00									
SO_4	0.06	0.31	0.33	−0.05	0.18	0.42	0.05	0.15	0.45	0.41	1.00								
PO_4	0.07	0.12	0.13	0.13	0.16	−0.06	0.00	0.15	−0.10	−0.06	−0.02	1.00							
F	0.12	−0.15	−0.15	−0.24	−0.28	0.15	−0.02	−0.16	−0.03	−0.07	0.05	−0.01	1.00						
Fe	−0.02	−0.13	−0.14	0.07	−0.12	−0.12	−0.07	−0.09	−0.05	−0.12	−0.11	−0.09	−0.03	1.00					
Cd	−0.58	−0.36	−0.30	−0.19	−0.36	−0.09	−0.12	−0.31	−0.10	0.07	0.13	−0.10	0.02	0.06	1.00				
Ni	−0.55	−0.38	−0.32	−0.24	−0.34	−0.10	−0.11	−0.33	−0.12	0.10	0.13	−0.10	0.07	0.02	**0.93**	1.00			
Co	−0.62	−0.34	−0.26	−0.09	−0.30	−0.18	−0.12	−0.28	−0.12	0.11	0.11	−0.10	0.00	0.05	**0.96**	**0.91**	1.00		
Pb	−0.53	−0.35	−0.31	−0.17	−0.36	−0.09	−0.08	−0.31	−0.07	0.11	0.11	−0.11	0.00	0.06	**0.95**	**0.86**	**0.93**	1.00	
Zn	−0.20	0.06	0.08	0.18	−0.09	0.02	0.07	0.06	−0.01	−0.02	0.11	−0.12	−0.03	0.17	0.36	0.24	0.37	0.42	1.00

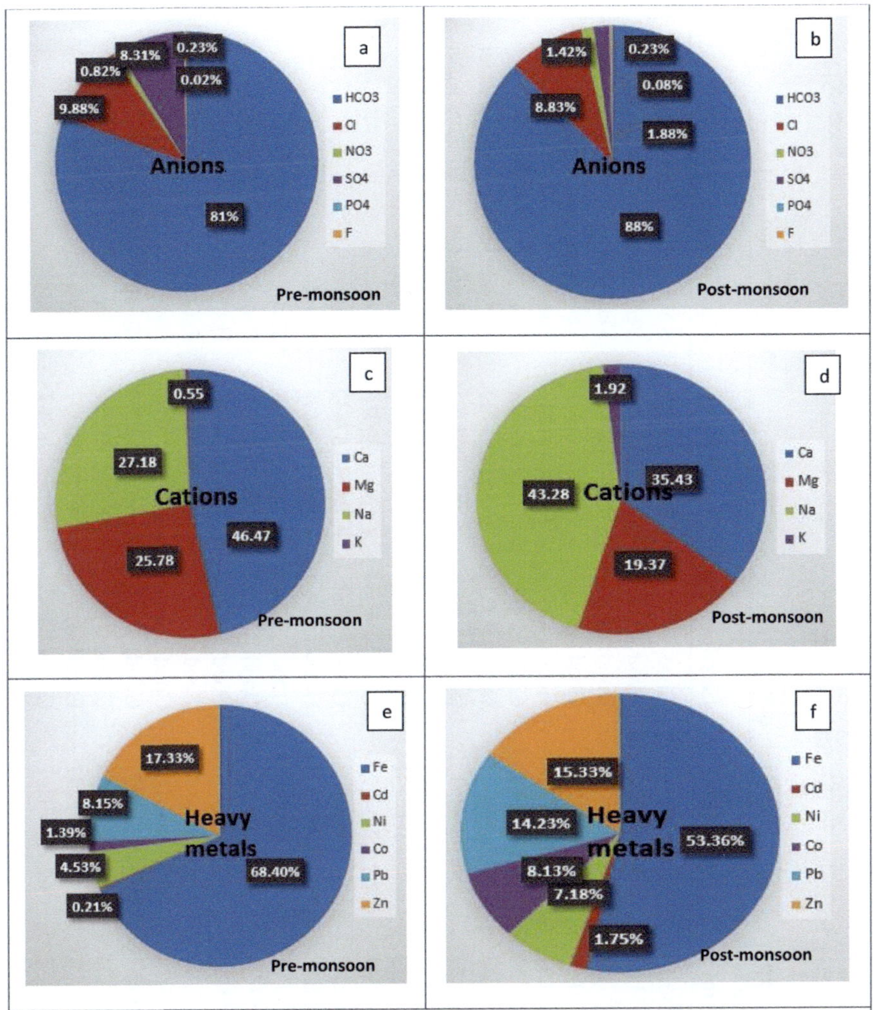

FIGURE 2.4 Distribution of cations, anions and heavy metals in pre- and post-monsoon groundwater samples.

TABLE 2.6
Classification of Groundwater Quality According to WQI Range (Prasad *et al.*, 2019)

WQI Range	Type of Water
<50	Excellent
51–100	Good
101–200	Poor
201–300	Very poor
>300	Not suitable for drinking water

TABLE 2.7
Water Quality Index Value for Groundwater Samples in the Study Area

S. No.	Sample Code	Type of GW Abstraction Structures	Depth of well (m)	Pre-Monsoon, 2018		Post-Monsoon, 2018	
				WQI Value	Classification	WQI Value	Classification
1.	GWS-1	Handpump	36.57	74.07882817	Good	522.4145047	Not suitable for drinking water
2.	GWS-2	Handpump	33.53	69.85865265	Good	249.4242053	Very poor
3.	GWS-3	Handpump	27.43	56.26780725	Good	298.2322422	Very poor
4.	GWS-4	Handpump	30.48	57.30294996	Good	262.6384331	Very poor
5.	GWS-5	Handpump	42.67	71.30496524	Good	121.649813	Poor
6.	GWS-6	Handpump	36.57	74.84819814	Good	108.5819788	Poor
7.	GWS-7	Handpump	27.43	72.5659871	Good	355.9253709	Not suitable for drinking water
8.	GWS-8	Handpump	30.48	56.7663451	Good	585.770821	Not suitable for drinking water
9.	GWS-9	Handpump	13.72	57.45679687	Good	585.2648142	Not suitable for drinking water
10.	GWS-10	Handpump	13.72	57.43119132	Good	440.4020792	Not suitable for drinking water
11.	GWS-11	Handpump	42.67	57.85904026	Good	108.0265026	Poor
12.	GWS-12	Handpump	**42.67**	**77.90896643**	**Good**	**88.30187632**	**Good**
13.	GWS-13	Handpump	**41.15**	**73.78995311**	**Good**	**88.18530699**	**Good**
14.	GWS-14	Handpump	39.62	56.05886698	Good	249.5349828	Very poor
15.	GWS-15	Handpump	53.34	75.26169958	Good	166.4736054	Poor
16.	GWS-16	Handpump	54.86	57.95845307	Good	141.0573827	Poor
17.	GWS-17	Handpump	32.00	66.35654828	Good	117.3219311	Poor
18.	GWS-18	Handpump	45.72	78.43880151	Good	118.4315234	Poor
19.	GWS-19	Handpump	**36.57**	**50.2151006**	**Good**	**60.49614797**	**Good**
20.	GWS-20	Handpump	45.72	51.55699505	Good	120.9263354	Poor
21.	GWS-21	Handpump	**48.77**	**48.50776297**	**Excellent**	**13.91120397**	**Excellent**
22.	GWS-22	Handpump	36.57	67.25470262	Good	106.3424867	Poor
23.	GWS-23	Handpump	**42.67**	**68.03376431**	**Good**	**97.36155944**	**Good**
24.	GWS-24	Handpump	42.67	83.57709978	Good	119.3710251	Poor
25.	GWS-25	Handpump	36.57	52.05868924	Good	113.6142773	Poor

26.	GWS-26	Handpump	**51.81**	**52.01496559**	**Good**	15.1154279	**Excellent**
27.	GWS-27	Handpump	**60.96**	**37.46723505**	**Excellent**	12.76052259	**Excellent**
28.	GWS-28	Handpump	15.24	37.07380478	Excellent	388.8547008	Not suitable for drinking water
29.	GWS-29	Handpump	18.29	57.89871714	Good	465.8185035	Not suitable for drinking water
30.	GWS-30	Handpump	38.10	53.1036884	Good	413.6146141	Not suitable for drinking water
31.	GWS-31	Handpump	21.33	51.05952038	Good	418.8953662	Not suitable for drinking water
32.	GWS-32	Handpump	25.91	69.83017738	Good	552.1734837	Not suitable for drinking water
33.	GWS-33	Handpump	21.33	**70.13537589**	**Good**	**32.10400078**	**Excellent**
34.	GWS-34	Handpump	13.72	61.02094465	Good	192.0628114	Poor
35.	GWS-35	Handpump	**27.43**	**30.27815356**	**Excellent**	**91.89419289**	**Good**
36.	GWS-36	Handpump	**22.86**	**27.63766732**	**Excellent**	**27.09159024**	**Excellent**
37.	GWS-37	Handpump	30.48	30.65614168	Excellent	585.8351389	Not suitable for drinking water
38.	GWS-38	Handpump	**30.48**	**29.56409158**	**Excellent**	**25.60903685**	**Excellent**
39.	GWS-39	Handpump	33.53	26.32956897	Excellent	560.4108834	Not suitable for drinking water
40.	GWS-40	Handpump	30.48	27.51368172	Excellent	544.2263011	Not suitable for drinking water
41.	GWS-41	Handpump	**36.57**	**26.67774431**	**Excellent**	**16.71557775**	**Excellent**
42.	GWS-42	Handpump	**18.29**	**30.65512013**	**Excellent**	**17.09823368**	**Excellent**
43.	GWS-43	Handpump	36.57	25.79434897	Excellent	397.6472508	Not suitable for drinking water
44.	GWS-44	Handpump	**27.43**	**24.71790086**	**Excellent**	**12.47201587**	**Excellent**
45.	GWS-45	Handpump	**36.57**	**27.74595438**	**Excellent**	**73.30070185**	**Good**
46.	GWS-46	Handpump	**10.67**	**29.61060375**	**Excellent**	**26.66011959**	**Excellent**
47.	GWS-47	Handpump	39.62	30.82850987	Excellent	132.9798123	Poor
48.	GWS-48	Handpump	**36.57**	**28.45862299**	**Excellent**	**83.48023718**	**Good**
49.	GWS-49	Handpump	33.53	25.98354649	Excellent	122.5316931	Poor
50.	GWS-50	Handpump	**21.33**	**30.77859409**	**Excellent**	**15.91478169**	**Excellent**
51.	GWS-51	Handpump	**36.57**	**24.22539721**	**Excellent**	**12.37697539**	**Excellent**
52.	GWS-52	Handpump	**61**	**23.63198006**	**Excellent**	**13.15387053**	**Excellent**

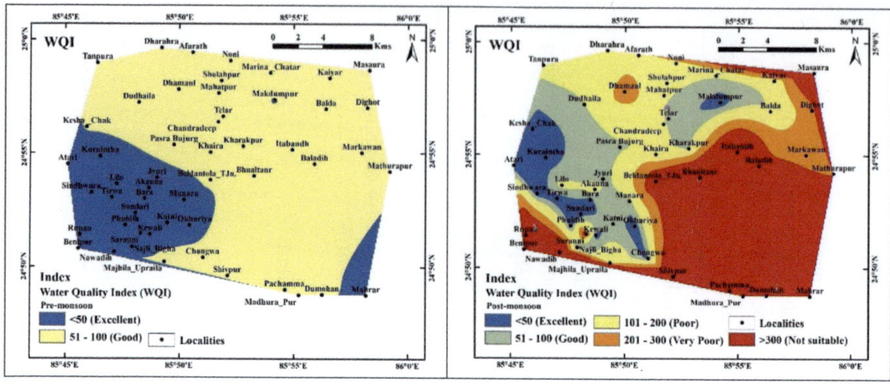

FIGURE 2.5 Water quality index map of the study area (pre-monsoon (a) and post-monsoon (b), 2018).

drinking category during the post-monsoon period. The degradation in quality is common to almost all samples, which appear to be affected. Not a single sample point observation indicates improvement in quality for those showing a change between seasons. This phenomenal contrast in water quality parameters across season thus warrant further site-specific and analytically aided study of existing media, rainfall conditions, role of prevalent geology vis a vis topography and anthropogenic interference if any.

ACKNOWLEDGMENTS

The authors are indebted to Smt. Sudipta Sarkar, Dy. Director General, SU: Bihar and Shri D.D. Bhattacharya, Former Dy. Director General (Retd.) SU: Bihar under whose direction the field assignment was done. The authors also acknowledge the Chemical Division, SU: Bihar, Patna; Chemical Division, Eastern Region, Kolkata and XRD laboratory, Mineral Physics Division, GSI, CHQ, Kolkata for facilitating the support for the chemical analyses of submitted samples. The authors are grateful to all their colleagues and those individuals, who helped them directly or indirectly during the work.

REFERENCES

APHA (1995) *Standard Method for the Examination of Water and Waste Water*, 14th ed. Washington: American Public Health Association, no.409A, pp. 316–317.

Ahmad S, David KS and Gerald S (2004) Environmental assessment: An innovation index for evaluation water quality in streams. *Environ Manage* 34:406–414.

Anitha P, Charmaine J and Nagaraja S (2011) Evaluation of groundwater quality in and around Peenya industrial area of Bangalore, South India using GIS techniques. *Environ Monit Assess*. doi: 10.1007/ s10661-011-2244-y

BIS (Bureau of Indian Standard) (2012) *10500, Indian Standard Drinking Water Specification*, second revision, pp.1–24.

Brown RM, McCleiland NJ, Deininger RA and O'Connor MF (1972) A water quality index – Crossing the psychological barrier (Jenkis SH, ed). *Proceedings of the International Conference on Water Pollution Research*, Jerusalem, Vol. 6, pp. 787–797.

Catroll D (1962) Rain water as a chemical agent of geological process – A view. *USGS Water Supply* 1533:18–20.

Comly HH (1945) Cyanosis in infants caused by nitrates in well water. *J Am Assoc* 129:12–114.

Dart FJ (1974) *The Hazard of Iron*. Canada: Ottawa Water and Pollution Control.

Davis SN and Dewiest RJM (1966) *Hydrogeology*. New York: John Wiley and Sons, Inc.

Faisal K, Tahir H and Ashok L (2003) Water quality evaluation and trend analysis in selected watersheds of the Atlantic region of Canada. *Environ Monit Assess* 88:221–248.

Freeze RA and Cherry JA (1979) *Groundwater*. Englewood Cliffs: Prentice Hall.

Ghosh BK (1980–81) Systematic geological studies of quaternary formations of Gaya district, Bihar (Unpublished GSI Report).

Ghosh BK (1982) Mapping of quaternary sediments in parts of Gaya, Nawada and Hazaribagh districts, Bihar. Unpub. Rep. Geol. Surv. Ind.: F.S. 1981–1982.

Gilly G, Corrao G and Favilli S (1984) Concentrations of nitrates in drinking water and incidence of gastric carcinomas, first descriptive study of the Piemonate region, Italy. *J Sci Total Environ* 34:35–37.

GSI (1968) Progress report on the geological mapping in parts of Monghyr, Gaya, and Hazaribagh districts, Bihar. Unpub. Rep. Geol. Surv. Ind.: F.S. 197–168.

Harrison TD, Cooper JAG and Ramm AEL (2000) *Water quality and aesthetics of South African estuaries*. South Africa: Department of Environmental Affairs and Tourism.

Hopfer SM, Fay WP and Sunderman FW Jr (1989) Serum nickel concentrations in hemodialysis patients with environmental exposure. *Ann Clin Lab Sci* 19:161–167.

Horton RK (1965) An index number system for rating water quality. *J Water Poll Control Fed* 37:300–305.

Janardhana Raju N, Dey S and Das K (2009) Fluoride contamination in groundwaters of Sonbhadra District, Uttar Pradesh, India. *Curr Sci* 96 (7): 979–985.

Krishna Kumar S, Chandrasekar N, Seralathan P, Godson PS and Magesh NS (2011) Hydrogeochemical study of shallow carbonate aquifers, Rameswaram Island, India. *Environ Monit Assess* 184 (7):4127–4139.

Kumar SK, Logeshkumaran A, Magesh NS, Godson PS and Chandrasekar N (2014) Hydro geochemistry and application of water quality index (WQI) for groundwater quality assessment, Anna Nagar, part of Chennai City, Tamil Nadu, India. *Appl Water Sci.* doi: 10.1007/s13201 014- 0196-4

McNeely MD, Nechay MW and Sunderman FW Jr (1972) Measurements of nickel in serum and urine as indices of environmental exposure to nickel. *Clin Chem* 18:992–995.

Moore CV (1973) *Iron: Modern Nutrition in Health and Disease*. Philadelphia: Lea and Fiibeger, p. 297.

National Science Foundation (2004) Water Quality Index, 2004.

Nives SG (1999) Water quality evaluation by index in Dalmatia. *Water Res* 33:3423–3440.

Palanisamy PN, Geetha A, Sujatha M, Sivakumar P and Karunakaran K (2007) Assessment of groundwater quality in and around Gobichettipalayam town Erode District, Tamil Nadu, India. *E-J Chem* 4:434–439.

Prasad M, Sunitha V, Reddy YS, Suvarna B, Reddy BM and ReddyM R (2019) Data on water quality index development for groundwater quality assessment from Obulavaripalli Mandal, YSR district, AP India, Data in brief, ISSN: 2352-3409, Vol. 24, pp. 103846.

Ranjan N, Kathal PK, Khatik J (2022) GIS Integrated evaluation of Groundwater Quality Index (GwQI) for granitic aquifers and its relation with zone of lineaments in parts of Bundelkhand Granite Complex (BGC), Central India. *Asian J Exp Sci* 36(1): 19–30.

Sadat-Noori SM, Ebrahimi K and Liaghat AM (2014) Groundwater quality assessment using the Water Quality Index and GIS in Saveh-Nobaran aquifer, Iran. *Environ Earth Sci* 71: 3827–3843. doi: 10.1007/ s12665-013-2770-8

Sawyer GN and McCartly DL (1967) *Chemistry of Sanitary Engineers*, 2nd ed. New York: McGraw Hill, p. 518.

Scottish Development Department (1975) Towards cleaner water. Report of a River Pollution Survey of Scotland. Edinburgh: HMSO.

Shiow-Mey L, Shang-Lien L and Shan-Hsien W (2004) A generalized water quality index for Taiwan. *Environ Monit Assess* 96:35–52.

Singh BB (1995) Pollution status of Rapti River at Gorakhpur. *J Environ Poll* 2(3):117–120.

Swamee PK and Tyagi A (2000) Describing water quality with aggregate index. *J Environ Eng* 126:451–455.

Umar A, Umar R and Ahmad MS *(2001)* Hydrogeological and hydrochemical framework of regional aquifer system in Kali-Ganga sub-basin, India. *Int J Environ Geol* 40 (4-5):602–611.

UN (2003) United Nations Meeting Coverage and Press release, SG/SM/8707-OBV/348 on 16 May 2003.

Verma AK and Saksena DN (2010) Assessment of water quality and pollution status of Kalpi (Morar) River, Gwalior, Madhya Pradesh: With special reference to conservation and Management plan. *Asian J Exp Biol Sci* 1(2):419–429.

Vigneswaran S, Viswanathan C (1995) *Water Treatment Process: Simple options*, New York: CRC Press, pp. 11.

WHO (2012) *Guidelines for Drinking Water, Recommendations*. Author.

3 Climate Change and Its Impact on Coldwater Fisheries

Darve Sabina Iqbal, Gohar Bilal Wani and Asra Imtiyaz Mattoo

3.1 INTRODUCTION

Climate change is one of the most important global environmental challenges of the twenty-first century. There are numerous reasons why climate change is concerning sincerural populations directly depend on climate-sensitive sectors, such asagriculture, forests and fisheries, which are based on natural resources for their subsistence and livelihoods. Any adverse impact on water availability due to recession of glaciers, decrease in rainfall and increased flooding in certain areas would threaten food security, cause further degradation of natural ecosystems and its resident species that sustain the livelihood of rural households.

Climate change is the variation in the Earth's global climate or in regional climates over time and it involves changes in the variability or average state of the atmosphere over durations ranging from decades to millions of years. The United Nations Framework Convention on Climate Change (UNFCCC) defines it as "a change of climate which is attributed directly or indirectly to human activity that alters the composition of the global atmosphere and which is, in addition to natural climate variability, observed over comparable time periods".

Climate change has both direct and indirect impacts on fish stocks. Direct effects act on the physiology and behavior and alter growth, reproductive capacity, mortality and distribution. Indirect effects alter the productivity, structure and composition of the ecosystems on which fish depend for food. However, many other factors, including fishing, biological interactions and non-climatic environmental factors, can also have similar effects.

Coldwater fishery resources comprise unique biodiversity with valuable indigenous germplasm and maintain environmental quality. The Indian coldwater fisheries include high- and mid-altitude lakes, rivers, streams, their tributaries and reservoirs dammed across such rivers. These resources are poorly developed, primarily due to paucity of financial resources and limited development efforts. In terms of species diversity in India, it is found that apart from the members of the family *Salmonidae*, which are commonly known as coldwater fishes, some Cyprinids, belonging to subfamily *Cyprininae*, which inhabit streams, lakes and reservoirs, receiving snow melt water directly from their watersheds, are also included in this definition. There are large numbers of indigenous and few exotic species of fishes that are present in rivers,

DOI: 10.1201/9781003485995-3

59

brooks, lakes, ponds of uplands. Of these trout, snow trout, mahseer, common carp and minor carps are important as sport and food fishes. These species are widely distributed both in the Himalayas and the Peninsular hill ranges (along the Western Ghats) of the country forming an entirely different ecogeographical entity. Indian coldwater fisheries harbour 258 species belonging to 21 families and 76 genera. Out of these, the maximum of 255 species are recorded from North-East Himalaya, 203 from the West and Central Himalaya and 91 from the Deccan plateau (Singh and Akhtar, 2015). The commercially important Indian coldwater species are as follows:*Tor tor, T. putiora, T. mosal, T. progeneius, T. khudree, T. mussullah, T. malabaricus, Naziritor chelynoides, Neolissochielus wynaadensis, N. hexagonolepis, Schizothoraichthys progastus, S. esocinus, Schizothorax richardsonii, S. plagiostomus, S. curvifrons, S. micropogon, S. kumaonensis, Barilius bendelisis, B. vagra, B. shacra, B. (Raiamas) bola, Banganadero, Crossocheilus periyarensis, Semiplotus semiplotus, Osteobrama belangeri, Garra lamta, Garra gotylagotyla, Glyptothorax pectinopterus, G. brevipinnis, G. stoliczkae, Labeo dero, L. dyochilus* and *Lepidopygopsis typus* (Sehgal, 1999; Sunder *et al.*, 1999; Jena and Gopalakrishnan, 2012; Sehgal, 2012).

3.2   EFFECT OF CLIMATE CHANGE WORLDWIDE

The Intergovernmental Panel on Climate Change (IPCC) has been offering frequent, fact-based updates on climate change and its effects on politics and the economy since 1988. These updates provide a thorough synopsis of the generally recognized body of knowledge regarding the causes, effects and science of climate change. Since the middle of the nineteenth century, the average surface temperature of the Earth has grown by more than 0.8 °C and it is currently warming at a rate of more than 0.1 °C every decade (Hansen *et al.*, 2010). The increase in the atmospheric concentration of (greenhouse gases) GHGs including carbon dioxide (CO_2), methane (CH_4) and nitrogen dioxide (NO_2) is thought to be the main cause of this warming. GHGs are in charge of sustaining life on Earth by acting as a heat blanket surrounding the planet (IPCC, 2014). Since the industrial revolution, GHG emissions have grown exponentially, resulting in the highest atmospheric concentrations of these gases in the past 800 000 years.

Climate change affects the sea also. The average annual expanse of Arctic sea ice shrank between 1979 and 2012 at a pace of 3.5 and 4.1 per cent per decade and the extent of permanent sea ice (summer minimum) shrank by 11.5 and 2.1 per cent, respectively. It is also quite likely that throughout the same time period, the average annual extent of Antarctic sea ice expanded by 1.2 per cent to 1.8 every decade (Vaughan *et al.*, 2013). Nearly all glaciers on Earth have declined; between 2003 and 2009, glaciers in Alaska, the Canadian Arctic, the Southern Andes, the Asian Mountains, and the edge of the Greenland ice sheet lost the most ice.

It is quite expected that Arctic sea ice cover will continue to diminish and thin in the near future as the global mean surface temperature rises, along with springtime snow cover in the northern high latitudes and near-surface permafrosts. Conversely, there is little optimism regarding anticipated short-term declines in the Antarctic sea ice extent and volume (Bindoff *et al.*, 2013). Water levels and flows in aquatic systems are influenced by melting ice and snow cover as well as shrinking mountain glaciers. While the loss of mountain glaciers will have an effect on river flow and lake levels during the short range, before they presumably disappear in the medium term, sea level rise, which is a direct result of ice melting, is projected to have effects in the long run.

Major zooplankton biogeographical shifts have been seen in the Northeast Atlantic as a result of global warming, including poleward expansion of warm-water species and a decline in the number of cold-water species in the same regions (Beaugrand *et al.*, 2009). Due to the Arrhenius relation, which states that chemical reaction rates double for every 10 °C increase in temperature, both abiotic and biotic conditions are sensitive to water temperature (Regier *et al.*, 1990). Temperature has an impact on almost all biological and chemical activities in freshwater habitats.

Glaciers retain 75 percent of the freshwater on Earth in a frozen condition and cover about 10 percent of the land surface (Milner *et al.*, 2009). There is general agreement that most glaciers (but not all) are receding as a result of recent warming and that the rate of recession has accelerated over the past three decades. As global temperatures have risen, the snow line the elevation above which precipitation falls in solid form has migrated upward, producing more rain and less snow, which reduces contributions to glacier mass. Glaciers will recede and disappear as the climate warms and precipitation patterns change. This will change how river systems receive their water supply.

3.3 IMPACT OF CLIMATE CHANGE ON COLDWATER FISH BIODIVERSITY

Climate variability is a key factor controlling the distribution and abundance of aquatic organisms and ecosystem structure (Jena and Gopalakrishnan, 2012). The climate change impact on fish depends on the magnitude of change and on the sensitivity of particular species or ecosystem. The impact of temperature shift due to climate change on aquatic organisms will affect their biological functions, as most of them are poikilothermic in nature. In temperate regions, coldwater fishes are important ecological indicators for climate change as they are very sensitive to changes in water temperature and other environmental conditions. The Himalayan coldwater fish species may be at the highest risk of global warming as many of them are endangered. In many areas, they are already living at the upper end of their thermal range and ecological models demonstrate that there would be significant losses of temperate fish species as climate change may lead to a reduction of fish habitat and diversity (Mohseni *et al.*, 2003).

Because coldwater fish species require coldwater with sufficient levels of dissolved oxygen (DO), they are especially vulnerable to climate change. In summer, their habitat is frequently restricted to the lower portion of the water columns of lakes, where suitably coldwater temperatures are maintained through the open water season. However, oxygen concentrations in those areas decline throughout the summer because decomposition of organic materials that settle into those areas consumes oxygen. Water temperatures in the upper layers of the water column become too warm for coldwater fish during the summer, thus limiting coldwater fish to a specific layer of habitat, where both water temperature and oxygen levels allow their survival. In especially warm years, this habitat may disappear altogether, leading to mortality events in the affected populations.

Due to temperature and oxygen requirements of coldwater fish, climate change may restrict available habitat. Increased solar radiation will expand the zone of warm surface water, while metabolic processes will eliminate oxygen in a larger portion near the bottom of the water column in a warmer climate, decreasing habitat suitable for coldwater fish.

3.4   CLIMATE CHANGE AND GROWTH OF FISHES

Global climate change is increasingly and profoundly threatening fishes, resulting in an uncertain future for fish diversity and global fisheries. Fish growth is a comprehensive consequence of synergistic interactions between gene-determined growth potential and environmental conditions and is fundamentally important for recruitment success (Brander, 1995; Rountrey *et al.*, 2014). Alterations in growth are the most direct, immediate and common way for fishes to respond to climate change. In addition, climate change is expected to change the life history of benthos and planktons (Thackeray *et al.*, 2013), which indirectly influence the energy intake and the growth of fish (Portner and Farrell, 2008; Prokesova *et al.*, 2020). Fish growth alterations are likely to have predictable long-lasting influences on population characteristics (e.g., age-size structure, egg size, reproduction phenology), dynamics (e.g., overwintering mortality) and recruitments (Jonsson and Jonsson, 2009; Murdoch and Power, 2013; Carozza *et al.*, 2019) and could then be transferred to higher organization levels through connecting ecological processes (e.g., inter-species relationships and food webs; Kroeker *et al.*, 2013; Thackeray *et al.*, 2013; Piliere *et al.*, 2014). Therefore, understanding how fish growth responds to changing environments is an essential and efficient step towards predicting the impacts of climate change on populations, communities and even ecosystems.

Fish growth could be affected by a number of environmental variables, of which temperature (e.g., water temperature and sea surface temperature) was most frequently considered (Li *et al.*, 2011). Other environmental variables, such as pH, salinity, rainfall, etc., could also contribute to the alteration of fish growth (Ding *et al.*, 2016). All kinds of environmental changes are likely to bring new physiological stress (thermal stress) to fishes, thus leading to extra energy expenditure and growth reduction. Prokesova *et al.*, (2020) evidenced that the growth of juvenile *Salvelinus fontinalis* and *Salvelinus alpinus* declined when the water temperature was higher than 13 °C, since the mortality rate increased to 30.5 percent with increasing temperature. Meanwhile, multiple physiological processes of fishes, including neurosensory processes, metabolic rates and otolith growth, were changed to adjust the pH and maintain acid-base balance in coping with ocean acidification. Under the future climate scenario, up to 40 percent of the fishes were expected to receive significant influence and 70 percent of energy intake at most is potentially distributed to adapt to the low pH environment (Heuer and Grosell, 2014). These variables may affect fish growth by changing the quantity and quality of fish food sources (Mo *et al.*, 2014). The growth rate is also slowed due to climate changes affecting the density of phytoplankton and zooplankton by increasing phosphorus and decreasing silicate concentrations (Thackeray *et al.*, 2013). The changes the feeding behavior or digestive efficiency of fishes could be another possible way for decrease in growth rate in fishes (Rogacki *et al.*, 2019).

3.5 CLIMATE CHANGE AND FISH REPRODUCTION

Any changes in the thermal regime occasioned by climate change will have major consequences for fish reproduction, and that these effects will be exercised across all stages of the reproductive process (Graham and Harrod, 2009; Jonsson and Jonsson, 2009). The extent of these effects will be determined by a range of factors including specific physiological tolerances, capacity for acclimation and adaptation, scope for behavioral avoidance, capacity to extend or shift ranges, and the timing of thermal challenges with respect to the reproductive cycle. As temperature is a fundamental physical regulatory factor in the lives of fishes and this effect is expressed particularly strongly in the control of all reproductive processes from gamete development to maturation, ovulation and spermiation, spawning, embryogenesis and hatching, to larval and juvenile development and survival. The effects of climate change on aquatic species will vary with latitude, habitat, water column characteristics. A reasonable prediction is that the effects of temperature change on reproduction will be most extreme in riverine systems. Riverine habitats are likely to experience elevated temperatures in association with decreased flow rates and increasing incidence of hypoxic conditions, but with quite marked regional differences in effect. Lacustrine habitats will experience a range of effects depending on lake inflows, water column structure and lake basin topography (Wahl and Loffler, 2009).

Temperature change has the capacity to affect the (hypothalamo-pitutary gonadal) HPG axis at multiple sites through its reaction-rate-determining effects on (follicle stimulating hormone (FSH = gonadotrophic hormone I or GtH I) and leutinizing hormone (LH = gonadotrophic hormone II or GtH II) hormone synthesis and action, and its effects on hormone structure. Inhibitory effects at higher temperature may arise from conformational changes in proteins (e.g., FSH, LH and their receptors, steroid-synthesizing enzymes), and also the increasing tendency for steroid hormones to form water-soluble conjugates at high temperatures (Van and Pankhurst, 1997). In cold temperate and sub-Arctic species, the inhibitory effects typically appear at temperatures of 11–12 °C, among cold-temperate species at around 18°C, temperate species at, 24°C and tropical species at 30°C and above. Exposure to elevated temperatures at the completion of vitellogenesis inhibits the synthesis of the maturational steroid 17,20βP and subsequent progression of oocytes through final oocyte maturation (the resumption of meiosis) and ovulation (Pankhurst *et al.*, 2010).

Thermal stress is known to have marked inhibitory effects on reproduction in fish (Pankhurst and Van, 1997; Leatherland *et al.*, 2010; Schreck, 2010). Stress stimulates activation of an acute catecholamine-mediated response that has the primary effect of rapidly increasing energy availability and the delivery of O_2 to the tissues, followed by a longer and more sustained activation of the hypothalamic–pituitary–interrenal

(HPI) axis, resulting in plasma elevations of the steroid cortisol in teleosts and chondrosteans, and 1α-hydroxycorticosterone in elasmobranchs (Pankhurst and Munday, 2011). Short-term increases in corticosteroids increase the availability of a variety of energy substrates, but longer-term exposure to elevated cortisol results in suppressive effects on a range of functions, including reproduction, growth and immune function.

REFERENCES

Beaugrand, G., Luczak, C., Edwards, M., 2009. Rapid biogeographical plankton shifts in the North Atlantic Ocean. *Global Change Biol.* 15(7), 1790–1803.

Bindoff, N.L., Stott, P.A., Achuta Rao, K.M., Allen, M.R., Gillett, N., Gutzler, D., Hansingo, K., 2013. Detection and attribution of climate change: From global to regional. In T.F. Stocker, D. Qin, G.-K. Plattner, M. Tignor, S.K. Allen, J. Boschung, A. Nauels, Y. Xia, V. Bex & P.M. Midgley, eds. *Climate Change 2013: The Physical Science Basis. Contribution of Working Group I to the Fifth Assessment Report of the Intergovernmental Panel on Climate Change*, pp. 867–952. Cambridge, UK and New York, USA, Cambridge University Press.

Brander, K.M., 1995. The effect of temperature on growth of Atlantic cod (*Gadus morhua* L.). *ICES J. Mar. Sci.* 52, 1–10.

Carozza, D.A., Bianchi, D., Galbraith, E.D., 2019. Metabolic impacts of climate change on marine ecosystems: Implications for fish communities and fisheries. *Glob. Ecol. Biogeogr.* 28, 158–169.

Ding, C.Z., Jiang, X.M., Chen, L.Q., Juan, T., Chen, Z.M., 2016. Growth variation of *Schizothorax dulongensis* Huang, 1985 along altitudinal gradients: Implications for the Tibetan Plateau fishes under climate change. *J. Appl. Ichthyol.* 32, 729–733.

Graham, C.T., Harrod, C., 2009. Implications of climate change for the fishes of the British Isles. *J. Fish Biol.* 74, 1143–1205.

Hansen, J., Ruedy, R., Sato, M., Lo, K., 2010. Global surface temperature change. *Rev. Geophys.* 48, 1–29.

Heuer, R.M., Grosell, M., 2014. Physiological impacts of elevated carbon dioxide and ocean acidification on fish. *Am. J. Physiol.-Regul. Integr. Comp.* 307, 1061–1084.

IPCC, 2014. Summary for policymakers. In C.B. Field, V.R. Barros, D.J. Dokken, K.J. Mach, M.D. Mastrandrea, T.E. Bilir, M. Chatterjee et al., eds. *Climate Change 2014: Impacts, Adaptation, and Vulnerability. Part A: Global and Sectoral Aspects. Contribution of Working Group II to the Fifth Assessment Report of the Intergovernmental Panel on Climate Change*, pp. 1–32. Cambridge, UK and New York, USA, Cambridge University Press.

Jena, J.K., Gopalakrishnan, G., 2012. Aquatic biodiversity management in India. *Proc. Natl. Acad. Sci. India Sect. B Biol. Sci.* 82, 363–379.

Jonsson, B., Jonsson, N., 2009. A review of the likely effects of climate change on anadromous Atlantic salmon *Salmo salar* and brown trout *Salmo trutta*, with particular reference to water temperature and flow. *J. Fish Biol.* 75, 2381–2447.

Kroeker, K.J., Kordas, R.L., Crim, R., Hendriks, I.E., Ramajo, L., Singh, G.S., Gattuso, J.P., 2013. Impacts of ocean acidification on marine organisms: Quantifying sensitivities and interaction with warming. *Glob. Change Biol.* 19, 1884–1896.

Leatherland, J.F., Li, M., Barkataki, S., 2010. Stressors, glucocorticoids and ovarian function in teleosts. *J. Fish Biol.* 76, 86–111.

Li, J., Wang, M.H., Ho, Y.S., 2011. Trends in research on global climate change: A science citation index expanded-based analysis. *Glob. Planet. Change* 77, 13–20.

Milner, A.M., Brown, L.E., Hannah, D.M., 2009. Hydroecological response of river systems to shrinking glaciers. *Hydrol. Process.* 23(1), 62–77.

Mo, W.Y., Cheng, Z., Choi, W.M., Man, Y.B., Liu, Y., Wong, M.H., 2014. Application of food waste based diets in polyculture of low trophic level fish: Effects on fish growth, water quality and plankton density. *Mar. Pollut. Bull.* 85, 803–809.

Mohseni, O., Stefan, H.G., Eaton, J.G., 2003. Global warming and potential changes in fish habitat in U.S. streams. *Clim. Change* 59, 289–409.

Murdoch, A., Power, M., 2013. The effect of lake morphometry on thermal habitat use and growth in Arctic charr populations: Implications for understanding climate change impacts. *Ecol. Freshwater Fish* 22, 453–466.

Pankhurst, N.W., King, H.R., Anderson, K., Elizur, A., Pankhurst, P.M., 2010. Thermal impairment of reproduction is differentially expressed in maiden and repeat spawning Atlantic salmon. *Aquaculture* 316, 77–87.

Pankhurst, N.W., Munday, P.L., 2011. Effects of climate change on fish reproduction and early life history stages. *Mar. Freshwater Res.* 62(9), 1015–1026.

Pankhurst, N.W., Van DerKraak, G., 1997. Effects of stress on growth and reproduction. In G.K. Iwama, A.D. Pickering, J.P. Sumpter & C.B. Schreck, eds. *Fish Stress and Health in Aquaculture*, pp. 73–93. Cambridge, UK and New York, USA, Cambridge University Press.

Piliere, A., Schipper, A.M., Breure, A.M., Posthuma, L., de Zwart, D., Dyer, S.D., Huijbregts, M.A., 2014. Comparing responses of freshwater fish and invertebrate community integrity along multiple environmental gradients. *Ecol. Indic.* 43, 215–226.

Portner, H.O., Farrell, A.P., 2008. Physiology and climate change. *Science* 322, 690–692.

Prokesova, M., Gebauer, T., Matousek, J., Lundova, K., Cejka, J., Zuskova, E., Stejskal, V., 2020. Effect of temperature and oxygen regime on growth and physiology of juvenile *Salvelinus fontinalis×Salvelinus alpinus* hybrids. *Aquaculture* 522, e735119.

Regier, H.A., Holmes, J.A., Pauly, D., 1990. Influence of temperature changes on aquatic ecosystem: An interpretation of empirical data. *Trans. Am. Fisheries Soc.* 119, 374–389.

Rogacki, T.C., Davie, A., Monroig, O., Migaud, H., 2019. Elevated temperature promotes growth and feed efficiency of farmed ballan wrasse juveniles (*Labrus bergylta*). *Aquaculture* 511, 734237.

Rountrey, A.N., Coulson, P.G., Meeuwig, J.J., Meekan, M., 2014. Water temperature and fish growth: Otoliths predict growth patterns of a marine fish in a changing climate. *Glob. Change Biol.* 20, 2450–2458.

Schreck, C.B., 2010. Stress and fish reproduction: The roles of allostasis and hormesis. *Gen. Comp. Endocrinol.* 165, 549–556.

Singh, A.K., Akhtar, M.S., 2015. Coldwater fish diversity of India and its sustainable development. International day for Biological Diversity. Biodiversity for Sustainable Development, Technical Report, pp. 97–105.

Sehgal, K.L., 1999. Coldwater fish and fisheries in the Indian Himalayas: Rivers and streams. In T. Petr, ed. *Fish and Fisheries at Higher Altitudes: Asia.* FAO Fisheries Technical Paper 385, pp. 41–63. Rome, FAO.

Sehgal, K.L., 2012. History of coldwater fisheries in India and some neighbouring countries. In D. Sarma, A. Pandey, S. Chandra & S.K. Gupta, eds. *DCFR Silver Jubilee Compendium on Coldwater Fisheries*, pp. 13–23.

Sunder, S., Raina, H.S., Joshi, C.B., 1999. Fishes of Indian uplands. Bull. No. 2, NRC on Coldwater Fisheries, Bhimtal, p. 64.

Thackeray, S.J., Henrys, P.A., Feuchtmayr, H., Jones, I.D., Maberly, S.C., Winfield, I.J., 2013. Food web de-synchronization in England's largest lake: An assessment based on multiple phenological metrics. *Glob. Change Biol.* 19, 3568–3580.

Van, D.K.G., Pankhurst, N.W., 1997. Temperature effects on the reproductive performance of fish. In C.M. Wood & D.G. McDonald, eds. *Global Warming: Implications for Freshwater and Marine Fish,* pp. 159–176.

Vaughan, D.G., Comiso, J.C., Allison, I., Carrasco, J., Kaser, G., Kwok, R., Mote, P., 2013. Observations: Cryosphere. In T.F. Stocker, D. Qin, G.-K. Plattner, M. Tignor, S.K. Allen, J. Boschung, A. Nauels, Y. Xia, V. Bex & P.M. Midgley, eds. *Climate Change 2013: The Physical Science Basis. Contribution of Working Group I to the Fifth Assessment Report of the Intergovernmental Panel on Climate Change*, pp. 317–382. Cambridge, UK and New York, USA, Cambridge University Press.

Wahl, B., Loffler, H., 2009. Influences on the natural reproduction of whitefish (*Coregonus lavaretus*) in Lake Constance. *Can. J. Fisheries Aqua. Sci.* 66, 547–556.

4 Implications of Climate Change on Agriculture and Adaptation Strategies in India

S.D. Attri and S. Naresh Kumar

4.1 INTRODUCTION

Implications of rapidly changing climate on different sectors of socio-economic activities have focused the attention of the global community to devise ways to address the negative consequences. The shifting weather patterns, rising sea levels and increasing extreme weather and climate events are being observed on unprecedented scale from the equator to poles and from developing to developed countries. The average global temperature has increased about 1.15 (1.02–1.28) °C in 2022 above the 1850–1900 pre-industrial period as per the World Meteorological Organization (WMO, 2023) statement on the "State of the Global Climate in 2022" (Figure 4.1). The most recent 7 years, i.e., 2015 to 2021 are the warmest years on record. The newer findings that warming is more pronounced than expected. Observed concentration of CO_2, CH_4 and N_2O were 415.7 ±0.2 ppm, 1908 ±2 ppb and 334.5 ±0.1 ppb, respectively, during 2020 showing an increase of 149%, 262% and 124%, respectively, from pre-industrial levels, i.e., year 1750 (WMO, 2022). The number of extreme climate-related disasters, including extreme heat, droughts, floods and storms, has doubled since the early 1990s. Over India, on average 213 of the climate extreme events occurred every year during 1990–2016. The annual mean global near-surface temperature for each year from 2022–2026 is projected to be between 1.1 °C and 1.7 °C higher than pre-industrial levels (1850–1900). The likelihood of the annual mean global near-surface temperature temporarily exceeding 1.5 °C above pre-industrial levels for at least one of the next 5 years is 48% and is increasing with time. However, there is only a small probability (10%) that the 5-year mean will exceed this threshold. The Paris Agreement level of 1.5 °C refers to long-term warming, but individual years above 1.5 °C are expected to occur with increasing regularity as global temperatures approach this long-term threshold. There is a 93% probability that at least 1 year in the next 5 will be warmer than the warmest year on record, 2016; and that the mean temperature for 2022–2026 will be higher than that of the last 5 years (WMO, 2022). Further, IPCC (2021) has projected that the global mean surface temperature may increase by 1.2 °C to 5.7 °C, under SSP-1.9, SSP1-2.6, SSP2-4.5, SSP3-7.0

DOI: 10.1201/9781003485995-4

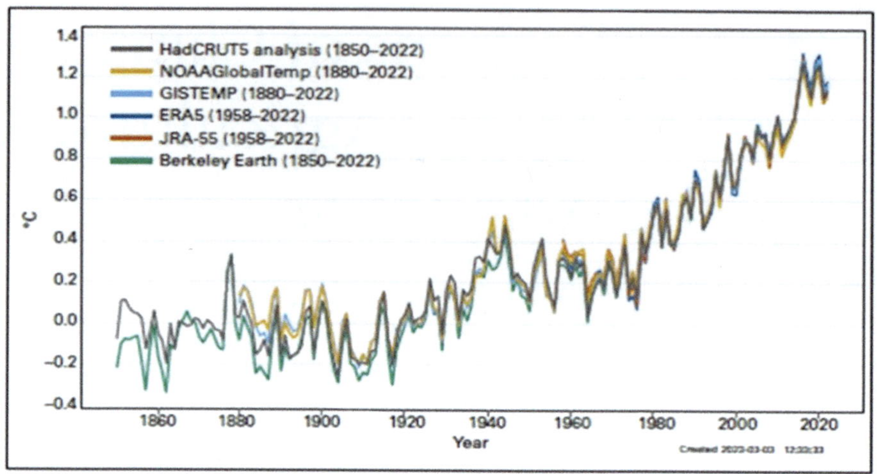

FIGURE 4.1 Global annual mean temperature anomalies with respect to the pre-industrial era (1850–1900) during 1850–2022 (WMO, 2023).

TABLE 4.1
Prediction of Temperature for the Current Century

Prediction of temperature for current century (IPCC 2021)

Scenario	Near term, 2021–2040		Mid-term, 2041–2060		Long term, 2081–2100	
	Best estimate (°C)	*Very likely* range (°C)	Best estimate (°C)	*Very likely* range (°C)	Best estimate (°C)	*Very likely* range (°C)
SSP1-1.9	1.5	1.2 to 1.7	1.6	1.2 to 2.0	1.4	1.0 to 1.8
SSP1-2.6	1.5	1.2 to 1.8	1.7	1.3 to 2.2	1.8	1.3 to 2.4
SSP2-4.5	1.5	1.2 to 1.8	2.0	1.6 to 2.5	2.7	2.1 to 3.5
SSP3-7.0	1.5	1.2 to 1.8	2.1	1.7 to 2.6	3.6	2.8 to 4.6
SSP5-8.5	1.6	1.3 to 1.9	2.4	1.9 to 3.0	4.4	3.3 to 5.7

Source: IPCC (2021).

and SSP-8.5 scenarios, respectively, during 2021–2100. Details of projections are depicted in Table 4.1 (IPCC, 2021).

The annual mean temperature during 1901–2021 in India showed an increasing trend of 0.63 °C/100 years with a significant increasing trend in maximum temperature (0.99 °C/100 years) and relatively lower increasing trend (0.26 °C/100 years) as records of IMD (Figure 4.2). All 10 warming years on record in order have occurred in the current century, viz., 2016 (anomaly +0.71 °C), 2009 (0.55 °C), 2017 (0.54 °C),

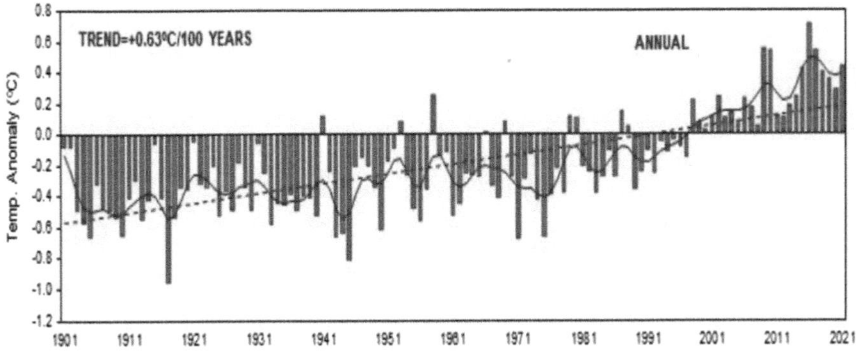

FIGURE 4.2   Observed temperature trends in India (IMD, 2022).

2010 (0.539 °C), 2021 (+0.44 °C), 2015 (0.42 °C), 2018 (0.40 °C), 2019 (+0.36 °C), 1958 (0.25 °C) and 2002 (0.246 °C). The past decade of 2011–2020 was warmest on record with anomalies of 0.37 °C above long-term mean (IMD, 2022). Annual monsoon rainfall over India does not show any trend during 1901–2021 but regional variations have been observed. A recent study by IMD (2021) has reported that five states, viz., Uttar Pradesh, Bihar, West Bengal, Meghalaya and Nagaland have shown significant decreasing trends in southwest monsoon rainfall and seven states in annual rainfall, viz., Uttar Pradesh, Bihar, West Bengal, Meghalaya, Nagaland, Himachal Pradesh and Arunachal Pradesh during the recent 30-year period (1989–2018). Increase in high extreme rainfall and decrease in low and moderate rainfall events have been observed in the country in recent decades. Further, significant increasing trends in frequency of heavy rainfall days are observed over Saurashtra and Kutch, south-eastern parts of Rajasthan, northern parts of Tamil Nadu, northern parts of Andhra Pradesh and adjoining areas of Southwest Odisha, many parts of Chhattisgarh, Southwest Madhya Pradesh, West Bengal, Manipur, Mizoram, Konkan & Goa and Uttarakhand (IMD 2020). Increase of temperatures by 2.7°C by 2099 (best case scenario) and 4.4°C (worst case scenario) has been projected for India (Krishnan *et al.*, 2020).

World Economic Forum (2023) has reported that 50 % and 60 % of global risks on 2-year and 10-year period, respectively, are related to environmental categories like extreme weather events and climate change. As such, there is a need to follow appropriate adaptation and mitigation strategies to address the challenges of a climate change by following low-carbon developmental pathway. Scenarios with very low or low GHG emissions (SSP1-1.9 and SSP1-2.6) may lead within years to discernible stabilizing effects on greenhouse gas, aerosol concentrations, and air quality, relative to high and very high GHG emission scenarios (SSP3-7.0 or SSP5-8.5). Under these contrasting scenarios, discernible differences in trends of global surface temperature would begin to emerge from natural variability within around 20 years, and over longer time periods for many other climatic impact-drivers do emerge and will have sectoral impacts. Global trends in yield of major crops like wheat, rice and maize during 50 years since 1960 has been presented in Figure 4.3 (FAOSTAT, 2018). Relatively few studies have considered impacts on

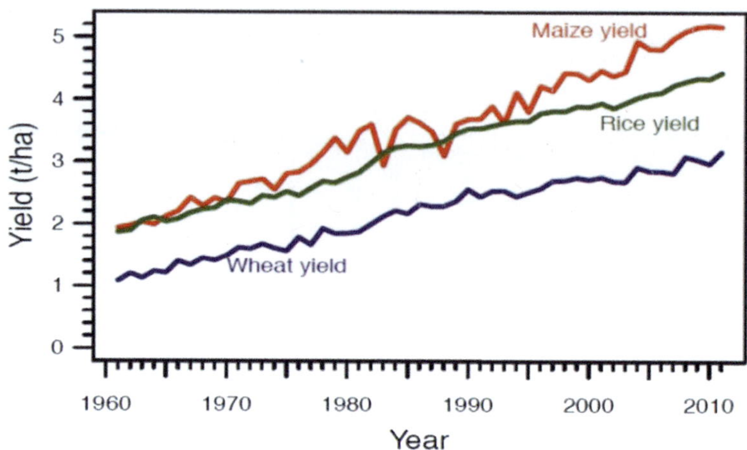

FIGURE 4.3   Trends in yield of major crops in the world (FAOSTAT, 2018).

cropping systems for scenarios where global mean temperatures increase by 4 °C or more.

IPCC (2022a) has reported that adaptation needs will reach $127 billion and $295 billion per year for developing countries alone by 2030 and 2050, respectively. Limiting warming to around 1.5 °C requires global greenhouse gas emissions to peak before 2025 at the latest, and be reduced by 43% by 2030. At the same time, methane would also need to be reduced by about a third. Achieving the target of net zero carbon dioxide emissions globally in the early 2050s and 2070s would be needed for limiting warming to 1.5 ° C and for 2 °C, respectively (IPCC, 2022b)

There is global consensus that adaptation action should follow a country-driven, gender-responsive, participatory and fully transparent approach, considering vulnerable groups, communities and ecosystems, and should be based on and guided by the best available science and, as appropriate, traditional knowledge, knowledge of indigenous peoples and local knowledge systems, with a view to integrating adaptation into relevant socio-economic and environmental policies and actions. To meet the food and nutritional security challenges, sustainably increasing agricultural productivity, helping food systems adapt and build resilience and reducing GHG emissions, promoting knowledge, good practices and available technologies and innovations are needed. As such, the strategies should revolve around technology, policy and decision support system driven initiatives.

4.2 OBSERVED IMPACTS OF CLIMATIC STRESSES ON AGRICULTURE IN INDIA

Climatic risks are affecting the Indian agriculture and monsoon is the major driver of agriculture performance. The rainfall amount and distribution, extreme rainfall events, dry spells, droughts and floods determine the crop yield in many regions in India. Even the irrigated agriculture is challenged by the quantity and quality of irrigation water availability. India is one of the three major countries with a significant water footprint of humanity with 1182 Gm³/yr and it is also one of the major gross virtual water exporters with 125 Gm³/yr (Hoekstra and Mekonnen, 2012). The Economic Survey (2017–2018) by the Government of India estimated the loss of farm revenue due to extreme temperatures and rainfall shocks for monsoon (kharif) and winter (rabi) crops. The analysis indicated a significantly higher loss in farm revenue due to extreme rainfall events during monsoon season. Whereas, the losses due to extreme temperature events are almost similar in monsoon and winter season crops. Changes in temperature and rainfall in the past (1969–2005) impacted the crop yields in India (Birthal *et al.*, 2014). For example, temperature extremes coinciding flowering and grain filling period is affecting the wheat yield in north parts of India (Lobell *et al.*, 2012; Duncan *et al.*, 2015); high daytime temperatures have larger negative impact on rice yield as compared to that of higher night time temperatures causing a loss of rice yield to the tune of 4.4 Mt/yr (Pattanayak and Kumar, 2014). A 10 percent decrease in rainfall over climate mean leads to significant reduction in rice production in India (Swaminathan, 1987). The production of kharif season (monsoon) crops (except sorghum) and rabi (post-monsoon) season crops (except rice and sorghum) were significantly correlated to monsoon rainfall over India (Krishna Kumar *et al.*, 2004). Negative anomalies of seasonal precipitation and number of rainy days are highly correlated with negative anomalies of kharif and rabi food grain yield (Prasanna, 2014). In addition to rainfall and temperature stresses, the extreme weather events such as flood, drought, cyclone, cold and heat wave have been adversely affecting agricultural productivity leading to increased risk of food insecurity.

Climatic extremes like drought in the year 2002 led to a reduction in rainy-season crop sown area to the tune of >15 Mha causing >10% loss in food production (Samra and Singh, 2002). High march temperatures in 2004 in the Indo-Gangetic plains caused early maturity of wheat crop resulting in about 4 Mt reduction in wheat production (Samra and Singh, 2004). During the 1970–2000 period, the Indo-Gangetic plains have experienced a declining trend in the rise in rice and wheat yields, which may partially be attributed to changes in weather conditions (Aggarwal *et al.*, 2000). High rainfall coinciding pod maturation caused severe yield loss in soybean in 2014 and 2015 in Madhya Pradesh. Soybean production has reduced by almost 45% from about 14 Mt in previous years during both the years due to such climatic risk. Hailstorm in 2014, 2015 in Maharashtra caused immense damage to many crops including horticultural crops. Crop loss due to heavy rains coinciding crop maturation is a recurring phenomenon in many districts of Andhra Pradesh and Telengana.

Several examples also exist in the horticulture sector. Apple productivity in Himachal Pradesh declined by up 40–50% due to warmer climate up to 1500 m amsl due to non-fulfilment of chilling requirement. This contributed to shifting of apple production areas to higher altitudes (2700 m amsl) from 10 m aml (Bhagat *et al.*, 2009). Cold waves during December 2002–January 2003 and frost events during 2004 have significantly impacted the horticultural crop production in northern India (Samra and Singh, 2003; 2004). Unseasonal rainfall during March 2008 has affected yields and quality of cashew nuts in northern districts of Kerala (Yadukumar *et al.*, 2010). In potato, tuber yield reduced by 10–50% due to frost damage, 10–20% due to high-temperature stress (Singh *et al.*, 2010). Analysis of the impact of climate change-related events like cyclones and droughts on coconut indicated that consecutive drought years in the Coimbatore district of Tamil Nadu and the Tumkur district of Karnataka caused not only severe reduction in yield but also left about two lakhs palms dead in both the districts (Naresh Kumar *et al.*, 2010). Length and frequency of dry spell has a negative impact on coconut yields in different agro-climatic zones (Naresh Kumar *et al.*, 2007).

The climatic risks for successful crop production have been on the rise and these are only a few examples of several such increasing incidences of crop loss due to climatic risks. All these stresses are increasing in frequency and intensity in recent years.

4.3　PROJECTED IMPACTS OF CLIMATE CHANGE ON CROPS

Different approaches for quantifying the impacts of climate change have been continuously evolving. Some commonly followed approaches include (i) past data analysis, (ii) multi-location, multi-seasonal experiments, (iii) surveys, (iv) controlled environment experiments, (v) meta-analysis and (vi) crop weather simulation modeling. All these are employed alone or in combinations to assess the impact of climate change on agriculture. Simulation models like InfoCrop, DSSAT and APSIM are often used for climate change impact assessments in India. Indian studies have projected climate change impacts on major crops, such as rice, wheat, sorghum, maize, mustard, ground nut, soybean, potato, cotton and coconut (Sinha, 1991, Sinha and Swaminathan, 1991; Aggarwal and Sinha, 1993, Mall et al., 2004; Naresh Kumar *et al*, 2011, 2012, 2013, 2014, 2015; Naresh Kumar and Aggarwal, 2013; NICRA Reports 2011–2017).

Climate change is projected to decrease the wheat yield in the range of 8–24% by 2050 in IGP and may constrain higher productivity of irrigated rice in Punjab, Haryana and Rajasthan. In the case of mustard, though yield may improve in Punjab and Haryana but in Uttar Pradesh, Bihar and Assam they are projected to reduce up to 6%. Climate change is projected to reduce potato yields by ~2.5, ~6 and ~11% in the IGP region in 2020 (2010–2039), 2050 (2040–2069) and 2080 (2070–2099) time periods. However, negative impacts can be overcome and substantial increase in yield can be realized by timely sowing and better management of improved varieties of wheat (~11%), rice (17-20%), mustard (~25%%) and potato (~8%) (Gupta and Pathak, 2016). Several studies have shown spatial variation in magnitude and direction of climate change impacts on crops.

4.3.1　CEREALS AND MILLETS

Negative impacts of climate change on rice yield are projected. About 4% loss in irrigated rice yield and 6% loss in rainfed rice yield are projected for the 2020 (2010–2039) scenario in India. But the projected yield loss in rainfed rice is <2.5 % in 2050 (2041–2070) and 2080 (2070–2099) scenarios. With adaptation, however, irrigated rice yield is to increase by ~17% while the increase in yield of rainfed rice is by ~20% (Naresh Kumar *et al.*, 2013).

Climate change is projected to affect wheat yield India ranging from 6% to 23% by 2050, without any specific adaptation. Yield would reduce in areas with current seasonal mean temperatures in excess of 27 °C and 13 °C for maximum and minimum, respectively. Sowing a suitable variety, adjusting sowing time, and fertilizer and irrigation management may be a low-cost and practical adaptation strategy that can significantly increase wheat yield in changing climates (Naresh Kumar *et al.*, 2014).

Initial studies from India have projected a reduction in wheat yield in the range of 28–68% due to climate change without considering the beneficial effects of CO_2 and a 2 °C increase in temperature would affect wheat yield in most parts of India (Agarwal and Sinha, 1993; Rao and Shina, 1994 a and b). Adaptation was projected to improve wheat yield (Attri and Rathore, 2003). A 1 °C increase in growing season temperature projected to result in a loss of 4–5 Mt of wheat yield (Rao and Sinha, 1994 a and b; Agarwal and Sinha, 1993; Aggarwal and Mall, 2002; Mall and Aggarwal, 2002; Aggarwal and Swaroopa Rani, 2009). Other studies have indicated a reduction of 450 kg/ha wheat yield with a 0.5 °C increase in temperature during winter (Sinha and Swaminathan, 1991; Lal *et al.*, 1998). Wheat potential is also projected to reduce by 5% with a warming of 0.5 to 1.5 °C (Aggarwal, 2003).

Climate change is projected to affect the yield of maize in the kharif season by about 18% in 2020 and 2050 scenarios, but about 21% increase in yield is possible in the 2020 scenario. In the 2050 scenario, increase in projected up to 10% in the 2050 scenario (Byjesh *et al.*, 2010; Naresh Kumar *et al.*, 2012). Projected effects of climate change on rainfed sorghum yield include a reduction of yield by about 2.5% in 2020 (2010–2039) and by 8% in the 2050 scenario. Simple adaptation options, however, can improve the productivity by ~8% in 2020 (Srivastava *et al.*, 2010; Naresh Kumar *et al.*, 2012). Climate change is projected to impact the pearl millet yield negatively (–1 and –14%) in parts of Maharashtra in future scenarios of 2030 and 2050, respectively (Singh *et al.*, 2017).

4.3.2 Oil Seeds

In India, climate change is projected to affect mustard seed yield by about 2% in the 2020 (2010–2039) scenario. The projected negative impacts on seed yield are about 7.9% in 2050 (2040–2069) and by about 15% in 2080 (2070–2099) scenarios. Areas with current mean seasonal temperatures (max/max) in excess of 25/10 °C will lose mustard yield due to temperature rise. It is also projected that there will be a reduction in a climatically suitable window for cultivation of mustard. Therefore, cultivars that are of short-duration (<130 days) having 63% of duration for pod filling can better adapt (Naresh Kumar *et al.*, 2015). Mustard in eastern parts of India is projected to be affected more (Boomiraj *et al.*, 2010; Naresh Kumar *et al.*, 2015) and a decrease in mustard seed yield up to 201 kg/ha for every 1°C increase in temperature in Haryana (*Kalra et al.*, 2008).

Seed yield of soybean is projected to increase by 8–13% in 2030 and 2080 climate scenarios, respectively (Naresh Kumar *et al.*, 2012).

Climate change is projected to reduce the seed yield of rainfed groundnut by ~4% in A1B 2080, while seed yield may increase in the range of 4–7% in other climate scenarios (Naresh Kumar *et al.*, 2012).

4.3.3 PULSES

The seed yield of chickpea is projected to increase by 17–25% in Haryana and central Madhya Pradesh, but to decrease by 7–16% in southern Andhra Pradesh in the 2050 climate scenario (Singh *et al.*, 2014).

Regression analysis indicated a plausible decrease in pigeon pea yield by ~–3.2 to –10.1% in the 2035 scenario, without considering the CO^2 and other factor effects. The decrease in yield may be even more in the 2065 scenario (Birthal *et al.*, 2014).

4.3.4 OTHER COMMERCIAL CROPS

Climate change is projected to marginally decline cotton productivity in northern India but in central India and in southern India, productivity may increase, but without any significant effect on all India cotton productivity in future climates (Hebbar *et al.*, 2013). Projected impacts on sugarcane yield in India are mainly attributed to increase in dry spells and water scarcity (Srivastava *et al.*, 2011). Climate change is projected to affect coconut productivity in the eastern region of India while improving in the western coast, parts of Tamil Nadu and Karnataka (Narsh Kumar and Aggarwal, 2013).

In addition to the impact on the yield, climate change is projected to affect grain/seed quality such as reduction in grain protein concentration, and some minerals (such as zinc and iron) due to elevated CO_2 (Porter *et al.*, 2014). On the other hand, high temperatures resulted in increase in their concentration in several cereal and pulse crops. In addition, climate change is projected to affect other factors that directly or indirectly influence the crop production. For instance, increase in the frequency of extreme rainfall events increase soil erosion and also affect water availability and quality (IPCC, 2014). However, temperature rise may decrease the occurrence of cold waves and frost events possibly reducing the damages to frost-sensitive field crops such as mustard, chickpea and potato, and also some vegetables.

4.3.5 CLIMATE CHANGE IMPACTS ON CROP-PESTS

Climate change driven temperature increase may enhance the abundance of crop-pests, at least some of them. A number of physiological processes including extension of temperature ranges, phenological changes, faster developmental rates of population, growth, over-wintering and migration will lead to differential incidences of pests (Prospero *et al.*, 2009; Chakraborty and Newton, 2011). Such changes are likely to alter the crop-pest–predator relationship, which may cause the emergence of new pests and increased or decreased incidence of current major pests. In addition, the emergence of current minor pests as major pests in changing climates is likely. The crop-pest interaction also depends on the response of host plants to climate change. For instance, increase in leaf tissue C:N ratio may drive insects to feed more foliage to meet their nitrogen requirement.

Impact of climate change on crop-pests was understood through an experimental approach and modeling. Experimental studies using the open top chambers (OTCs) indicated that elevated CO_2 levels (550 ppm and 700 ppm) increase the C:N ratio of foliage, leading to higher consumption rate, lower digestive efficiency and slower growth of the castor semi-looper, *Achaea janata* (Rao *et al.,* 2009) and tobacco caterpillar *Spodoptera litura* (Rao *et al.,* 2012). Honey dew excretion was also more under elevated CO_2 (Prasannakumar *et al.,* 2012). Under elevated CO_2 conditions, susceptibility of wheat to spot blotch was related to stomatal density and involvement of PR protein (Shukla *et al.,* 2017).

Crop-pest simulation-model-based studies on rice-brown plant hopper (BPH) projected that the BPH population may decrease by 3.5 and 9.3–14 % by 2020 and 2050 scenarios, respectively, during kharif in New Delhi, while a marginal decline (2.1–3.5 %) in pest population during winter at Aduthurai (Tamil Nadu) by 2050 is projected (Sujithra and Chander, 2013). Higher yield loss of rice under elevated CO_2 (29.9–34.9%) than the ambient conditions (17–23.1%) is projected due to increased BPH population and higher sucking rate because of reduced nitrogen level of plant sap (Guru-Pirasanna-Pandi *et al.,* 2018).

Similarly, rice pink stem borer, *Sesamia inferens* projected a decline in the range of 5.82–22.8 % by 2020 and 19.0–42.7 % by 2050 in the Delhi region (Selvaraj and Chander, 2015). The population of *P. solenopsis* increased within a temperature range of 15–40 ºC, wherein 25–35 ºC was the optimum. Increased pest activity due to climate change may decrease cotton yield (Fand *et al.,* 2014). Spot blotch (*Bipolaris sorokinnina*), an emerging disease in wheat, is projected to occur more in future climates in Indo-Gangetic plains (Viani *et al.,* 2013). For IGP region, the future risk of leaf blast disease in rice is reported to be higher during winter season than during monsoon season (Viswanath *et al.,* 2017). Population generation rates are projected for Harmigera in pigeon pea growing areas of India in future climates (Rao *et al.,* 2016).

From the above studies, it is clear that climate change directly or indirectly affects the productivity and quality of various crops in India. Significant spatial and temporal variations exist for possible impacts on crops making adaptation in agriculture important.

4.4 ADAPTATION TO CLIMATE CHANGE: STRATEGIES AND EVIDENCES

4.4.1 ADAPTATION TO CLIMATE CHANGE: APPROACHES AND MECHANISMS

Adaptation to climate change should address both mean change in climatic conditions and increased climatic variability. Adaptation approaches and options are highly dependent on type and intensity of impact, exposure and vulnerability as well as adaptive capacity of the systems. Several approaches are followed, but they generally fall into a few categories. Adaptation may be to address (i) short-term, (ii) medium and/or (iii) long-term. The short-term adaptations generally address climatic risk issues while the long-term adaptation generally addresses the mean change in climate. Adaptation approach can also be (i) agro-ecosystem component specific (ex. crop, livestock) or (ii) integrated adaption (ecosystem-based) or both. It is imperative to choose low-GHG emitting adaptation technologies as an alternative.

4.4.2 Adaptation Strategies and Projected Adaptation Gains

Developing the adaptation strategy includes either a "top-down" or "bottom-up" approach and/or a combination of both. For instance, in the "top-down" approach includes insurance safety nets, neem oil coated urea among many. The bottom-up approach includes several location-specific and locally preferred interventions. However, a combination of both is preferable where convergence of local needs with national goals is achieved through several mechanisms to (i) reduce vulnerability, (ii) achieve resilience and (iii) sustain food production systems. The FAO (2010) has given the concept of a climate-smart agricultural (CSA) system, which includes a sustainable increase in the productivity and income, building resilience and reducing and/or removing of GHG emissions.

Since climate change impacts are location- and situation-specific and adaptation options need to be modified to suit agro-ecological conditions, socio-economic contexts and farmers' priorities (Rosenstock *et al.*, 2015). So the adaptation options and resultant gains will also have significant spatial and temporal variations. The contextualizing adaptation approach involves participatory adaptation planning (Naresh Kumar *et al.*, 2014). Under this approach, firstly the needs, technologies and aspirations of the farming community as desired by them are listed. Then, the best technological alternatives to optimize the production systems while minimizing the GHG emissions are identified using experimental or simulation approaches in current and future climate scenarios. Thus shortlisted technological options are matched with the farmers' needs/aspirations before they are implemented in their field for performance testing in the pursuit of developing "climate-resilient agriculture". Thereafter, the scope for out-scaling and up-scaling of successful technologies is analyzed. Agrometeorlogical services are also significant tools in managing climate change impacts (Attri and Mohapatra, 2021).

Simulation studies have indicated that significant improvement of up to 40% in the yield of several crops can be achieved by adopting simple and low-cost adaptation technologies like adjusting sowing time, change in variety, and improved water and nitrogen management with national level out-scaling (Naresh Kumar *et al.*, 2012, 2013, 2014, 2015).

4.4.3 Adaptation Evidences from Farmers' Fields

In India, several evidences on farmers' adaptation to climatic risks and climate change began to emerge. Under National Network Project on Climate Change (2004–2013), National Initiative on Climate Change (2011–2015), National Innovations in Climate Resilient Agriculture (2016 onwards), World Bank-GEF project (2009–2014) and DST-MEDP (2018 onwards) projects several adaptation technologies are demonstrated in farmers' fields. Under these projects, the "climate-resilient villages (CRVs)" are being developed (NICRA Reports, 2011–2017; Naresh Kumar *et al.*, 2014; Srinivasa Rao, Gopinath, *et al.*, 2016). The concept of CRVs and the details of adaptation options, their performance in various agroecosystems in different parts of India are extensively dealt with in earlier publication (Srinivasa Rao, Prabhakar, *et al.*, 2016). Apart from these, the CCAFS programme also has contributions in developing CSVs (CCAFS reports at https://ccafs.cgiar.org/flagships/climate-smart-technologies-and-practices). These concepts use the basic approach of managing the resources and knowledge for reducing the climatic risks to agriculture and farm income. Some of these components are being used for developing CSAs and CRVs.

The adaptation interventions in farmer's fields are based on the analysis of past and projected climatic risks and resource analysis. The interventions have been based on crop varieties, crops agronomical management and use of weather information apart from income augmentation activities (Naresh Kumar *et al.*, 2014, CCAFS reports). In addition, convergence of line departments plays an important role in developing climate-resilient villages. Adaptation interventions that include conservation agricultural practices indicate huge potential in improving the yields from current level despite climatic risks. The summary of results indicates the availability of technologies and their potential in minimizing climatic risks in different agricultural systems in various parts of India (Table 4.2).

Several technologies have shown promise to substantially increase the yields of many crops. Many of the existing technologies may fall short of climatic challenges in the future. So, innovations in adaptation need to be continued to find technologies for climate smart agriculture.

4.4.4 INNOVATIONS IN ADAPTATION APPROACHES

The adaptation to climate change demands innovations due to uncertainty in occurrence of climatic risk-events, in terms of space, time and magnitude. In agriculture the adaptation to climate change must have innovative paths. These can be of different types, viz., technological innovations, managerial, policy and institutional innovations.

Currently, several adaptation technologies are being developed and tested in the villages under major projects like the WB-GEF project, NICRA, CCAFS and DST projects. For example, under NICRA, technology demonstration is being carried out in about 151 village clusters in as many districts, covering climatically sensitive districts. Along with promotion of climate-resilient technologies, behavioral and institutional interventions have been made for developing climate-resilient villages across India.

Some of the innovations already in place to address climatic risks in India include contingency plans being disseminated by ICAR Institutes and SAUs in collaboration with IMD for almost all districts in India; increased use of ICT tools for dissemination of weather information and other farm advisories including pest forewarning; use of neem oil-coated urea across India;

farm level soil health cards, etc.; the research initiatives/projects such as the National Network Project on Climate Change (Planning Commission); National Initiatives on Climate Resilient Agriculture (under NMSA); National Innovations in Climate Resilient Agriculture (under NMSA); National projects under DST on climate change (under NMSKCC); National Mission for Sustaining Himalayan Ecosystem-Task Force-Agriculture Network project (under NMHE) have contributed and continue to contribute to innovation in adaptation.

TABLE 4.2

Adaptation Interventions and their Potential in Minimizing Climatic Risks

Type of Stress	Adaptation Options and Observed Improvements in Yield			
	Crop	Adaptation option	Improvement in yield over farmers practice (%)	Details (indicative)
Drought stress	Rice	Tolerant/	20–35	Short duration drought tolerant varieties
	Wheat	improved	20–28	
	Maize	varieties and	25–41	
	Mustard	increase in seed	8–25	
	Ground nut	replacement	5–10	
	Soybean	rate	22–40	
	Chick pea		25–35	
	Pigeon pea		30–40	
	Green gram		15–25	
	Black gram		25–30	
	Cotton		5–10	
Submergence <1 m for <10 days	Rice	Tolerant varieties	30–35	Swarna-sub 1, MTU-1010, 1100, 1064, 1140, 7029, BPT-5204; Dehangi, Gitsh, Sahasrang
Submergence >1 m for >10 days	Rice	Tolerant varieties	20–25	Jalashree, Rajashree, Karjat-2, GAR-13 Lalat,
Drought prone areas with areas <750 mm	Intercrops	Altering the intercrops	4–47	Soybean+pigeon pea Pigeon pea+pearl millet/ sunflower Rabi sorghum+safflower/ chickpea Maize + balckgram
Drought prone areas with areas >750 mm	Intercrops	Altering the intercrops	11–>50	Groundnut + pigeon pea/ castor Maize+ pigeon pea/ cow pea/ black gram/ green gram Pigeon pea+ black gram/ sorghum/maize/soybean/okra Soybean+ pigeon pea Ground nut + castor Finger millet +pigeon pea Cotton+ pigeon pea
Regions with <500 mm rainfall	Different crops	*In situ* water conservation	11–61	Bunding and leveling Conservation furrow Land leveling

TABLE 4.2 (Continued)

Type of Stress	Adaptation Options and Observed Improvements in Yield Improvements in Yield			
	Crop	Adaptation option	Improvement in yield over farmers practice (%)	Details (indicative)
Regions with more than 500 mm rainfall	Different crops	*In situ* water conservation	8–>50	Compartmental bunding, mulching, conservation furrows Trench cum bunding, bunding, land leveling, contour sowing ridge and furrow, summer ploughing, set furrows
Regions with <500 mm rainfall	Different crops	Water harvesting and irrigation	5–>50	Farm ponds, furrow, sprinkler, drip
	Different crops	Planting methods	4–43	Broadbed and furrow, raised bed
	Wheat	Tillage method	23–34	Zero tillage
Drought prone areas	Different crops	Improved crop management	12–15 25% & 30% reduction in CO_2 emission / ha & water saving respectively	Recommended seed rate, sowing method, dose and scheduling fertilizer and irrigation
	Different crops	Water management	~40% water saving, 25% irrigation-time saving and 25% reduction in CO_2 gas emission.; water user groups	Laser leveling, underground pipeline, sprinkler/ rain gun, drip irrigation; open-well deepening, de-silting and renovation of community ponds

Source: NICRA Annual Reports 2011–2017; Bandyopadhyay *et al.* (2013); Naresh Kumar *et al.* (2014).

In addition, the policy initiatives that are directly related to agriculture include National Environment Policy (NEP), National Action Plan on Climate Change; National Action Plan on Adaptation and State Action Plan on Climate Change.

The research and technological innovations should include germplasm exploration, widening the species base used for food, and data-driven technological interventions including simulation models, remote sensing, IoT, AI, ML and block-chain technologies. A portion of adaptation funds must be delineated for research and climate services.

4.4.5 ADAPTATION COSTS

Indian farmers have been adapting to climatic risks. Though several efforts have been going on to test the adaptation options at farm level, only a very few studies have analyzed the cost of adaptation at farm and at household level. A detailed household level analysis in a village indicated that the cost of adaptation and gains vary with type of agriculture and farm size (*Naresh Kumar et al.*, 2014, 2016). In this study, adaptation options that included crops (such as improved varieties, crop diversification, crop water, pest management), farm (includes livestock management), off-farm (value additions, etc.), non-farm (income augmentation) and knowledge (weather forecast) have improved resilience of crop, farm, farmer and household to climatic risks. The analysis indicated that income from agriculture alone could not support the small and marginal farm families. However, adaptation to climatic risks can improve the income sustainability. Though farmers incur adaptation cost, it is not always the case particularly for middle- and large-farm holdings. Science- and technology-based crop management can rationalize costs of adaptation to climatic risks (Naresh Kumar *et al.*, 2014; 2016). The study further shows that though additional cost is not imperative for adaptation, the cost of adaptation is not directly proportional to gains across different strata of farmers.

4.5 ADAPTATION TO CLIMATE CHANGE: PRIORITIZATION OF INVESTMENTS AND POLICY

4.5.1 PRIORITIZING ADAPTATION AREAS AND OPTIONS

Climate change impacts on agriculture cut across several sectors, such as water, energy, governance at different levels, finances, markets and industry. Under the NICRA project, the district-level vulnerability atlas was prepared (Rao *et al.*, 2016). Based on the climatically challenged districts as identified by the Planning Commission, GoI, the technological demonstrations under the NICRA project are implemented in 151 districts (NICRA Reports 2011–2017). Farmers' priorities for available adaptation technologies generally are based on the prevailing climatic conditions, socio-economic conditions and their willingness-to-pay (Khatri-Chhetria *et al.*, 2017). Further, the willingness-to-pay is related to the evidence-based adaptation gains/profits and also the cost of technology. For instance, farmers in the <600 mm annual rainfall zone preferred crop insurance, rainwater harvesting, fodder management, weather-based agro-advisory and contingent crop planning over others. In the 600–800 mm rainfall zone, farmers' preference was on weather-based agro-advisory, crop insurance, laser leveling of land, rainwater harvest and agro-forestry to minimize climatic risks. On the other hand, farmers in the >800 mm rainfall zone preferred crop insurance, weather-based agro-advisory, weather smart shelters for livestock, agro-forestry and site-specific nutrient management (Khatri-Chhetria *et al.*, 2017).

Technologies for minimizing climatic risk have been evolving and being demonstrated in farmers' fields with a significant success. However, knowledge-based interventions ably supported by the decision support systems linked to ICTs will immensely improve the timely adaptation actions. A well-informed farmer can take timely action to minimize the negative impacts or/and to optimize resource utilization that reduces the cost of cultivation. Current information generation and dissemination systems such as contingency plans are inadequate to meet the requirements of an individual farmer. Thus there is a need for enhanced use of technology in agriculture for improving the productivity and profitability (Aggarwal *et al.*, 2016). Use of crop simulation models in combination with remote sensing is one such important step towards developing knowledge-based agriculture in India. There is a need to develop and use the prioritization tools such as the CSA-RA tool (Mwongera *et al.*, 2017).

4.5.2 INVESTMENT NEEDS AND OPPORTUNITIES

Adaptation requires huge investments at almost all levels of adaptation, right from research to farm level and to the infrastructure development level. Since the cost of adaptation and gains significantly vary within a community, policies for incentivizing adaptation costs for small and marginal farmers would be required (Naresh Kumar *et al.*, 2014; 2016). Mainstreaming of climate change adaptation help in the flow of huge investments to the needy. There exists several opportunities for investments by government/private sectors in the field of weather-based products, artificial intelligence and ICT tools, input delivery and output management, market intelligence, agri-residue management, value addition, small-scale farm machinery, sensor-based agriculture, financial sector, safety nets, and ecosystem-based adaptation, and many more.

4.6 WAY FORWARD

Climate change affects all systems of Earth and thus demands collective actions to understand impacts, causes of vulnerability to climatic risks, develop context-specific forewarning and decision support systems, use of emerging technologies, develop adaptation technologies and to find synergistic technologies. Developing integrated adaptation strategies and convergence of local institutions, resource mobilization and capacity building of all stakeholders for adaptation is very important to make India agriculture resilient to climate change.

ACKNOWLEDGEMENTS

Authors are grateful to DGM, IMD for encouragement and sincerely acknowledge the financial support provided for various studies under Indian Council of Agricultural Research (ICAR) Network Project on Climate Change; ICAR-National Innovations on Climate Resilient Agriculture; ICAR-World-Bank GEF project; MoEF & CC, DST project, etc.

REFERENCES

Aggarwal, P.K., 2003, Impact of climate change on Indian agriculture. *J Plant Bio*, 30(2), 189–98.

Aggarwal, P.K., Chand, R., Bhutani, A., Kumar, V., Goe, S.K., Rao, K.N., Poddar, M.K., Sud, U.C., Lochan, R., 2016, Report of the Task Force on 'Enhancing Technology Use in Agriculture Insurance' NITI (National Institute for Transforming India) Aayog, Government of India, 24.

Aggarwal, P.K., Mall, R.K., 2002, Climate change and rice yields in diverse agro-environments of India. II. Effect of uncertainties in scenarios and crop models on impact assessment. *Climate Change*, 52(3), 31–43.

Aggarwal, P.K., Naresh Kumar, S., Pathak, H., 2010, Impacts of climate change on growth and yield of rice and wheat in the Upper Ganga Basin. WWF Report, 1–44.

Aggarwal, P.K., Sinha, S.K., 1993, Effect of probable increase in carbon dioxide and temperature on wheat yields in India. *J Agric Meteorol*, 48(5), 811–814.

Aggarwal, P.K., Swaroopa Rani, D.N., 2009, Assessment of climate change impacts on wheat production in India. *Global Climate Change and Indian Agriculture: Case Studies From the ICAR Network Project*, 5–12.

Aggarwal, P.K., Talukdar, K.K., Mall, R.K., 2000, Potential yields of rice-wheat system in the Indo-Gangetic plains of India. *Rice-Wheat Consortium Paper Series*, 10, 16.

Attri, S.D., Mohapatra, M., 2021, Agrometeorological services for climate resilient agriculture. In: *Climate Resilience and Environmental Sustainability Approaches – Global Lessons and Local Challenges*. Springer, pp. 127–139. https://doi.org/10.1007/978-981-16-0902-2

Attri, S.D., Rathore, L.S., 2003, Simulation of impact of projected climate change on wheat in India. *Int J Climatol*, 23, 693–705.

Bassu, S., Brisson, N., Durand, J.L., Boote, K., Lizaso, J., 2014, How do various maize crop models vary in their responses to climate change factors? *Global Change Biol*. https://doi.org/10.1111/gcb.12520

Bhagat, R.M., Rana, R.S., Kalia, V., 2009, Weather changes related shift of apple belt in Himachal Pradesh. Impact of increased temperature on quality of rice and wheat grains. In: Aggarwal, P.K. (ed), *Global Climate Change and Indian Agriculture-Case Studies From the ICAR Network Project*, ICAR Pub, pp. 48–53.

Birthal, P.S., Digvijay N., Shiv. K., Shaily A., Suresh M.K., 2014(a), How sensitive is Indian agriculture to climate change. *Ind J Agric,* 69(4), 474–487.

Birthal, P.S., Khan, M.T., Negi, D.S., Agarwal, S., 2014(b), Impact of climate change on yields of major food crops in India: Implications for food security. *Agric Econ Res Assoc*, 27(2), 145–155.

Boomiraj, K., Chakrabarti, B., Aggarwal, P., Choudhary, R., Chander, S., 2010, Assessing the vulnerability of Indian mustard to climate change. *Agric Ecosyst Environ*, 138(3-4), 265–273.

Byjesh, K., Naresh, K.S., Aggarwal, P.K., 2010, Simulating impacts, potential adaptation and vulnerability of maize to climate change in India. *Mitigation Adapt Strateg Glob Change*, 15, 413–431.

Chakraborty, S., Newton, A.C, 2011, Climate change, plant diseases and food security: An overview. *Plant Pathol*, 60(1), 2–14.

Duncan, J., Dash, J., Atkinson, P.M., 2015, Elucidating the impact of temperature variability and extremes on cereal croplands through remote sensing. *Glob Change Biol,* 21(4), 1541–1551.

FAO, 2010, Climate-Smart Agriculture. Sourcebook. www.fao.org/4/i3325e/i3325e.pdf

FAOSTAT, 2018, FAOSTAT, Food and Agriculture Organization Corporate Statistical Database. www.fao.org/faostat/en/#home

Fand, B.B., Tonnang, H.E., Kumar, M., Bal, S.K., Singh, N.P., Rao, D.V.K.N., Kamble, A.L., Nangare, D.D., Minhas, P.S., 2014, Predicting the impact of climate change on regional and seasonal abundance of the mealybug *Phenacoccus solenopsis* Tinsley (Hemiptera: Pseudococcidae) using temperature-driven phenology model linked to GIS. *Ecol Modell*, 288, 62–78.

Gupta, A., Pathak, H., 2016, Climate change and agriculture in India. A Thematic Report of National Mission on Strategic Knowledge for Climate Change (NMSKCC) under National Action Plan on Climate Change (NAPCC), 1–51.

Guru-Pirasanna-Pandi, G., Chander, S., Pal, M., Soumia, P.S., 2018, Impact of elevated CO_2 on *Oryza sativa* phenology and brown planthopper, *Nilaparvata lugens* (Hemiptera: Delphacidae) population. *Curr Sci*, 114(8), 1767–1777.

Hebbar, K.B., Venugopalan, M.V., Prakash, A.H., Aggarwal, P.K., 2013, Simulating the impacts of climate change on cotton production in India. *Climatic Change*, 118(3–4), 701–713.

Hoekstra, A.Y., Mekonnen, M.M., 2012, The water footprint of humanity. *Proc Natl Acad Sci*, 109(9), 3232–3237.

IMD, 2020, Observed Monsoon Rainfall Variability and Changes During Recent 30 years (1989–2018), 1–10.

IMD, 2021, Changing Rainfall Pattern in the Country. https://pib.gov.in/Pressreleaseshare. aspx?PRID=1696515...pl

IMD, 2022, Statement on Climate of India During 2021, 1–9.

INCCA Report, 2010, Climate Change and India: A 4x4 Assessment – A Sartorial and Regional Analysis for 2030s: Chapter on Agriculture, Ministry of Environment and Forestry. Under Indian Network for Climate Change Assessment, 67–88.

IPCC, 2001, Climate Change 2001: Impacts, Adaptation, and Vulnerability. *Contribution of Working Group II to the Third Assessment Report of the Intergovernmental Panel on Climate Change.*

IPCC, 2013, Climate Change 2013: The Physical Science Basis. *Working Group I Contribution to the Fifth Assessment Report of the Intergovernmental Panel on Climate Change.*

IPCC, 2014, Climate Change 2014: Impacts, Adaptation, and Vulnerability. Part A: Global and Sectoral Aspects. *Contribution of Working Group II to the Fifth Assessment Report of the Intergovernmental Panel on Climate Change.*

IPCC, 2021, Climate Change 2021: The Physical Science Basis. In: Masson-Delmotte, V., Zhai, P., Pirani, A., Connors, S.L., Péan, C., Berger, S., Caud, N., Chen, Y., Goldfarb, L., Gomis, M.I., Huang, M., Leitzell, K., Lonnoy, E., Matthews, J.B.R., Maycock, T.K., Waterfield, T., Yelekçi, O., Yu, R., Zhou, B. (eds.), *Contribution of Working Group I to the Sixth Assessment Report of the Intergovernmental Panel on Climate Change.* Cambridge University Press, Cambridge, United Kingdom and New York, NY, USA, In press, https://doi.org/10.1017/9781009157896

IPCC, 2022a, Climate Change 2022: Impacts, Adaptation, and Vulnerability. In: Pörtner, H.O., Roberts, D.C., Tignor, M., Poloczanska, E.S., Mintenbeck, K., Alegría, A., Craig, M., Langsdorf, S., Löschke, S., Möller, V., Okem, A., Rama, B. (eds.), *Contribution of Working Group II to the Sixth Assessment Report of the Intergovernmental Panel on Climate Change.* Cambridge University Press, Cambridge, UK and New York, NY, USA, 3056. https://doi.org/10.1017/9781009325844

IPCC, 2022b, Climate Change 2022, Mitigation of Climate Change. In: Shukla, P.R., Skea, J., Slade, R., Al Khourdajie, A., van Diemen, R., McCollum, D., Pathak, M., Some, S., Vyas, P., Fradera, R., Belkacemi, M., Hasija, A., Lisboa, G., Luz, S., Malley, J. (eds.), *Contribution of Working Group III to the Sixth Assessment Report of the*

Intergovernmental Panel on Climate Change. Cambridge University Press, Cambridge, UK and New York, NY, USA. https://doi.org/10.1017/978100915792

Kalra, N., Chakraborty, D., Sharma, A., Rai, H.K., Jolly, M., Chander, S., Kumar, P.R., Bhadraray, S., Barman, D., Mittal, R.B., Lal, M., Sehgal M., 2008, Effect of increasing temperature on yield of some winter crops in northwest India. *Curr Sci*, 94(1), 82–88.

Krishna Kumar, K., Rupa Kumar, K., Ashrit R.G., Deshpande N.R., Hansen J.W., 2004, Climate impacts on Indian agriculture. *Int J Climatol*, 24(11), 1375–1393.

Krishnan, R., Sanjay, J., Gnanaseelan, C., Mujumdar, M., Kulkarni, A., Chakraborty, S., eds., 2020, *Assessment of Climate Change over the Indian Region: A Report of the Ministry of Earth Sciences (MoES)*. Government of India Springer Open, p. 226, https://doi.org/ 10.1007/978-981-15-4327-2

Lal, M., Singh, K.K., Rathore, L.S., Srinivasan, G., Saseendran, S.A., 1998, Vulnerability of rice and wheat yields in NW-India to future changes in climate. *Agric Meteorol*, 89, 101–114.

Lobell, D.B., Sibley, A., Ortiz-Monasterio, J.I., 2012, Extreme heat effects on wheat senescence in India. *Nat Climate Change*, 2(3), 186–189.

Mall, R.K., Aggarwal, P.K., 2002, Climate change and rice yields in diverse agro-environments of India. I. Evaluation of impact assessment models. *Climate Change*, 52, 331–343.

Mall, R.K., Lal, M., Bhatia V.S., Rathore, L.S., Singh, R., 2004, Mitigating climate change impact on soybean productivity in India: A simulation study. *Agric Meteorol*, 121, 113–125.

Mwongera, C., Kelvin, M., Shikuku, T.J., Läderach, P., Ampaire, E., Asten, P.V., Twomlow, S., Winowiecki, L.A., 2017, Climate smart agriculture rapid appraisal (CSA-RA): A tool for prioritizing context-specific climate smart agriculture technologies. *Agric Syst*, 151, 192–203.

Naresh Kumar, S., 2009, Climate change and Indian agriculture: Current understanding on impacts, adaptation, vulnerability and mitigation. *J Plant Biol*, 37(2), 1–16.

Naresh Kumar, S., Aggarwal, P.K., 2013, Climate change and coconut plantations in India: Impacts and potential adaptation gains. *Agric Syst*, 117. DOI:10.1016/ j.agsy.2013.01.001

Naresh Kumar, S., Aggarwal, P.K., Rani, D.S., Saxena, R., Chauhan, N., Jain, S., 2014, Vulnerability of wheat production to climate change in India, *Climate Res*, 59, 173–187.

Naresh Kumar, S., Aggarwal, P.K., Rani, S., Jain, S., Saxena, R., Chauhan, N., 2011, Impact of climate change on crop productivity in Western Ghats, coastal and north-eastern regions of India. *Curr Sci*, 101(3), 33–42.

Naresh Kumar, S., Aggarwal, P.K., Saxena, R., Rani, D.S., Jain, S., Chauhan, N., 2013, An assessment of regional vulnerability of rice to climate change in India. *Climate Change*, 118(3–4), 683–699.

Naresh Kumar, S., Anuja, Md. Rashid, A.M., Bandyopadhyay, S.K., Padaria, R., Khanna, M., 2016, Adaptation of farming community to climatic risk: Does adaptation cost for sustaining agricultural profitability? *Curr Sci*, 110(7), 1216–1224.

Naresh Kumar, S., Bandyopadhyay, S.K., Padaria, R.N., Singh, A.K., Rashid, M., Wasim, M., Prasad, S., 2014, *Climatic Risks and Strategizing Agricultural Adaptation in Climatically Challenged Regions*. IARI, New Delhi Publication. TB-ICN: 136/2014, 106.

Naresh Kumar, S., Govindakrishnan, P.M., Swarooparani, D.N., Nitin, C., Surabhi, J., Aggarwal, P.K., 2015, Assessment of impact of climate change on potato and potential adaptation gains in the Indo-Gangetic Plains of India. *Int J Plant Prod*, 9(1), 151–170.

Naresh Kumar, S., Kasturi Bai, K.V., Thomas, G.V., 2010, Climatic change and plantation crops: Impact, adaptation and mitigation with special reference to coconut. In: Singh,

H.P., Singh, J.P., Lal, S.S. (eds.), *Challenges of Climate Change – India Horticulture.* Westville Publishing House, New Delhi, pp. 9–22.

Naresh Kumar, S., Rajagopal, V., Thomas, T., Cherian, S., Narayanan, V.K., Ananda, M.K.R., Nagawekar, K.S., Hanumanthappa, D.D., Vincent, M.S., Srinivasulu, B., 2007, Variations in nut yield of coconut (*Cocos nucifera* L) and dry spell in different agroclimatic zones of India. *Indian J Hortic*, 64(3), 309–313.

Naresh Kumar, S., Singh, A.K., Aggarwal, P.K., Rao, V.U.M., Venkateswarlu, B., 2012, Climate Change and Indian Agriculture: Salient Achievements From ICAR Network Project. IARI Publication, p. 32.

National Initiative on Climate Resilient Agriculture (NICRA), Research Highlights (2011-17). Central Research Institute for Dryland Agriculture, Hyderabad. www.nicra-icar.in/nicra revised/index.php/photogallery?layout=edit&id=104

Pattanayak, A., Kumar, K.S.K., 2014, Weather sensitivity of rice yield: Evidence From India. *Climate Change Econ*, 5(4), 1–24. https://doi.org/10.1142/S2010007814500110

Porter, J.R., Xie, L., Challinor, A.J., Cochrane, K., Howden, S.M., Iqbal, M.M., Lobell, D.B., Travasso, M.I., 2014, Food security and food production systems. In: Field, et al. (ed.), *Climate Change 2014: Impacts, Adaptation, and Vulnerability. Part A: Global and Sectoral Aspects. Contribution of Working Group II to the Fifth Assessment Report of the Intergovernmental Panel on Climate Change.* Cambridge University Press, Cambridge, UK and New York, USA, pp. 485–533.

Prasanna, V., 2014, Impact of monsoon rainfall on the total food grain yield over. *Indian J Earth Syst Sci*, 123(5), 1129–1145.

Prasannakumar, N.R., Chander, S., Pal, M., 2012, Assessment of impact of climate change with reference to elevated CO_2 on rice brown planthopper, *Nilaparvata lugens* (Stal.) and crop yield. *Curr Sci*, 103(10), 1201–1205.

Prospero, S., Grunwald, N.J., Winton, L.M., Hansen, E.D.M, 2009, Migration patterns of the emerging plant pathogen *Phytophthora ramorum* on the west coast of the United States of America. *Phytopathology*, 99, 739–749.

Rao, B.R., Raju, M.K., Subba Rao, A.V.M, Rao, K.V., Rao, V.U.M, Rao, C.S., 2016, A district level assessment of vulnerability of Indian agriculture to climate change. *Cur Sci*, 110.

Rao, G.D., Sinha, S.K., 1994(a), Climate changes and agriculture. *Nature*, 437, 102–109.

Rao, G.D., Sinha, S.K., 1994(b), Impact of climate change on simulated wheat production in India. Implications of climate change for international agriculture. *Crop Modelling Study*, 2(3), 4–10.

Rao, M.S., Manimanjari, D., Vanaja, M., Rama Rao, C.A., Srinivas, K., Rao, V., Venkateswarlu, B., 2012, Impact of elevated CO_2 on tobacco caterpillar, *Spodoptera litura* on peanut, *Arachis hypogea. J Insect Sci*, 12(103), 1–10. doi: 10.1673/031.012.10301

Rao, M.S., Srinivas, K., Vanaja, M., Rao, G.G.S.N, Venkateswarlu, B., Ramakrishna, Y.S., 2009, Host plant (*Ricinus communis* Linn.) mediated effects of elevated CO_2 on growth performance of two insect folivores. *Curr Sci*, 1047–1054.

Rosenstock, T., Lamanna, S., Chesterman, C., Bell, S., Arslan, P., Richards, A., Rioux, M., Akinleye, J., Champalle, A.O., Cheng, C., Corner-Dolloff, Z., Dohn, C., English, J., Eyrich, W., Girvetz, A.S., Kerr, E.H., Lizarazo, A., Madalinska, M., McFatridge, A., Morris, S., Namoi, K.S., Poultouchidou, N., Ravina, N., Silva, D., Rayess, M., Ström, S., Tully, H., Zhou, K.L.W, 2015, The Scientific Basis of Climate-Smart Agriculture: A Systematic Review Protocol. https://ccafs.cgiar.org/publications/scientific-basis-climate-smart-agriculture-systematic-review-protocol#.W-VAw9IzaUk

Samra, J.S., Singh, G., 2002, Drought Management Strategies. Indian Council of Agricultural Research, New Delhi, 68.

Samra, J.S., Singh, G., 2004, Heat Wave of March 2004: Impact on Agriculture. Indian Council of Agricultural Research, New Delhi, 32.

Samra, J.S., Singh, G., Ramakrishna, Y.S., 2003, Cold Wave of 2002-03: Impact on Agriculture. Natural Resource Management Division. Indian Council of Agricultural Research, 49.

Selvaraj, K., Chander, S., 2015, Simulation of climatic change impact on crop-pest interactions: A case study of rice pink stem borer *Sesamia inferens* (Walker). *Climate Change*, 131(2), 259–272. DOI 10.1007/s 10584-015-1385-3

Shrivastava, A.K., Srivastava, A.K., Solomon, S., 2011, Sustaining sugarcane productivity under depleting water resources. *Curr Sci*,1(6), 748–754.

Shukla, K.K., Parimal, S., Vivek, S., Ali, V., Amit, S., 2017, Repercussion of elevated CO_2 concentration on wheat susceptibility to *Bipolaris sorokiniana*. *Environ Econ*, 35(2), 768–773.

Singh, P., Boote, K.J., Kadiyala, M.D.M., Swamikannu, N., Gupta, S.K., Srinivas, K., Bantilan, M.C.S., 2017, An assessment of yield gains under climate change due to genetic modification of pearl millet. *Sci Total Environ*, 601–602, 1226–1237.

Singh, P., Swamikannu, N., Boote, K.J., Gaur, P.M., 2014, Climate change impacts and potential benefits of drought and heat tolerance in chickpea in South Asia and East Africa. *Eur J Agron*, 52, 123–137.

Singh, S., Gupta, A.K., Gupta, S.K., Kaur, N., 2010, Effect of sowing time on protein quality and starch pasting characteristics in wheat (*Triticum aestivum* L.) genotypes grown under irrigated and rain-fed conditions. *Food Chem*, 122(3), 559–565.

Sinha, S.K., Swaminathan, M.S., 1991, Deforestation, climate change and sustainable nutrition security. *Climate Change*, 16, 33–45.

Srinivasa Rao, Ch., Gopinath, K.A., Prasad, J.V.N.S., Channalli, P., Singh, A.K., 2016, Climate resilient villages for sustainable food security in tropical India: Concept, process, technologies, institutions, and impacts. *Adv Agron*, 140, 101–214, doi.org/10.1016/bs.agron.2016.06.003

Srinivasa Rao, Ch., Prabhakar, M., Maheswari, M., Srinivasa Rao, M., Sharma, K.L., Srinivas, K., Prasad, J.V.N.S., Rama Rao, C.A., Vanaja, M., Ramana, D.B.V., Gopinath, K.A., SubbaRao, A.V.M., Rejani, R., Bhaskar, S., Sikka, A.K., Alagusundaram, K., 2016, National Innovations in Climate Resilient Agriculture (NICRA), Research highlights 2015–16. Central Research Institute for DrylandAgriculture, Hyderabad, p. 112.

Srivastava, A., Soora, N.K., Aggarwal, P.K., 2010, Assessment on vulnerability of sorghum to climate change in India. *Agric Ecosyst Environ*, 138, 160–169.

Sujithra, M., Chander, S., 2013, Simulation of rice brown planthopper, *Nilaparvata lugens* population and crop-pest interactions to assess climate change impact. *Climate Change*, 121, 331–347. https://doi.org/10.1007/s 10584-013-0878-1

Summerfield, R.J., Hadley, P., Roberts, E.H., Minchin, F.R., Rawsthorne, S., 1984, Sensitivity of chickpeas (*Cicer arietinum*) to hot temperatures during the reproductive period. *Exp Agric*, 20(1), 77–93.

Swaminathan, M.S., 1987, Abnormal monsoons and economic consequences: The Indian experiment. In: Fein, J.S., Stephens, P.L. (eds.), *Monsoons*. Wiley, New York, pp. 121–134.

Viani, A., Sinha, P., Singh, R., Singh, V., 2013, Simulation of spot blotch in wheat under elevated temperature. *Ann Pl Protec Sci*, 21(2), 368–376.

Viswanath, K., Sinha, P., Kumar, S.N., Sharma, T., Saxena, S., Panjwani, S., Pathak, H., Shukla, S.M., 2017, Simulation of leaf blast infection in tropical rice agro-ecology under climate change scenario. *Climate Change*, 142, 155–167.

World Economic Forum, 2023, *The Global Risks Report 2023*, 18th Edition, pp. 1–97.

WMO, 2022, *WMO Provisional State of the Global Climate 2022*, pp. 1–26.

WMO, 2023, *State of the Global Climate 2022*, WMO No. 1316, pp. 1–49.

Yadav, S.B., Patel, H.R., Lunagaria, M.M., Parmar, P.K., Chaudhari, N.J., Karande, B.I., Pandey, V., 2013, Impact assessment of projected climate change on pearl millet in Gujarat. *National Seminar on Climate Change Impacts on Water Resources Systems.* International Crops Research Institute for the Semi-arid Tropics India, Hyderabad, pp. 33–38.

Yadukumar, N., Raviprasad, T.N., Bhat M.G., 2010, Effect of climate change on yield and insect pests incidence on cashew. In: Singh, H.P., Singh, J.P., Lat, S.S., (eds.), *Challenges of Climate Change – Indian Horticulture*, Westville Publishing House, New Delhi, p. 224.

5 Impact of Climate Change in India
Challenges to Human Health

Vandana Prajapati, Shailender Kumar Verma,
Anwar Alam and Neloy Khare

Climate change is an alteration in the composition of global atmosphere, weather conditions, including environment factors like temperature, humidity, precipitation, cloudiness and wind patterns over comparable time periods. Climate change is a natural process whereby the Earth's climate undergoes change in periodic cycles over very long periods of time. The shifts in climate change occur naturally due to Earth's axial tilt angle, orbital eccentricity and variations in the solar activity, including sun spots, solar flares and solar storms. In addition, volcanic eruptions and temperature distribution due to ocean currents and phenomenon like El Nino, La Nina and El Nino-Southern Oscillations (ENSO) also contribute to climate change.

Scientific studies suggest that long-term natural climate change has been accelerated owing to anthropogenic factors. Currently, "climate change" refers to rapid changes in the global climate induced by human activities such as emissions of greenhouse gases (GHG) by burning fossil fuels, deforestation and urbanization.

As per the Climate Transparency Report 2022, India lost \$159 billion, 5.4% of its gross domestic product (GDP) due to extreme heat in 2021. It is further estimated that India may lose upto US\$6 trillion by 2050 that is 6% of the GDP by 2050 and US\$35 trillion, i.e., 12.6% of the GDP by 2070 (Phillip *et al.*, 2021).

5.1 ENVIRONMENTAL FACTORS AFFECTING HUMAN HEALTH

1. <u>Floods and heavy rainfall</u>
   - Storm runoff and floods help in mobilization and transportation of pathogens causing water-borne diseases.
   - Floods damage vital water supply and sanitation infrastructure by overtaxing containment systems and releasing untreated effluent. The effectiveness of water treatment is compromised by overburdened or degraded infrastructure.
   - Floods force people to relocate to unhygienic places that increases their risk towards communicable diseases.
2. <u>Drought</u>
   - Water availability and quality are declined owing to increased demand for and sharing of limited water resources, such as with cattle.

DOI: 10.1201/9781003485995-5

- To overcome water scarcity, water is stored in home water storage tanks. Uncovered home water storage tanks are a breeding ground for insects that act as vectors for transmission of numerous infectious diseases.
- Poor hygiene due to reduction in water availability leads to a high risk of exposure to pathogens breeding in animal manure and stored human waste, thereby resulting in disease.

3. <u>Temperature rise</u>
   - Provides an extended period for survival, proliferation and transmission of opportunistic pathogens and increases pathogen load in hosts.
   - Causes wildfires in the forests during heat waves, which leads to degradation of air and water quality.
   - Exposure to polluted air and water affects health and enhances morbidity and mortality.
   - Extended summer season leads to behavioral changes in humans and wildlife.

4. <u>Sea level rise</u>
   - Strong storm surges lead to population eviction at coastal areas.
   - Coastal flooding causes disruption in drinking water supplies and damages sanitation infrastructure.
   - Saline seawater intrusion into coastal aquifers causes a decline in the quality of the soil and drinking water.

The effect of environmental factors on health is illustrated in Figure 5.1.

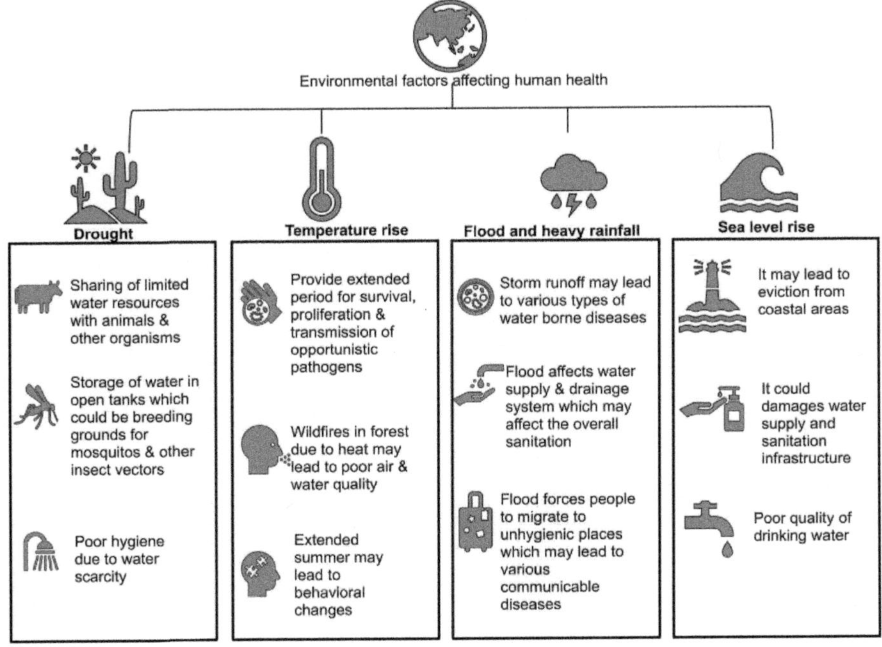

FIGURE 5.1 Environmental factors affecting human health.

The adverse consequences of climate change are alarmingly pervasive for the environment which also affects the components of the entire ecosystem, including human health. Climate change can be catastrophic as it can not only disturb the equilibrium and pattern of the season but also the transition between seasons. Changes in the pattern of the season primarily affect the temperature. A change of 1 °C in temperature can increase the height of the ocean and water bodies by 6 m due to melting of glaciers, which may lead to the submergence of land areas near the ocean and large water bodies. Changes in the pattern of seasons affect plants, which can have an indirect effect on human health. Globally, climate change poses challenges for health organizations to maintain the supply of adequate food. Some of the prominent effects of climate change include asthma exacerbations due to extended pollen seasons and increase in the number of cases of emergence or re-emergence of various vector-borne diseases (Rocklöv and Dubrow, 2020). Moreover, many people are affected by heat-related illnesses due to prolonged and frequent heat waves (Vicedo-Cabrera *et al.*, 2021). On a global scale, a study conducted on 652 cities concluded that pollution caused by burning fossil fuels may be linked with elevated risks of cardiovascular and respiratory disorders due to air particulate matter, thereby causing increase in human mortality and aging. The special report "Global warming of 1.5°C" (IPCC, 2018) concludes that the development of climate-smart technologies may still require more than 10 years to minimize the negative impacts of climate change on human health. As a result, significant increase in morbidity and mortality are anticipated over the next decades. Additionally, the productivity of working class people is expected to decline due to poor quality of food and vector-borne diseases.

Drought, flood and heavy precipitation, high temperatures and sea level rise due to climate change adversely affects human health and may lead to communicable and non-communicable diseases.

India stands at a high risk from climate change due to its sizable population, which depends on climate conditions for agricultural practices and availability of natural resources for survival. The pressure on natural resources in water, air, soil and forests has already reached its threshold in India. The strategic location of India makes it a hub for economic activity, which has a toll on the environment and climate. Development of infrastructure for creating jobs and to provide food security in India has led to indiscriminate non-compliance of regulatory environmental norms, thereby causing pollution of air, water and soil. In 2019, India ranked seventh in terms of economic losses (66,182 million US dollars) and deaths (2,267 individuals) due to climate change-related extreme weather conditions (UNICEF India, 2022). Heat exposure in India in 2022 caused a loss of 167 billion potential labor hours. The labor productivity in the country is projected to further decline if global temperatures increase by 1.5°C.

It has been estimated that 90 percent of children across the globe are breathing polluted air. Africa and Asia (including India) accounts for highest number of deaths caused due to $PM_{2.5}$ pollutants. Between 1901 and 2018, the average temperature in India increased by around 0.7°C (PIB Delhi, 2022) due to emission of greenhouse gases (GHGs). India ranks 26 out of 163 countries in the index that assesses climate risk for children. In total, t21 out of 30 of the most polluted cities in the world are located in India and in 2017 nearly 1.24 million people died due to air-pollution-related illnesses (Balakrishnan *et al.*, 2019).

Human health has been impacted by evolution of various pathogens due to changes in weather conditions causing transmission of epidemics including Spanish flu, plague and cholera. Currently, SARS-CoV-2 caused by coronavirus has severely impacted the global economy and affected millions of people worldwide. The distribution of various kinds of insects that act as disease vectors is directly influenced by climate. In recent times, diseases including West Nile Fever, dengue, and chikungunya viruses have become more prevalent in India (Dash *et al.*, 2013). Many people experience viral flu, conjunctivitis, gastrointestinal distress, and other ailments. People living in flood water areas are exposed to contaminated water and have a high risk of acquiring skin disorders, and helminthiasis. Under-nutrition, poor mental health and infections transmitted via contaminated food are additional health concerns. It has been observed that the climate change linked mortality rate among children below the age of five is quite high in low-income countries (Perin *et al.*, 2022).

5.2   DIRECT AND INDIRECT IMPACTS ON HUMAN HEALTH DUE TO CLIMATE CHANGE

5.2.1   HEALTH ISSUES DUE TO TEMPERATURE-RELATED STRESS

The temperature of the human body is managed and controlled by the hypothalamus in the brain. The heat escapes from the body through the processes of convection, radiation and perspiration. There is often no radiant heat gain or loss when the ambient temperature is equal to that of a person's skin. As the body heats up, the hypothalamus and neurological system activate the sweat glands to release sweat, which evaporates and cools the skin. The pace of heat acquisition exceeds the rate of heat loss under hot conditions resulting in a rise of body temperature (Ahima, 2020). Excessive exposure to heat may cause weariness, dehydration, exhaustion and heat stroke. Extreme situations, among the elderly or youngsters, or both, might result in abrupt stress that has an impact on mental health. It may result in sleep disturbances, anxiety attacks and neurological and psychiatric issues. When the body temperature rises over 41°C, heat stress occurs, which results in skin rashes, muscular cramps, body fatigue and heat stroke, which may result in death. Heat waves caused large-scale hospitalizations and fatalities in Europe, Russia, Japan and India (Vicedo-Cabrera *et al.*, 2021). Heat stress conditions are already reaching the maximum limits of worker productivity and human survival in certain several parts of southAsia including the Indian sub-continent; and may severely affect the working population. From 1980 to 2017, there has been a 500% rise in heat waves in 150 of the most populous cities across the globe (Li *et al.*, 2021). In low- and middle-income nations, heat has mainly been disregarded as a health risk factor. According to the global burden of disease (GBD) research, the cost of disability adjusted life years (DALYs) associated with low temperatures in 2019 was 2.2 times more than the cost associated with high temperatures (Abbafati *et al.*, 2020a). The cardiovascular health of an individual is affected by stress caused due to climate change. Diuretics and beta-blockers, which are used to treat cardiovascular disorders (CVDs), may make people more susceptible to heat stress (Stewart *et al.*, 2017). During heat waves, there are additional cases of hospitalizations due to electrolytic imbalance and deaths due to respiratory illnesses.

Atmospheric pollution and global warming resulting from climate change causes asthma, respiratory tract illnesses and cardiovascular diseases. Environmental change had drastic effects on children and maternal health and the elderly population.

Climate change adversely affects human health by causing temperature-related stress, vector-borne, water-borne and food-borne diseases, respiratory tract infections, non-communicable diseases and affects mental health. Environment change also impacts the livelihoods by affecting food security and nutrition.

Outdoor workers who experience occupational heat stress have dehydration, decline in renal function, weariness, drowsiness, disorientation and loss of focus. There is evidence that suggests that the necessity to wear personal protective equipment during the COVID-19 epidemic increased the risk of occupational heat stress for healthcare personnel (Bose-O'Reilly *et al.*, 2021). The cost of heat-related loss in work time increased over the years. The maximum loss of work due to heat stress was observed in lower- and middle-income nations.

It has been observed that in India about a 0.5% increase in the mean temperature from 1960 to 2009 has led to 146% more mortality due to heat-related stress (Mazdiyasni *et al.*, 2017). It has also been observed that the global temperature will rise upto 5.5 % by the end of thetwenty-first century, which may lead to drastic increase in heat-stress-related deaths (IPCC Working Group I et al., 2013).

5.2.2 Allergic Responses and Asthma Due to Decline in Air Quality

The extensive effects of climate change may lead to increased cases of respiratory and allergy illnesses worldwide. The World Health Organization (WHO) predicts that by the end of year 2030 the annual cost of healthcare for the management of diseases associated with climate change would be in the range of 2 to 4 billion. A major proportion of these expenses shall be related to allergic diseases, which include allergic rhinitis, asthma, atopic dermatitis, and food allergies that affect approximately10–20% of the global population (Ober and Yao, 2011) and about 37.5 million people in India (Krishna *et al.*, 2020). The prevalence of asthma and common allergic rhinitis has significantly increased during the last 40 years. There is growing evidence that suggests that the prevalence of allergic symptoms is due to changes in climate.

The concentration and global distribution of air-borne pollutants are significantly influenced by climate and weather. Pollen seasons last longer and produce more pollen due to increasing air temperatures and CO_2 concentration in the atmosphere. Therefore, allergy-related respiratory diseases occur more often and more severe climate change is predicted to affect food sources' nutritional value, which might have an impact on asthma and other related conditions. For instance, climate change modifications have been linked to changes in the composition of peanuts, which become more immunogenic owing to increasing CO_2 and atmospheric temperature (Beggs and Walczyk, 2008).

Globally, over seven million people die each year owing to illnesses related to $PM_{2.5}$ pollution, as stated in a study published by the WHO. Nine out of ten of the most polluted cities in the world are in India. In India, air pollution led to 1.24 million fatalaties in 2014, which rose to about 1.67 million by 2019. Most of these deaths were due to occupational pollution from particulate matter followed by residential pollution. Moreover, deaths due to particulate matter pollution and ozone pollution increased by 11.5% and 139%, respectively, between 1990 to 2019 in India. This in turn led to substantial economic costs nearly 5.7% of India's GDP (Maji *et al.*, 2016).

5.2.3 Vector-Borne Diseases

Blood-sucking insects are the primary agents for disease transmission from animals to humans and from humans to humans. Mosquitoes, ticks, flies, sandflies, fleas, bed bugs, and a few freshwater aquatic snails are sensitive to climate and their local population is dependent on environmental factors such as temperature and humidity. Steady rise in vector-borne diseases (VBDs) transmitting dengue, West Nile disease, chikungunya, zika and nipah viruses, especially in regions located within the tropics have been attributed to changes in climate (Thomson and Stanberry, 2022). Infectious organisms, vectors and secondary hosts that serve as reservoirs for pathogens are affected by local climatic conditions. Sensitivity towards the climate, changes in the rate of proliferation or reproduction of VBDs at higher temperatures, cause a prolonged transmission season (Semenza and Paz, 2021). For several VBDs, age-standardized DALY rates have fallen during the last 10 years. The vulnerable groups, especially children, old-aged people and pregnant women, are at a higher risk to VBDs. Other factors like socio-economic status, standard of living, occupation and access to healthcare affects the vulnerability towards VBDs (Semenza and Paz, 2021). India is already affected by at least six major VBDs namely chikungunya, dengue, malaria, filariasis, encephalitis and leishmaniasis. It has been predicted that kala-azar may re-emerge in north India and chikungunya in south India due to climate change (Dhiman *et al.*, 2010).

5.2.4 Mosquito-Borne diseases

5.2.4.1 Malaria

The protozoan *Plasmodium* spp., which is spread by female Anopheles mosquitoes, is the primary cause of malaria. There are five species known to transmit malaria to humans. They are *P. knowlesi, P. vivax, P. ovale, P. falciparum* and *P. malariae*. Malaria has been observed to be intimately linked with environmental conditions, including excessive rain and high humidity that contribute to an increase in mosquito reproduction and survival. Climate factors have a direct impact on vector reproduction, survival and parasite incubation time. India targets to eliminate malaria by the year 2030; however malaria infection in forest areas is still a significant health threat. In the year 2022, about 1.73 lakh cases and 64 deaths due of Malaria were reported in India (NCVBDC, 2022).

5.2.4.2 Dengue

Temperature, relative humidity, and rainfall can affect the transmission of dengue disease. Numerous dengue epidemics have occurred in the Asia-Pacific area because of early monsoons and exceptionally humid and warm weather. Dengue virus is spread by *Aedes aegypti* and *Aedes albopictus* mosquitoes. Increased temperatures shorten the extrinsic incubation time, i.e., the time needed for the virus to replicate and propagate within the mosquito. This takes place before the virus gets into the mosquitoes' salivary glands and infects people. The probability that a mosquito may infect a man during its life-cycle increases when it becomes contagious due to rise in temperature. A large number of fatalities from dengue, which has the second-highest incidence of VBDs, occurs in Asia (Bhatt *et al.*, 2013). There have been 1.1 lakh cases and 86 deaths due to dengue in 2022 in India (NCVBDC, 2022). Since 1950, there has been an increase in the worldwide burden of dengue for the past several decades, which may be attributed to expansion of the vector species, transportation, urbanization and poor vector control methods. Globally, higher dengue case incidence and/or transmission rates are strongly correlated with temperature, relative humidity, and rainfall. Increased dengue incidence has been correlated to variations in winds patterns, sea surface temperatures and precipitation across the tropical regions (Pramanik *et al.*, 2020).

5.2.4.3 Chikungunya

This viral disease is transmitted by the *Aedes albopictus* and *Aedes aegypti* mosquitoes. Chikungunya is caused by arbovirus, which results in significant post-stress hair loss and nutrient shortages. More than a million people in India and on the Réunion Islands in the southwest Indian Ocean were impacted in 2005. There have been 5320 cases chikungunya cases in 2022 in India (NCVBDC, 2022). Many individuals have been impacted by this virus since the sickness has reemerged in South India. It indicates that a variety of variables, including environment, have a role in the re-emergence of the chikungunya virus.

5.2.4.4 Zika Virus

Altering climatic trends have made it easier for zika fever and Japanese encephalitis to spread globally. The zika virus, which is a flavivirus also spreads through mosquitoes. The zika virus outbreaks have been observed to occur following extreme temperatures and drought conditions (Tesla *et al.*, 2018). Usage of domestic water storage containers is also connected to increasing household exposure to the Aedes vector. Similarly, the transmission of Japanese encephalitis to higher elevations seems to be related to changing environment (Ghimire and Dhakal, 2015). There have been 942 cases of JEV and 117 deaths in 2022 in India (NCVBDC, 2022).

5.2.5 NON-VECTOR-BORNE DISEASES (NVBDs)

5.2.5.1 Avian Influenza

Avian influenza is also called bird flu, is caused by influenza virus strain, and birds serve as its hosts. According to the World Health Organization, there were over 900 human cases of H7N9 from 2013 to 2017. The majority of those who were affected had either eaten tainted chicken items or visited poultry markets. Climate change not only alters the bird migratory patterns but also modifies the environment for the replication and transmission of influenza virus. Thus, climate change plays a significant role in the spread of avian influenza.

5.2.6 WATER-BORNE DISEASES (WBDs)

Water-borne diseases (WBDs) include chistosomiasis, leptospirosis, hepatitis A and E, poliomyelitis, and diarrheal diseases include cholera, shigella, cryptosporidiosis and typhoid. The key factor contributing to the burden of WBDs is drinking water contaminated with pathogenic bacteria. Intestinal disease outbreaks, in particular, are caused by contamination of local drinking water systems with various pathogens, such as bacteria, protozoa, viruses or parasites. WBDs and high temperatures have been shown to be positively correlated, particularly in regions with severe water, sanitation and hygiene (WASH) inadequacies (Levy *et al.*, 2016;2018). WBD outbreaks including diarrheal illness have regularly been linked to heavy precipitation episodes, water scarcity and drought (Boithias *et al.*, 2016).

There is a correlation between higher risks of various gastro-intestinal (GI) diseases and heavy rainfall, warmer temperatures, and drought. The bacterial causes of GI infections seem to grow with temperature, while humidity and rainfall have varying effects on bacterial-linked GI infections (Levy *et al.*, 2016). Pathogens in the environment can enter water treatment and distribution systems when intense or prolonged precipitation flushes them from open fields into groundwater and surface water resources like rivers, and lakes.

Cholera risk is raised in areas where there has been a lot of rain and temperature is warmer than usual environment. The bacterium *Vibrio cholera* causes acute diarrheal illness resulting in cholera, which has high morbidity and fatality rate. Cholera cases show a correlation with maximum and minimum temperatures as well as precipitation.

5.2.7 FOOD-BORNE DISEASES (FBDs)

Sickness caused by consuming food that has been contaminated by pathogens, viruses, parasites, toxins, pesticides or medications is referred to as food-borne disease (FBD). The risk of FBDs is prevalent throughout the food chain right from sources of food production to consumption. They result from contamination of food source and inappropriate handling, preparation and storage of food items (Semenza and Paz, 2021). Similar to WBDs, FBD outbreaks can have multiple root causes resulting from interaction of climatic risk factors with food production and distribution of system, urbanization, scarcity of resource and energy due to population and overutilization of resources, declining agricultural yield, and changes in dietary trends. Malnutrition also enhances the risk of FBDs due to low immunity and increased susceptibility towards food-borne infections and toxins.

Increases in FBDs are strongly correlated with a prolonged hot summer season and high temperatures in the air and water. The food-borne pathogens with low infectious dose, high stress tolerance to temperature change, and extensive environmental persistence (e.g., enteric viruses, Campylobacter spp., *E. coli* strains, Mycobacterium, Salmonella and parasitic protozoa) pose major challenge to health (Semenza and Paz, 2021). Increased usage of chemicals to enhance food production (fertilizers, pesticide, herbicides antibiotics and other veterinary medications), and their possible residues in food are a major concern for health due to FBDs.

Increases in Salmonella infections have been shown to be strongly correlated with increase in the average ambient temperature. Salmonella infections often result in salmonellosis, but certain strains, such as *Salmonella typhi*, may also cause typhoid fever. One of the most common FBDs, non-typhoidal Salmonella infection, is often transferred to people by consuming food that has been contaminated with animal feces. Cryptococcus outbreaks in humans and animals have been linked to confluence of climate variables and changes in host and vector populations. In the tropics and sub-tropics, the prevalence of cryptosporidiosis, the second most common cause of diarrhoea in infants, is correlated with high rainfall. It has been noticed that contamination of food source from *Cryptosporidium* spp. is 2.61 times more common during and after precipitation episodes (Khalil *et al.*, 2018).

5.2.8 INFECTIONS OF THE RESPIRATORY TRACT

Extremes in temperature and humidity conditions, dust storms, severe rainfall and enhanced climate variability contribute as risk factors for respiratory tract infections (RTIs) caused by numerous pathogens. Pneumonia and influenza are two RTIs that have a large disease burden and may be caused by both climatic and non-climate variables. Chronic diseases such as chronic obstructive pulmonary disease (COPD) and asthma, comorbidities, compromised immune system, smoking and air pollution may further complicate the relationship between climate change and pneumonia.

According to reports, pneumonia incidence is greater in tropical and sub-tropical regions during the rainy season, indicating a correlation between pneumonia patterns and climatic conditions (Lim and Siow, 2018). Flu outbreaks have generally been associated with periods of low temperatures and high humidity in temperate regions and increased precipitation in the regions within tropics (Lam *et al.*, 2020). The dynamics of the influenza pandemic may be impacted by large-scale climatic variability factors like ENSO La Nina and the Indian Ocean Dipole that greatly limit the frequency of weather patterns in particular regions of the globe (Oluwole, 2017).

5.2.9 NON-COMMUNICABLE DISEASES (NCDS)

Non-communicable diseases (NCDs) are those that cannot be transferred directly from one person to another. Whereas in high-income nations, NCDs account for around 80% of the disease burden, in low- and middle-income nations, the burden is lower but is predicted to increase (Bollyky *et al.*, 2017). Non-communicable respiratory illness, cardiovascular disease (CVD), endocrine disorders, such as diabetes and cancer, are among the diseases classified as NCDs that are primarily caused by environmental, lifestyle and other factors (Figure 5.2).

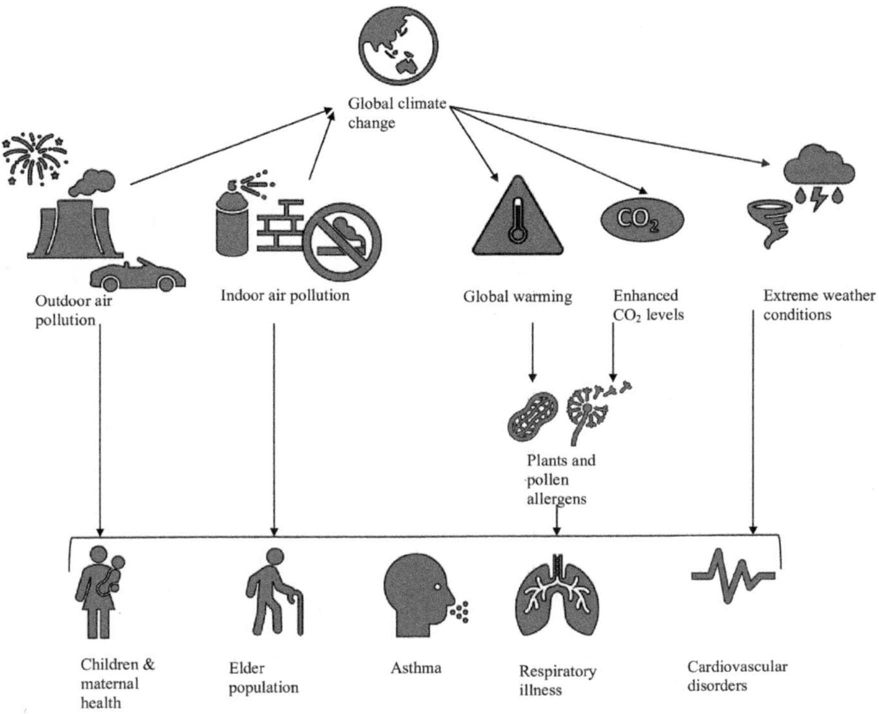

FIGURE 5.2 Climate change causes non-communicable diseases.

5.2.10 CARDIOVASCULAR CONDITIONS

Coronary heart disease, cerebrovascular disease, peripheral arterial disease, rheumatic heart disease, congenital heart disease, deep vein thrombosis and pulmonary embolism are among the heart and blood vessel illnesses collectively known as cardiovascular diseases (CVDs). The main cause of mortality worldwide is CVD, and low- and middle-income nations currently account for more than 75 percent of all CVD fatalities (Roth *et al.*, 2020). In 2016 India reported 63% of total deaths due to NCDs, of which 27% were attributed to CVDs. Of these deaths, 85% were due to heart attack and stroke. The results of the Global Burden of Disease Study 2021 stated age-standardized CVD death rate of 272 per lakh population in India, which is much higher than that of global average of 235 (Sreeniwas *et al.*, 2020). High temperatures, intense heat and other climate change factors like air pollutants enhance the risk of CVDs. For instance, exposure to air pollutants such as particulate matter (PM 10, PM 2.5), ozone, black carbon, oxides of nitrogen and sulfur, hydrocarbons and traces of metals may cause prothrombotic conditions, endothelial dysfunction, pro-inflammatory and hypertensive reactions (Giorgini *et al.*, 2017).

Climate-change-induced wildfires result in smoke-related CVD morbidity and mortality risks; and significantly higher rates of cardiovascular conditions (e.g., cardiac arrests). Reduced physical activity due to hot weather, sleep disturbances, and dehydration are other results of climate change that raise CVD risk. Moreover, saline groundwater intrusion due to sea level rise may lead to consumption of more salt, leading to raise blood pressure and ultimately CVD risk (Talukder *et al.*, 2017).

5.2.11 Respiratory Conditions That Are Non-transmittable

The most prevalent non-communicable respiratory pulmonary diseases include asthma, COPD and lung cancer. The burden of all chronic lung diseases including lung cancer, is significant worldwide and accounted for 10.6% of fatalities and 5.9% of DALYs in the year 2019 (Abbafati *et al.*, 2020b). Non-communicable respiratory disease is caused by a variety of exposure pathways, some of which are related to climate change such as the movement and transport of dust particles and air pollutants like smoke particles, PM 2.5 PM 10 and ozone.

5.2.12 Cancer

Climate change enhances the risk of developing cancer in people exposed to carcinogens. The estimated number of incident cases of cancer in India for the year 2022 was found to be 14.61 lakh (crude rate:100.4 per 100,000) and there were around 2.7 million cancer cases and 0.85 million deaths due to cancer in India in 2020 (Sathishkumar *et al.* 2022). In India, one in nine people are likely to develop cancer in their lifetime. GLOBOCAN predicted that cancer cases in India would increase to 2.08 million, accounting for a rise of 57.5 percent in 2040 from 2020 (Sung *et al.*, 2020).

Climate change may affect transportation and fate of carcinogens like polyaromatic hydrocarbons as well as the mobilization of persistent organic pollutants (POPs) like polychlorinated-biphenyls and bromide from areas contaminated by industrial runoff. Increased floods due to climate change may cause mobilization of sediments with accumulated carcinogens and expose people to the known carcinogens.

Environmental pollutants and chlroflorocarbons (CFCs) can deplete the ozone layer, which may lead to higher exposure to UV radiations. UV light exposure may enhance the prevalence of skin cancers like malignant melanoma, especially in outdoor workers. Additional cancer causing risks include the spread of infectious diseases like schistosomiasis, which raises cancer risk owing to migration associated with climate change, and greater exposure to liver flukes resulting in hepatobiliary cancer. Cyanobacteria pose challenges as they are sources of carcinogens and are expected to become more frequent and widely dispersed due to climate change (Prueksapanich *et al.*, 2018).

5.2.13 Diabetes

It has been shown that extreme weather conditions and increasing temperatures increase morbidity and mortality in diabetic patients, particularly those with cardiovascular comorbidity (Zilbermint, 2020). The estimates in 2019 showed that 77 million individuals had diabetes in India, which is expected to rise to over 134 million by 2045. The prevalence of diabetes in India has risen from 7.1% in 2009 to 8.9% in

2019. Total deaths in India, attributable directly to diabetes were about 1 million (Pradeepa *et al.*, 2021). Evidence shows that diabetes-related deficiencies affect the flow of blood under the skin in response to local heat loss or pharmacological stimulation. Type-2 diabetes may also decrease thermoregulatory perspiration, which affects the body's capacity to lose heat to the surrounding environment resulting in electrolytic imbalance (Xu *et al.*, 2019).

Due to treatment disruptions and drug access issues, people with chronic diseases are more vulnerable during and after severe weather events. The effects of severe weather events on the health of persons with chronic illnesses are caused by a variety of issues, such as transportation problems, compromised healthcare systems, including medication supply chains, power outages, and population migration.

5.2.14　Effects on the Health of the Maternal, Fetal and Newborn

According to the available research, heat is linked to greater rates of preterm delivery, low birthweight, stillbirth, neonatal stress, and poor child health (Wang *et al.*, 2020). Extreme weather events are linked to decreased access to pediatric healthcare, unattended births and prenatal care.

5.2.15　Mental Wellness

Climate change may influence the food, social and economic status of an individual, which can have a toll on mental health. Farmers in particular have been found to be more likely to commit suicide as a result of the economic effects of droughts. Moreover, the occupations that are likely to be impacted by climate change are a prominent source of substance abuse and suicidal tendencies among people (Kabir, 2018). In the year 2021, the number of suicides among agricultural workers in India increased by 9 percent from 2020 and by around 29 percent from 2019 (NCRB, 2021). The risk of mental illness increases when family members pass away, for instance because of severe weather disasters, such as floods, droughts and wildfires. The population in low and middle income nations has more severe effects because they have less access to mental health treatments and less money to deal with the adverse effects of climate-induced disasters than people in high-income countries (Keyes *et al.*, 2014).

High temperatures have been linked to a visible decline in mental health. Suicide, psychiatric hospital admissions, trips to the emergency department for mental illnesses, feelings of worry, sadness and acute stress are only a few of the mental health consequences linked to various climatic hazards, such as heatwaves (Liu *et al.*, 2021), wildfires (Eisenman and Galway, 2022), drought (OBrien *et al.*, 2014), floods (Hrabok *et al.*, 2020) and storms (Obradovich *et al.*, 2018), which are strongly associated with climate change.

According to previous studies from South America and Asia, depression and post-traumatic stress disorder (PTSD) strike 20–30% of storm survivors within a few months after the catastrophic incident (Rataj *et al.*, 2016). These percentages are comparable for individuals who have experienced floods. After severe weather occurrences, children and adolescents are more prone to developing depression and PTSD.

There have been harmful effects of wildfires on mental health due to the shock of what happened right away or following relocation and evacuation after the disaster. In reaction to wildfires and harsh weather, sub-clinical consequences, such as increases in anxiety, insomnia or drug misuse, have been recorded.

5.3 OTHER HEALTH EFFECTS

Certain populations, such as children, the elderly, and Indigenous people, are more vulnerable to a variety of negative effects of climate change. Multiple injuries caused by floods, fires and storms are clearly climate sensitive, but more data are needed to determine the present injury burden associated with climate change. There is a serious concern about coastal flooding becoming worse due to climate change since it affects 120 million people yearly (Nicholls *et al.*, 2007) resulting in 12000 fatalities due to cyclones (Shultz *et al.*, 2005).

5.3.1 Hunger, Malnutrition and Food Insecurity

Climate change has a direct impact on fisheries, forests, and agriculture. The metabolism, biodiversity, and nutrient cycle of plants and animals are all impacted by climate change. Food security may be at danger because of the advent of new diseases and pests that damage plants and animals. Crops, cattle, fish and ultimately people will all face new dangers because of these changes. Climate change will have an impact on the stability of the food system as well as the accessibility, usage and availability of food. Additionally, it may influence food production, livelihoods and human health. Food chains will be affected, market prices may rise, livelihood chances may be endangered and health may also be at risk. The most vulnerable groups are those whose livelihoods are reliant on agriculture. They are initially affected by the possibility of crop failure, insect and disease assaults, a shortage of seeds and animal losses. Thus, the climate-change-linked food insecurity has a strong relation with hunger and malnutrition.

The worst cases of hunger are seen in Africa, India, certain regions of north Asia, and the Western Pacific. More people worldwide are undernourished and food insecure. There may be a drop in the production of basic foods because of rising temperatures and erratic precipitation risking a large population to starvation and malnutrition. By the middle of the twenty-first century, crop yields in Central and South Asia may decrease (Aryal *et al.*, 2020). Additionally, nutrient concentrations and greenhouse gas emissions might have an impact on plant development. The productivity and economic progress of developing countries are hampered by their large populations, inadequate nutrition and bad health. The WHO estimates that 2.5 billion people worldwide need access to better sanitation and more than 700 million people lack access to clean drinking water (WHO, 2008). Children under the age of three experience at least three bouts of diarrhea annually in poor nations. Due to the loss of essential nutrients, during diarrheal diseases children show stunted development, vitamin D deficiency, Beriberi, Kwashiorkor, hypoproteinemia, goiter, etc.

In low- and middle-income nations, climate variability and change leads to, food insecurity thereby, resulting in malnutrition and disease susceptibility. Moreover, it is observed that undernutrition, obesity, and susceptibility to NCDs are all impacted by food insecurity and limited availability to nutrient-rich food in low- and middle-income countries including India (FAO United Nations, 2020).

Globally, about 690 million people are undernourished, 144 million children suffer from chronic undernutrition that results in stunting, 47 million children suffer from acute undernutrition that results in wastedness, and over 2 billion people have micro-nutrient deficiencies (FAO United Nations, 2020). Increased temperatures may cause people to be less active, pay more for produce, or change their eating habits to include more processed foods, all of which may raise the risk of obesity (Cole *et al.*, 2018).

Thus, all the four facets of food security that isproduction and availability of food, stability of food supply, access to food, and food useare impacted by climate change (FAO United Nations, 2019). Extreme weather and climatic events may cause people to consume too little food thereby increasing their vulnerability to infectious and chronic diseases.

5.3.2 IMPACTS OF CLIMATE CHANGE ON LIVELIHOOD, MIGRATION AND HUMAN SETTLEMENTS

People are forced to look for better economic circumstances and jobs, either temporarily or permanently, due to drought and other climate catastrophes happening suddenly. Migration results from climate change both inside countries and beyond borders. Certain food items may become scarce at certain periods of the year due to climate fluctuation. Livelihoods may be impacted by such seasonal fluctuations in food availability, as well as vulnerability to floods and other hazards. Due to the proximity of people to the poverty line, these effects are crucial for people belonging to marginal groups who depend heavily on food for their survival. Agriculture, cattle raising and fishing are the marginal communities' primary sources of income. The low-income population impacted by drought and flooding, as well as those with inadequate food distribution and emergency response systems, are among the livelihood categories that are of concern in the context of climate change. Therefore, such abrupt changes in temperature and precipitation might destroy the way of life.

The exposure, vulnerability and adaptability to a wide variety of climatic and non-climatic factors like social, economic, cultural and political factors affect the migration and displacement decision among population. Indirect causes of migration and displacement include things like rural income losses and/or food insecurity caused by crop failures brought on by extreme heat or drought, which in turn lead to fresh population shifts. Direct drivers include catastrophe such as tropical cyclones that destroy dwellings. Droughts, excessive heat and changes in precipitation patterns are more likely to cause long term changes in migratory patterns. Besides this, major storms, floods and wildfires are significantly linked to both short and long term displacements (Hoffmann *et al.*, 2020).

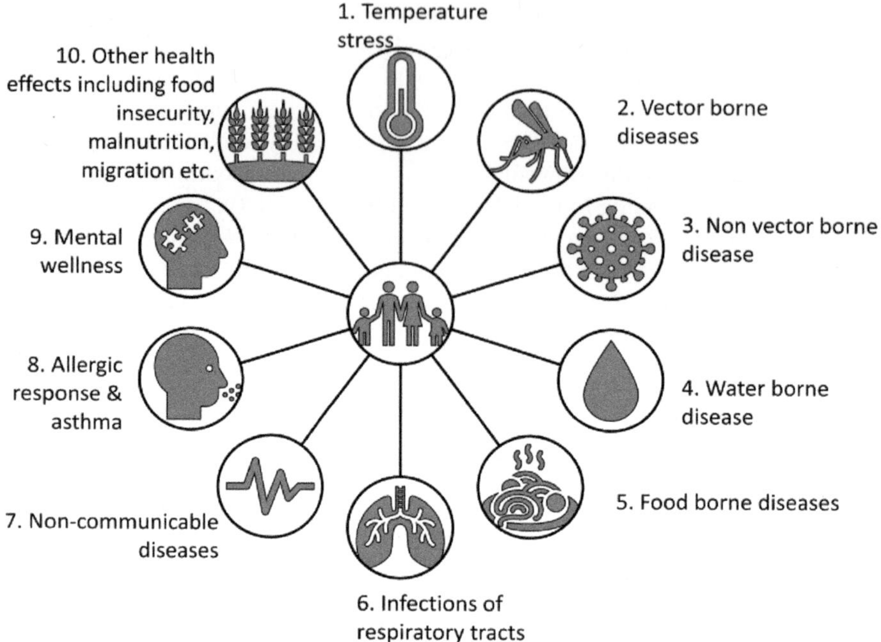

FIGURE 5.3   Effects of climate change on human health.

The effect of climate change on various aspects of human health is illustrated in Figure 5.3.

5.4 CONCLUSIONS

While industrial and economic development is required for sustaining human growth, it must not be at the cost of misusing the natural resources, which may have a negative impact on the environment. A conscious effort is required not to create environmental pollutants that will skew the balance of the components of the environment, including air, water and soil. Changes in the natural resources will inevitably have an impact on the climate of our planet. Changes in climate variability result in extreme weather events and changes in the ecosystems. These changes modify the vector habitats and favor the spread of diseases. Therefore, it is important to address climate change and raise awareness in order to address human health.

Earth has a very balanced climatic condition that sustains life as compared to other planets where extremes of temperature, gases and radiations make them unworthy of existence of life. It is therefore required that societal awareness be created about the value of climate in sustaining life. Policies made worldwide for economic development must prioritize ways that will have minimum impact on environment and in sustaining the natural resources. It is imperative that the effect of climate on human health must be at the forefront for any human development index so that the future of life on Earth is at a dynamic balance with nature.

REFERENCES

Abbafati C, Abbas K M, Abbasi-Kangevari M, et al. (2020a) Global burden of 87 risk factors in 204 countries and territories, 1990–2019: a systematic analysis for the Global Burden of Disease Study 2019. *The Lancet* 396: 1223–1249. https://doi.org/10.1016/S0140-6736(20)30752-2

Abbafati C, Abbas K M, Abbasi-Kangevari M, et al. (2020b) Global burden of 369 diseases and injuries in 204 countries and territories, 1990–2019: a systematic analysis for the Global Burden of Disease Study 2019. *The Lancet* 396: 1204–1222. https://doi.org/10.1016/S0140-6736(20)30925-9

Ahima R S (2020) Global warming threatens human thermoregulation and survival. *J Clin Invest* 130 : 559–561. https://doi.org/10.1172/JCI135006

Aryal J P, Sapkota T B, Khurana R, et al. (2020) Climate change and agriculture in South Asia: adaptation options in smallholder production systems. *Environ Dev Sustain* 22: 43777–43791. https://doi.org/10.1007/s10668-019-00414-4

Balakrishnan K, Dey S, Gupta T, et al. (2019) The impact of air pollution on deaths, disease burden, and life expectancy across the states of India: the Global Burden of Disease Study 2017. *Lancet Planet Health* 3: e26–e39. https://doi.org/10.1016/S2542-5196(18)30261-4

Beggs P J, Walczyk N E (2008) Impacts of climate change on plant food allergens: a previously unrecognized threat to human health. *Air Qual Atmos Health* 1: 119–123. https://doi.org/10.1007/s11869-008-0013-z

Bhatt S, Gething P W, Brady O J, et al. (2013) The global distribution and burden of dengue. *Nature* 496: 504–507. https://doi.org/10.1038/nature12060

Boithias L, Choisy M, SoELKuliyaseng N, et al. (2016) Hydrological regime and water shortage as drivers of the seasonal incidence of diarrheal diseases in a tropical montane environment. *PLoS Negl Trop Dis* 10: e0005195. https://doi.org/10.1371/journal.pntd.0005195

Bollyky T J, Templin T, Cohen M, Dieleman J L (2017) Lower-income countries that face the most rapid shift in noncommunicable disease burden are also the least prepared. *Health Aff* 36: 1866–1875. https://doi.org/10.1377/hlthaff.2017.0708

Bose-O'Reilly S, Daanen H, Deering K, et al. (2021) COVID-19 and heat waves: new challenges for healthcare systems. *Environ Res* 198: 111153. https://doi.org/10.1016/j.envres.2021.111153

Climate Transparency Report (2022). www.climate-transparency.org/wp-content/uploads/2022/10/CT2022-Summary-report.pdf. Accessed 15 June 2024.

Cole M B, Augustin M A, Robertson M J, Manners J M (2018) The science of food security. *NPJ Sci Food* 2. https://doi.org/10.1038/s41538-018-0021-9

Dash A P, Bhatia R, Sunyoto T, Mourya D T (2013) Emerging and re-emerging arboviral diseases in Southeast Asia. *J Vector Borne Dis* 50: 77–84.

Dhiman R C, Pahwa S, Dhillon G P S, Dash A P (2010) Climate change and threat of vector-borne diseases in India: are we prepared? *Parasitol Res* 106: 763–773. https://doi.org/10.1007/s00436-010-1767-4

Eisenman D P, Galway L P (2022) The mental health and well-being effects of wildfire smoke: a scoping review. *BMC Public Health* 22: 2274. https://doi.org/10.1186/s12889-022-14662-z

FAO United Nations (2019) Food security and right to food. www.unicef.org/media/55921/file/SOFI-2019-full-report.pdf. Accessed 5 June 2024.

FAO United Nations (2020) The state of food security and nutrition in the world. www.unicef.org/media/72676/file/SOFI-2020-full-report.pdf. Accessed 10 June 2024.

Ghimire S, Dhakal S (2015) Japanese encephalitis: challenges and intervention opportunities in Nepal. *Vet World* 8: 61-65. https://doi.org/10.14202/vetworld.2015.61-65

Giorgini P, di Giosia P, Petrarca M, et al. (2017) Climate changes and human health: a review of the effect of environmental stressors on cardiovascular diseases across epidemiology and biological mechanisms. *Curr Pharm Des* 23: 3247–3261. https://doi.org/10.2174/1381612823666170317143248

Hoffmann R, Dimitrova A, Muttarak R, et al. (2020) A meta-analysis of country-level studies on environmental change and migration. *Nat Clim Chang* 10: 904–912. https://doi.org/10.1038/s41558-020-0898-6

Hrabok M, Delorme A, Agyapong V I O (2020) Threats to mental health and well-being associated with climate change. *J Anxiety Disord* 76. https://doi.org/10.1016/j.janxdis.2020.102295

IPCC (2018) Global warming of 1.5°C. An IPCC Special Report on the impacts of global warming of 15°C above pre-industrial levels and related global greenhouse gas emission pathways, in the context of strengthening the global response to the threat of climate change, 2: www.ipcc.ch/site/assets/uploads/sites/2/2019/06/SR15_Full_Report_High_Res.pdf. Accessed 10 June 2024.

IPCC Working Group I, Stocker T F, Qin D, et al. (2013) IPCC, 2013: climate change 2013: the physical science basis. Contribution of Working Group I to the Fifth Assessment Report of the Intergovernmental Panel on Climate Change. IPCC AR5. www.ipcc.ch/site/assets/uploads/2017/09/WG1AR5_Frontmatter_FINAL.pdf. Accessed 10 June 2024.

Kabir S M S (2018) Psychological health challenges of the hill-tracts region for climate change in Bangladesh. *Asian J Psychiatr* 34: 74–77. https://doi.org/10.1016/j.ajp.2018.04.001

Keyes K M, Pratt C, Galea S, et al. (2014) The burden of loss: unexpected death of a loved one and psychiatric disorders across the life course in a national study. *Am J Psychiatr* 171: 864–871. https://doi.org/10.1176/appi.ajp.2014.13081132

Khalil I A, Troeger C, Rao P C, et al. (2018) Morbidity, mortality, and long-term consequences associated with diarrhoea from Cryptosporidium infection in children younger than 5 years: a meta-analyses study. *Lancet Glob Health* 6: e758–e768. https://doi.org/10.1016/S2214-109X(18)30283-3

Krishna M T, Mahesh P A, Vedanthan P, et al. (2020) An appraisal of allergic disorders in India and an urgent call for action. *World Allergy Organ J* 13: 100446. https://doi.org/10.1016/j.waojou.2020.100446

Levy K, Smith S M, Carlton E J (2018) Climate change impacts on waterborne diseases: moving toward designing interventions. *Curr Environ Health Rep* 5: 272–282. https://doi.org/10.1007/s40572-018-0199-7

Levy K, Woster A P, Goldstein R S, Carlton E J (2016) Untangling the impacts of climate change on waterborne diseases: a systematic review of relationships between diarrheal diseases and temperature, rainfall, flooding, and drought. *Environ Sci Technol* 50: 4905–4922. https://doi.org/10.1021/acs.est.5b06186

Li L, Jiang C, Murtugudde R, et al. (2021) Global population exposed to extreme events in the 150 most populated cities of the world: implications for public health. *Int J Environ Res Public Health* 18: 1293. https://doi.org/10.3390/ijerph18031293

Lim T K, Siow W T (2018) Pneumonia in the tropics. *Respirology* 23: 28–35. https://doi.org/10.1111/resp.13137

Liu J, Varghese B M, Hansen A, et al. (2021) Is there an association between hot weather and poor mental health outcomes? A systematic review and meta-analysis. *Environ Int* 153: 106533.

Maji K J, Dikshit A K, Deshpande A (2016) Human health risk assessment due to air pollution in 10 urban cities in Maharashtra, India. *Cogent Environ Sci* 2: e1700066. https://doi.org/10.1080/23311843.2016.1193110

Mazdiyasni O, AghaKouchak A, Davis S J, et al. (2017) Increasing probability of mortality during Indian heat waves. *SciAdv* 3. https://doi.org/10.1126/sciadv.1700066

National Centre for Vector Borne Disease Centre (NCVBDC). Annual Report 2022. ncvbdc.mohfw.gov.in

National Crime Records Bureau India (NCRB). Crime in India 2021 Statistics. https://ncrb.gov.in/en/Crime-in-India-2021. Accessed 10 June 2024.

Nicholls R J, Wong P P, Burket V R, et al. (2007) Coastal systems and low-lying areas. Climate change 2007: impacts, adaptation and vulnerability. www.ipcc.ch/site/assets/uploads/2018/02/ar4-wg2-chapter6-1.pdf. Accessed 10 June 2024.

Ober C, Yao T C (2011) The genetics of asthma and allergic disease: a 21st century perspective. *Immunol Rev* 242: 10–30. https://doi.org/10.1111/j.1600-065X.2011.01029.x

Obradovich N, Migliorini R, Paulus M P, Rahwan I (2018) Empirical evidence of mental health risks posed by climate change. *Proc Natl Acad Sci USA* 115: 10953–10958. https://doi.org/10.1073/pnas.1801528115

OBrien L V., Berry H L, Coleman C, Hanigan I C (2014) Drought as a mental health exposure. *Environ Res* 131: 181–187. https://doi.org/10.1016/j.envres.2014.03.014

Oluwole O S A (2017) Dynamic regimes of El Niño southern oscillation and influenza pandemic timing. *Front Public Health* 5: 301. https://doi.org/10.3389/fpubh.2017.00301

Perin J, Mulick A, Yeung D, et al. (2022) Global, regional, and national causes of under-5 mortality in 2000–19: an updated systematic analysis with implications for the sustainable development goals. *Lancet Child Adolesc Health* 6: 106–115. https://doi.org/10.1016/S2352-4642(21)00311-4

Phillip P, Symons W, Ibrahim C, Hodges C, McGrath M. (2021) India's turning point: how climate action can drive our economic future, pp. 1–46. www2.deloitte.com/content/dam/ Deloitte/in/Documents/about-deloitte/in-indiaturning-point-noexp.pdf. Accessed 25 May 2024.

PIB Delhi (2022) Unusual rise in temperature due to climate change. Ministry of Environment, Forest and Climate Change. https://pib.gov.in/PressReleasePage.aspx?PRID=1809123. Accessed 15 June 2024.

Pradeepa R, Mohan V (2021). Epidemiology of type 2 diabetes in India. *Indian J Ophthalmol* 69(11): 2932–2938. https://doi.org/10.4103/ijo.IJO_1627_21

Pramanik M, Singh P, Kumar G, et al. (2020) El Niño Southern Oscillation as an early warning tool for dengue outbreak in India. *BMC Public Health* 20: 1498. https://doi.org/10.1186/s12889-020-09609-1

Prueksapanich P, Piyachaturawat P, Aumpansub P, et al. (2018) Liver fluke-associated biliary tract cancer. *Gut Liver* 12: 236–245. https://doi.org/10.5009/gnl17102

Rataj E, Kunzweiler K, Garthus-Niegel S (2016) Extreme weather events in developing countries and related injuries and mental health disorders – a systematic review. *BMC Public Health* 16: 1020. https://doi.org/10.1186/s12889-016-3692-7

Rocklöv J, Dubrow R (2020) Climate change: an enduring challenge for vector-borne disease prevention and control. *Nat Immunol* 21: 479–483. https://doi.org/10.1038/s41590-020-0648-y

Roth G A, Mensah G A, Johnson C O, et al. (2020) Global burden of cardiovascular diseases and risk factors, 1990-2019: update from the GBD 2019 study. *J Am Coll Cardiol* 76: 2982–3021. https://doi.org/10.1016/j.jacc.2020.11.010

Sathishkumar K, Chaturvedi M, Das P, Stephen S, Mathur P (2022). Cancer incidence estimates for 2022 & projection for 2025: result from National Cancer Registry Programme, India. *Indian J Med Res.* 156(4–5): 598–607. https://doi.org/10.4103/ijmr.ijmr_1821_22

Semenza J C, Paz S (2021) Climate change and infectious disease in Europe: impact, projection and adaptation. *Lancet Reg Health* 9: 100230. https://doi.org/10.1016/j.lanepe.2021.100230

Shultz J M, Russell J, Espinel Z (2005) Epidemiology of tropical cyclones: the dynamics of disaster, disease, and development. *Epidemiol Rev* 27: 21–35. https://doi.org/10.1093/epirev/mxi011

SreeniwasKumar A, Sinha N (2020) Cardiovascular disease in India: a 360 degree overview. *Med J Armed Forces India* 76(1): 1–3. https://doi.org/10.1016/j.mjafi.2019.12.005

Stewart S, Keates A K, Redfern A, McMurray J J V (2017) Seasonal variations in cardiovascular disease. *Nat Rev Cardiol* 14: 654–664. https://doi.org/10.1038/nrcardio.2017.76

Sung H, Ferlay J, Siegel RL, et al. (2020) GLOBOCAN estimates of incidence and mortality worldwide for 36 cancers in 185 countries. *CA Cancer J Clin.* 71: 209–249. https://doi.org/10.3322/caac.21660

Talukder M R R, Rutherford S, Huang C, et al. (2017) Drinking water salinity and risk of hypertension: a systematic review and meta-analysis. *Arch Environ Occup Health* 72: 126–138. https://doi.org/10.1080/19338244.2016.1175413

Tesla B, Demakovsky L R, Mordecai E A, et al. (2018) Temperature drives Zika virus transmission: evidence from empirical and mathematical models. *Proc R Soc B Biol Sci* 285: 20180795. https://doi.org/10.1098/rspb.2018.0795

Thomson M C, Stanberry L R (2022) Climate change and vectorborne diseases. *N Engl J Med* 387 : 1969–1978. https://doi.org/10.1056/NEJMra2200092

Unicef India (2022) Climate change and environmental sustainability. *Unicef India.* www.unicef.org/india/what-we-do/climate-change. Accessed 6 January 2023.

United Nations (2019) World urbanization prospects – population division. United Nations.

Vicedo-Cabrera A M, Scovronick N, Sera F, et al. (2021) The burden of heat-related mortality attributable to recent human-induced climate change. *Nat Clim Chang* 11: 492–500. https://doi.org/10.1038/s41558-021-01058-x

Wang Q, Li B, Benmarhnia T, et al. (2020) Independent and combined effects of heatwaves and PM2:5 on preterm birth in Guangzhou, China: a survival analysis. *Environ Health Perspect* 128: 17006. https://doi.org/10.1289/EHP5117

WHO (2008) 2.5 billion live with poor sanitation facilities. www.who.int/news/item/17-07-2008-2.5-billion-live-with-poor-sanitation-facilities. Accessed 10 June 2024.

Xu Z, Tong S, Cheng J, et al. (2019) Heatwaves and diabetes in Brisbane, Australia: a population-based retrospective cohort study. *Int J Epidemiol* 48: 1091–1100. https://doi.org/10.1093/ije/dyz048

Zilbermint M (2020) Diabetes and climate change. *J Community Hosp Intern Med Perspect* 10: 409–412. https://doi.org/10.1080/20009666.2020.1791027

6 Use of Biotechnological Tools in Assessing Climate Impact on Mariculture

An Overview of Likely Scenarios over the Indian Subcontinent

*Dilip Kumar Jha, Josephine Anthony,
G. Dharani and R. Kirubagaran*

6.1 INTRODUCTION

The Indian subcontinent is a physiographic area in Southern Asia located on the Indian Plate, which extends southward from the Himalayas into the Indian Ocean. Geographically, it encompasses Bangladesh, Bhutan, India, the Maldives, Nepal, Pakistan and Sri Lanka (Figure 6.1). The Indian subcontinent evolved from Insular India, an isolated landmass that rifted from the Cretaceous supercontinent of Gondwana and fused with the mainland of Eurasia roughly 55 million years ago, generating the Himalayas (Jones, 2011). The Tibetan Plateau to the north, the Indochinese Peninsula to the east, the Iranian Plateau to the west, and the Indian Ocean to the south are the geographical regions surrounding the subcontinent. The zone where the Eurasian and Indian subcontinent plates meet remains geologically active and prone to major earthquakes. The Indian subcontinent occupies the major landmass of South Asia (Lukacs, 2013). The Indian Ocean, the Bay of Bengal, and the Arabian Sea form the boundary of the Indian subcontinent in the south, southeast, and southwest respectively (Mukherjee, 2001).

The growing human population puts pressure on coastal regions to use resources, which leads to habitat deterioration, fragmentation and destruction (Gray, 1997). The anthropogenic activity along the Indian subcontinent is high; the maritime ecosystem remains generally pristine in the Islands (Mahajan *et al.*, 1996; Sahu *et al.*, 2013) due to its distance from neighboring nations such as Myanmar, Thailand, Malaysia, Indonesia and India's mainland. Mariculture has a long history in Southern Asia, particularly the Indian subcontinent, and it has a significant impact on the seafood

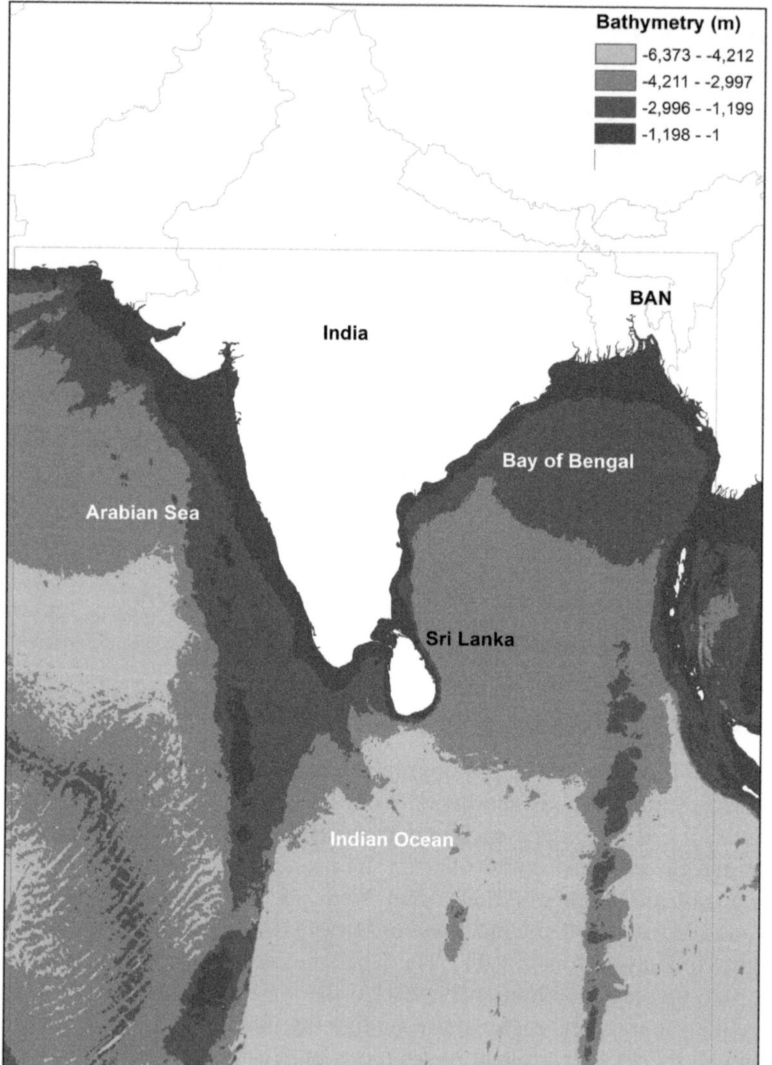

FIGURE 6.1   Study area map.

industry (FAO, 2016). While certain wild fisheries may benefit from climate change by 2050, the total landings are predicted to fall by 10% globally (Barange *et al.*, 2014).

The oceans have a key role in global climate dynamics because they absorb 90% of the heat accumulated in the atmosphere and also absorb 25% of the additional carbon dioxide since the industrial revolution, the latter causing ocean acidification (Tiedje *et al.*, 2022). Deoxygenation is also caused by a warmer, more stratified ocean, and all three risks are caused by excess carbon dioxide emissions from human activity (Gruber, 2011). Ocean plays a vital role in the absorption of carbon dioxide and converting it to carbonic acid, which combines with seawater. Carbonic acid dissociates very instantly to generate bicarbonate and hydrogen ions; nevertheless, seawater becomes more acidic as the concentration of hydrogen ions increases. However, some of the excess hydrogen ions mix with carbonates ions to generate more bicarbonate, which subsequently reduces the abundance of carbonate ions and affects many marine organisms that absorb carbonate from seawater to build calcium carbonate shells and skeletons (corals and clams) in a process called calcification. Ocean acidification affects the biological process such as building shells, maintaining metabolic processes, photosynthesis, and difficulty in obtaining the essential minerals and nutrients from seawater.

Ecosystem services and the surrounding environment are used to varied degrees by the majority of mariculture systems across the world. This suggests a possible vulnerability to climate change's consequences. To benefit from this scenario, adaptive capacity must be built by recognizing and adapting to climate change's effects (Cinner *et al.*, 2018). Understanding how climate change impacts biological responses, supplies, and economics in mariculture is therefore essential. However, the mariculture process is widely connected through a chain to other organisms such as microbes, algae and corals. It is known that microbes produce and consume primary greenhouse gases such as carbon dioxide, methane and nitrous oxide, whereas certain microbes cause diseases in humans, animals and plants that can be aggravated by climate change (Tiedje *et al.*, 2022). Further, macroalgae cultivation using CO_2 has been an important aspect, and the ocean is considered a carbon sink, which could support mitigating the climate impact (Barati *et al.*, 2021). Corals and fish are important entities in the mariculture cycle and hence any change in the climatic temperature could impact its growth and productivity. Under this circumstance, biotechnology is one of the essential climate mitigation tools and has already offered significant solutions such as environmental, physiological and genetic manipulation, etc., to protect mariculture against climatic calamities. However, recent techniques such as transgenesis technology (modifies or improves the genetic traits of commercially important fishes, molluscs and crustaceans for aquaculture) and nutrigenomics (effects of nutrients on genomes, proteome and metabolome) find indispensable applications (Müller and Kersten, 2003). In line with transgenesis, remodeling of chromosomes, improvement of disease tolerance, vaccine productions, gamete cryopreservation or gene banking, and other bioinformatics coupled tools piqued greater interest as climate mitigation tools in the aquaculture sector. To facilitate effective mariculture adaptation and understanding of the role of the biotechnological tool in assessing climate change, temperature fluctuation, sea-level rise, changing weather, sea surface salinity and ocean acidification research implementation is reviewed in context with the Indian subcontinent.

6.2 GLOBAL CLIMATE IMPACT OF TEMPERATURE

The global ocean covers 70% of the Earth and reaches an average depth of 4,000 meters. Oceanic ecosystems are physically and chemically varied, with over 50 biomes ranging from the tropics to the poles and from the bright surface layer to the dark abyss (Longhurst, 2010). Each biome supports a distinct microbe-based eco-system that functions as a complex adaptive system (Hagstrom and Levin, 2017), with emerging processes and services intricately related to habitat variability. Controlling the oceanic process temperature is a very crucial factor in the growth, metabolism and development of aquatic species (Ngoan, 2018). In the increasing global climatic impact, microbes play a vital role because of their ability in fixing atmospheric carbon. Microbes, on the other hand, play an important role in regulating the longevity and stability of carbon, as well as whether or not it is released into the atmosphere as a greenhouse gas, therefore mediating the carbon cycle processes (Weiman, 2015). It also slows down global warming, which has ramifications for critical ecological processes like the nitrogen cycle, which relies on microbial activity. It also plays an important role in the breakdown and transformation of decaying organic material into forms that may be utilized by other organisms and as a result, the microbial enzyme systems involved are regarded as critical "engines" that power the Earth's biogeo-chemical cycles (Abatenh *et al.*, 2018). Similarly, macroalgae and corals are also affected due to temperature changes, which could influence the ecosystem function. Considering the impact of global climate change, genetically modified microalgae are very important for enhanced carbon dioxide sequestration and enriched biomass productivity (Priyadharsini *et al.*, 2022). Table 6.1 provides a detailed account of organisms and genetically modified organisms in mitigating global climate change.

Since fish are poikilothermic, they may be especially vulnerable to temperature fluctuations caused by climate change (Adhikari *et al.*, 2018). Most fish, particularly cold-water species such as Atlantic halibut, salmon, cod and intertidal shellfish, are expected to die as a result of the anticipated 1.5°C rise in mean world temperature (Hamdan *et al.*, 2012; Gubbins *et al.*, 2013). Further, prolonged temperature stress may have an impact on aquaculture production in a variety of ways, with a focus on decreased output. Chronic stress, for example, has been shown to alter cardio-respiratory performance and aerobic scope, as well as immune responses in several economically important species (Brodie *et al.*, 2014; Gazeau *et al.*, 2014; Paukert *et al.*, 2016; Stewart *et al.*, 2019; Zhang *et al.*, 2019; Maulu *et al.*, 2021).

Furthermore, most finfish and shellfish species' metabolism and physiology, as well as dietary habits and growth performance, are likely to be impacted (Lemasson *et al.*, 2018). Meanwhile, rising ocean temperatures and subsequent ocean acid-ification gradually erode the ocean's carbon sink capacity, causing changes in the hydrology and hydrography of water systems (Cochrane *et al.*, 2009). Thermal strati-fication in deep water bodies caused by temperature variations may also have an impact on the distribution and abundance of nutrients in the water and in the event of upwelling, aquaculture producers operating in open waters will suffer severe eco-nomic losses (Seggel and De Young, 2016).

The global scenario has a great impact on climate change and related risks to health, food security, livelihood, water supply, human security and economic devel-opment. The increase in temperature due to global warming is projected to be 1.5°C (IPCC, 2018). The increment in global temperature, which is evident in the Indian subcontinent during mass coral bleaching events on various occasions is already reported (Krishnan *et al.*, 2011). An increment in mean temperature in the last 20 years, which triggered coral bleaching, has also been well explained (Jha *et al.*,

TABLE 6.1
Organisms and Genetically Modified Group's Role in Mitigating Global Climate Change

Sl. No.	Name of the Organism	Remarks
1	*Escherichia coli*	It produces its biomass from atmospheric CO_2, negating its need for sugar (Gleizer *et al.*, 2019).
2	*Chlamydomonas reinhardtii* and cyanobacterium *Anabaena variabilis*	Genetic improvement of microalgae for enhanced carbon dioxide sequestration and enriched biomass productivity (Barati *et al.*, 2021; Priyadharsini *et al.*, 2022)
3	Genetically modified heat-resistant coral and zooxanthellae	Genetically modified corals and algae (zooxanthellae) can provide better heat tolerance towards the warming oceans (Hobman *et al.*, 2022).
4	*Pleuronectes americanus* (Atlantic Ocean Bream)	Fishes undergo adverse stress upon low temperatures and it stimulates a signal to secrete antifreeze proteins or glycoproteins, which helps the fishes to prevent ice crystal formation by lowering the subzero temperature (Hew *et al.*, 1995).
5	*Halomonas sp. NIOT-EQR_ J251*	Efficient in mercury detoxification mechanisms by secreting Isooctyl thioglycolate compound. This metal-resistant species can create a suitable steady cellular environment for survival under mercuric stress conditions (Joshi *et al.*, 2022).
6	*Alicyclobacillus disulfidooxidans*	Tolerate extreme cold and hot temperature conditions, 4–40°C (Dopson *et al.*, 2023).
7	*Bacillus horneckiae*	These radio-resistant microbes could tolerate ultraviolet radiation up to 1000 J m^{-2} and hence could survive the environment, where even nuclear power plant wastes are discharged into seawater (Vaishampayan *et al.*, 2010).
8	*Symbiodinium* sp.	Genetically engineered microalgal species *Symbiodinium* sp. enhances the stress tolerance of *Symbiodinium* sp.,which reduces coral bleaching due to the rise in ocean temperature. This genetically modified microalga could produce molecules that feed the corals, which is indispensable for the corals to grow and form coral reefs (Levin *et al.*, 2017).

2018). Moreover, it is understood that the sea surface temperature (SST) lower value ranges from 23 to 24.1°C (2002–2021) whereas the higher SST value ranges from 31.1 to 32.6°C (Figure 6.2a–d). This model was prepared to analyze the change in the temperature for the past 20 years in the Indian subcontinent and results indicated that there is a change in the lower and higher temperature, which could be attributed to the

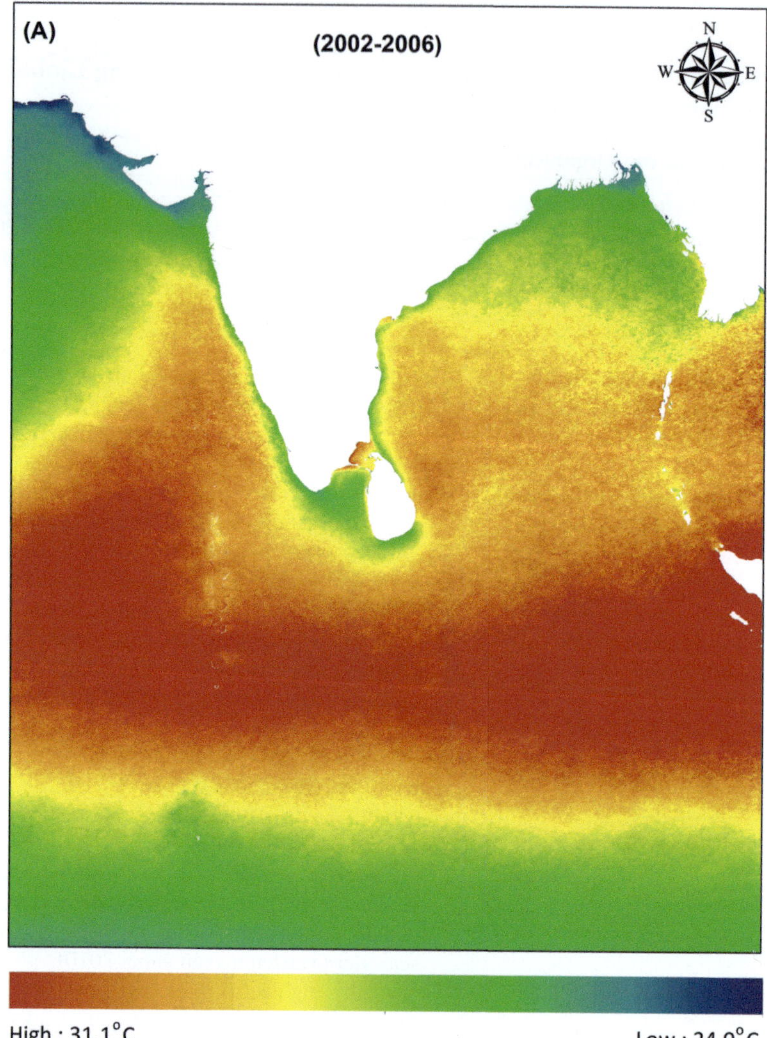

FIGURE 6.2　Temporal variations in mean SST in the Indian subcontinent. [(a) 2002–2006; (b) 2007–2011; (c) 2012–2016 and (d) 2017–2021].

global climate impact in this region. As global temperatures increase, the ice sheets in the Arctic and Antarctica are rapidly melting, which might have an impact on ocean water levels especially in the low-lying coastal zones. The temperature has an impact on the physiology and ecology of marine fish and shellfish (Pörtner and Farrell, 2008; Pörtner and Peck, 2010). It also affects oxygen consumption (Fry and Hart, 1948), hypoxia tolerance (Remen *et al.*, 2013), reproductive performance (Pankhurst and Munday, 2011), growth rate (Reid *et al.*, 2015), maturation (Wilkinson *et al.*, 2010), nutrition and feeding of fish (Britz *et al.*, 1997, Siikavuopio *et al.*, 2012).

To minimize the influence of temperature shifts on fish, biotechnological tools offer potent advantages. Generally, fishes undergo adverse stress upon low temperatures and it stimulates a signal to secrete antifreeze proteins or glycoproteins, which helps the fishes to prevent ice crystal formation by lowering the sub-zero temperature. Numerous studies have reported on the nature of these antifreeze proteins, particularly in the Atlantic Ocean Bream, *Pleuronectes americanus*. The genes encoding for the liver antifreeze protein of Winter Flounder fish were transformed into the genomic sequence of Atlantic Salmon and further incorporated into the germ line to develop the progeny with the potential to produce the antifreeze protein in the liver (Hew *et al.*, 1995). Such mutated traits were observed to be phenomenally efficient in overcoming the low-temperature effects on fish and improving production. However, a similar study for tropical fish needs to be strengthened and adopted for the betterment of fish resilience to temperature variation due to climate impact on mariculture production. Climate variability alters the spatial distribution pattern of fish and the catching potential, which has a substantial effect on fish landings especially tuna catch in the Indian Ocean (Kumar *et al.*, 2014). According to Sumaila *et al.* (2011), the relocation of fishing grounds and redistribution of tuna fisheries are advantageous among maritime countries and are anticipated effects of climate change that could have a significant impact on developing countries that rely on fishing in one way or another (Barange *et al.*, 2014). The productivity of fisheries will be impacted by the relocation of fishing grounds because fish will seek out habitat in an area where fishing is less intensive. This could support the areas where fishing activity is less such as Maldives, Lakshadweep, Andaman & Nicobar Islands, etc.

Further, according to models with a "business as usual" scenario, skipjack tuna biomass in the Indian Ocean is predicted to move to higher latitudes in the first half of the twenty-first century as equatorial waters are becoming warm due to global climatic impact (Dueri *et al.*, 2014). Small island economies that rely heavily on fishing and tourism like Maldives, Lakshadweep, etc., may be the first to feel the effects of the biomass shift as well as sea-level rise in the coastal vicinity.

6.3 CHANGING WEATHER AND SEA-LEVEL RISE

It is well understood that sea-level rise could potentially result in the loss of cultivable areas (Hargreaves, 2014), salt intrusion into coastal groundwater (Smajgl *et al.*, 2015), and in some areas, an increase in seasonal or episodic flooding due to storms surges (IPCC, 2013). The typical rise of sea level from 1901 to 1990 was about 1.2 mm/year, but it increased to about 3.0 mm/year from 1993 to 2010 (Hay *et al.*, 2015). By the end of the twenty-first century, it is quite likely that sea levels will increase in more than 95% of the world's oceans (Stocker *et al.*, 2013). The rise in the sea level could impact coastal cities and islands, which are located in the low-lying areas along the Indian subcontinent as well as in global scenarios. Most land-based aquaculture (such as ponds and hatcheries) located close to the coastal area is susceptible to flooding, which can result in escapes, the introduction of invasive species, and the pollution of culture water (Oyebola and Olatunde, 2019). Moreover, sea-level rise will also impact the coastal and marine water quality because of the higher mixing of fresh water in the marine environment. This could be attributed to a change in the fish's

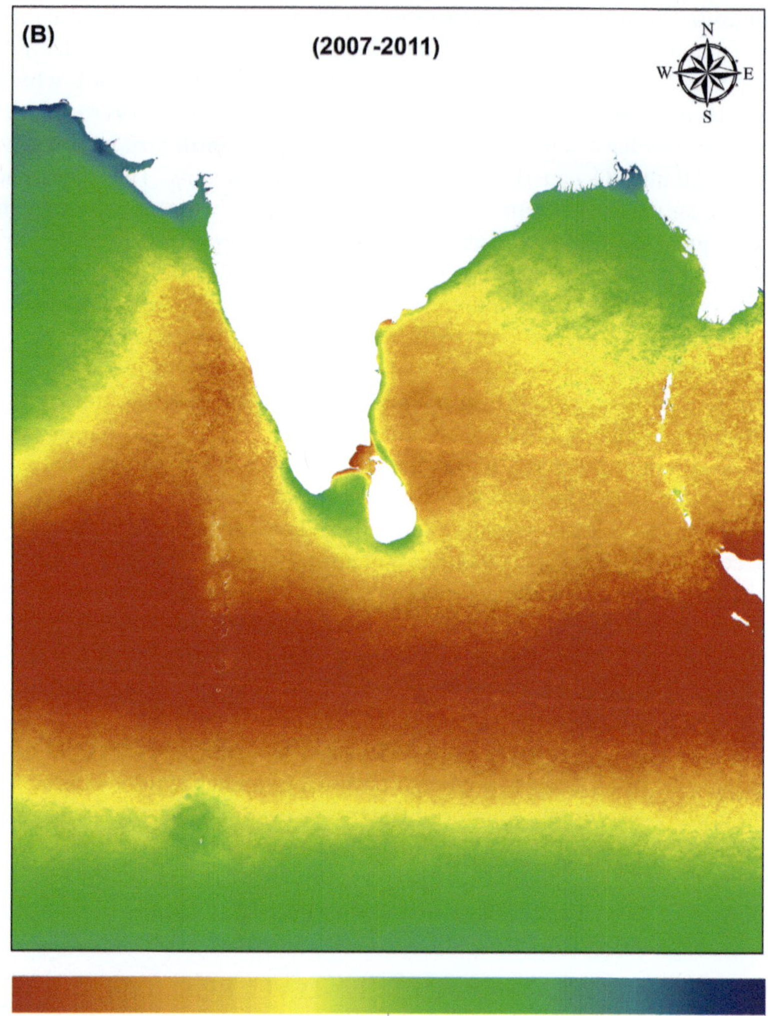

FIGURE 6.2   Continued

behavior or physiology or its osmoregulatory pattern due to continuous exposure to the varied salinity condition.

Due to impending El Niño/Southern Oscillation (ENSO) events, significant regional temperature fluctuations are predicted to increase annually. Because of sea-level rise, hypoxia is a prevalent condition in marine coastal regions and not only affects fish movement and metabolism but also harms several anti-predator behaviors (Domenici *et al.*, 2017). Significant changes in gene expression were also found in fishes under hypoxic circumstances, prompting the creation of a catalog of hypoxia-responsive genes in fishes (HRGFish) using bioinformatics tools such as Linux Apache MySQL PHP and Perl (LAMPP) technology (Rashid *et al.*, 2017).

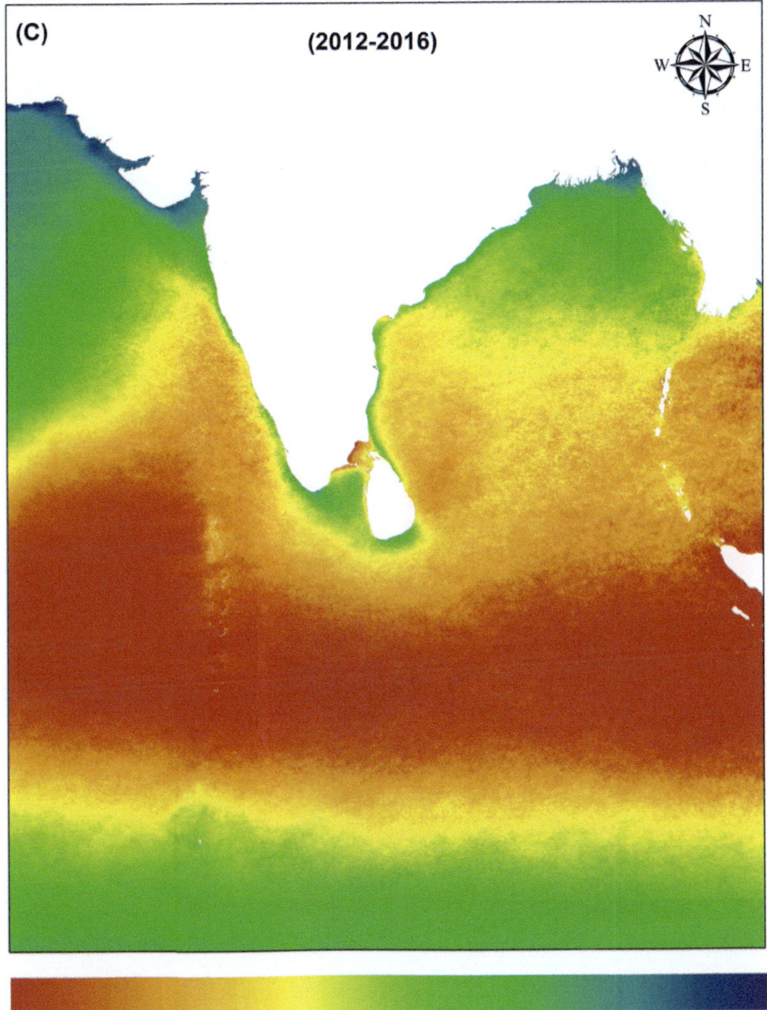

FIGURE 6.2 Continued

According to Wells *et al.* (2015), it is unclear how climate change may affect the frequency and intensity of harmful algal blooms in an aquatic ecosystem. Regardless of regional differences, it is predicted that general global warming will cause terrestrial areas to dry out due to increased evaporation (Sherwood and Fu, 2014). Increased eutrophication is expected in the twenty-first century as a consequence of changing precipitation trends and expanding farmland (Tilman *et al.*, 2001; Sinha *et al.*, 2017). Increased land runoff has the potential to intensify algal blooms, which lower dissolved oxygen levels and finally harm fish population. Microalgae offer a key contribution in minimizing the impact of climate change and pollution in the sea. Recently, carbon capture and storage methods have gained

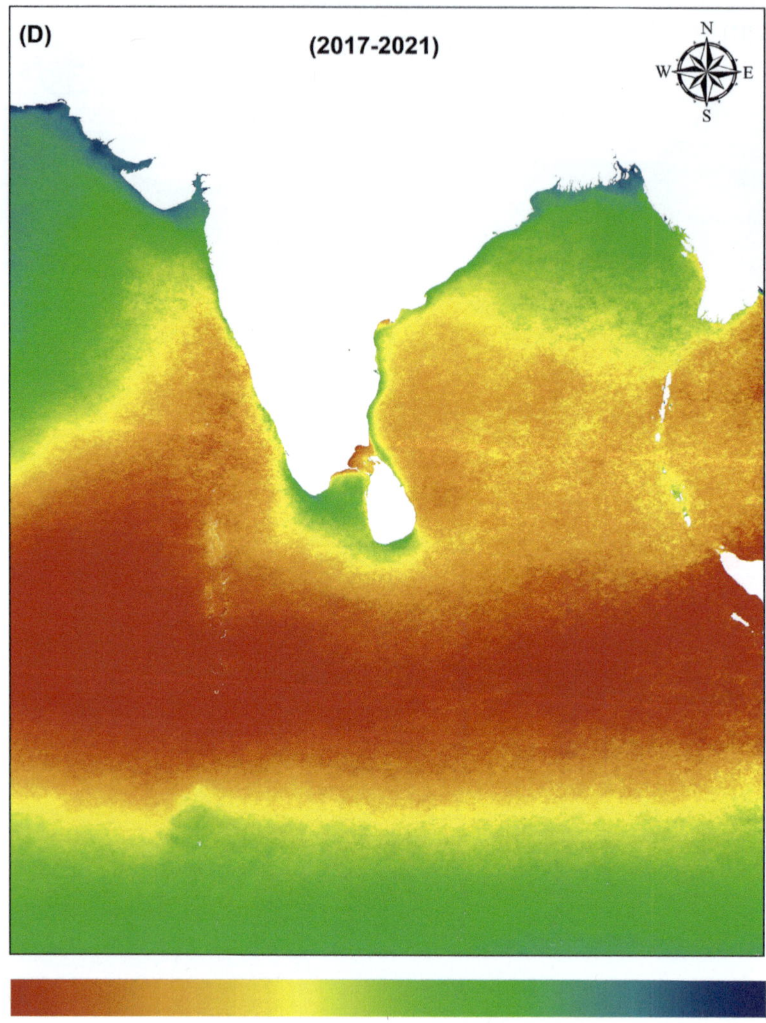

FIGURE 6.2   Continued

less attention, as they are economically not feasible, although it is one of the most promising technologies to alleviate the threatening scenario of climate change. On the other hand, the biological capture of carbon dioxide by microalgae, and utilizing it for generating algal biomass is considered an alternative feasible option for recycling the excess carbon dioxide generated from pollution and natural calamities (Helen *et al.*, 2021).

As per the recent research, changes in storminess may have serious consequences for worldwide fisheries, particularly if they result in the loss of ships, facilities and communities (Salisbury *et al.*, 2008). Such concerns can be extended to coastal

TABLE 6.2
Cities and Islands Predicted to be Impacted by Sea-Level Rise

Sl. No.	Cities	Country	Remarks
1	Mumbai, Kochi, Mangalore, Chennai, Vishakapatnam, and Thiruvananthapuram	India	Coastal roads will be affected (www.livemint.com/news/india/mumbai-to-chennai-these-coastal-cities-that-may-drown-by-2050-11649384760837.html)
2	Kadmat and Kavaratti would suffer land loss along the shoreline. Androth is expected to be impacted the least while Minicoy, the southernmost island, and Agatti, which houses the archipelago's only airport, would face heavy flooding, the study suggests.	India	Minor land loss along the shoreline (www.zuccess.in/uploads/news/MARCH-2023/1679112914678.pdf)

aquaculture, where farms are extremely vulnerable to storm events (Luening, 2013), as evidenced by the storm-driven destruction of nearly complete land-based aquaculture sectors (Kais and Islam, 2018). Open-water aquaculture is also vulnerable to extreme weather, with large-scale escapes from marine nets linked to storms (Jensen *et al.*, 2010). Apart from the direct impact on mariculture, sea-level rise could also impact the coastal area by inundating the low-lying areas, and some of the important locations are listed in Table 6.2.

6.4 SEA SURFACE SALINITY CHANGE AND ITS IMPACT

Salinity is one of the important environmental parameters, which control and impact the stenohaline (can survive in a narrow range of salinity) and euryhaline (can survive in a higher range of salinity) organisms. The change in sea surface salinity (SSS) is attributed to the freshwater intake from precipitation, glacier melting, river runoff, water loss through evaporation, and the mixing and circulation of ocean surface water with subsurface water (Koblinsky *et al.*, 2003; Cochrane *et al.*, 2009). The change in SSS could be attributed to the greater evaporation caused by higher temperatures and changes in ocean circulation, or they might be caused directly by climate change (Robinson *et al.*, 2005). These changes could be attributed to climate change and hence environmental sustainability of aquaculture will get impacted because SSS variation will greatly impact the stenohaline fish and its productivity in the oceanic environment. Most aquatic organisms have SSS range especially stenohaline fishes and any changes in the SSS range may result in the mortality of the fish (Jahan *et al.*, 2019). The change in salinity levels beyond the permissible range has been shown to reduce survival, growth and red blood cell production in catfish, which ultimately affects the immune system of fish (Jahan *et al.*, 2019). The SSS variations are likely to have a detrimental impact on the economic benefits of some aquaculture species, thus affecting the socio-economic factors of sustainable production in aquaculture. However, a greater SSS effect has been significantly observed in aquaculture production systems in coastal areas (Nguyen *et al.*, 2018). Greater mortality was observed in immature clams with reduced seawater salinity (Baker *et al.*, 2013). Rodrick (2008) revealed that reduced seawater salinity makes oysters vulnerable to bacterial invasion because of a reduction in the ability of hemocytes (blood cells) to resist foreign bacterial invasion. Although clams have a higher tolerance to salinity than other marine mollusks, Baker *et al.* (2013) found that clams cannot tolerate extended exposure to either high or low salinity. In general, changes in water salinity would result in increasing mortalities for various species, threatening the sector's economic and social sustainability through greater species losses and higher management costs. The majority of existing knowledge on the influence of climate-related changes on SSS in aquaculture, however, has been biased toward reporting the effect of greater salinity. There is a need for research into the effects of salinity levels below an ideal need on finfish and shellfish. Furthermore, the response of some commercially important species to climate-induced salinity variations is unknown and needs focused research in this sector. This information is particularly valuable for aquaculture adaptation, since variations in salinity may favor the development of resistant species (Jahan *et al.*, 2019).

6.5   OCEAN ACIDIFICATION

Ocean acidification is a very important aspect and is defined as the reduction of ocean pH over time (typically decades) as a result of atmospheric CO_2 accumulation (Richards *et al.*, 2015; Bahri *et al.*, 2018). It is a well-known fact that seas are thought to contain around 50 times higher CO_2 than the atmosphere (Seggel and De Young, 2016). At 1.5°C or higher temperature rise, the expected increase in CO_2 absorption by seas would harm the growth, development, calcification, survival and abundance of various aquatic organisms (IPCC, 2018). The effects of ocean acidification on tuna fishing are challenging to measure because our understanding of the involved processes and possible synergistic interactions (e.g., between pH and temperature changes) is presently very limited. Studies on the impacts of ocean acidification on fishing, in general, are rare, and practically non-existent in the Indian Ocean (Hilmi *et al.*, 2013; Sumaila *et al.*, 2011). Ocean acidification is occurring more quickly in coastal and estuarine areas (Waldbusser *et al.*, 2011), which are important aquaculture production zones, with more variable pH and CO_2 levels than in the open ocean, as local inputs and dual cycles of photosynthesis and respiration can drive changes in pH beyond the influence of ocean acidification. Natural variability, eutrophication, net heterotrophy (Kemp *et al.*, 2005; Borges and Gypens, 2010), acid-forming chemical deposition (Doney *et al.*, 2012), regional changes in sediment saturation (Green *et al.*, 2009), and watershed sources can all cause variations in coastal and estuarine pH (Dove and Sammut, 2007; Salisbury *et al.*, 2008), which could be correlated to mortality for juvenile bivalves. Many factors influence the experimental outcomes of ocean acidification exposure studies on shellfish, including species, acute *versus* chronic exposure, animal size, parental exposure, the driver investigated (i.e., CO_2, calcium carbonate saturation, pH), the magnitude of change, diet availability, and the presence of other stressors (Ries *et al.*, 2009; Thomsen *et al.*, 2010; Gobler *et al.*, 2014). The bulk of shelled mollusc larval research on aquaculture and industrial fishing species has found unfavorable reactions to ocean acidification (Gazeau *et al.*, 2013). Because most shellfish farming depends on naturally occurring wild or hatchery-derived juveniles as the initial stage of production, declines in seed availability will significantly impact industry viability. The effects of ocean acidification on physiology and reproduction are expected to have a significant influence on the long-term viability of cultivated or wild shellfish populations. In wild finfish, impairments in olfactory responses under elevated pCO_2 have been linked to receptor impairment of GABA-A, the primary inhibitory neurotransmitter receptor in the vertebrate brain (Reid *et al.*, 2019).

By contrasting simulations from ten different Earth system models (ESMs), it was possible to determine the robustness of anticipated trends under the "business as usual" scenario (RCP 8.5) (Bopp *et al.*, 2013). The comparison demonstrates the robustness of the Indian Ocean Dipole (IOD)-like warming pattern that was previously highlighted. Between models, surface pH trends also reveal a gradual, very uniform decline in the Indian Ocean. Strong projections show a drop in the southeast and the central Indian Ocean (from the southern border of Indonesia to the eastern boundary of Madagascar) and a rise in the western Indian Ocean (north of 5°S) subsurface dissolved oxygen content. Models reliably reproduce the decline in net primary production (NPP) in the tropical Indian Ocean's western upwelling zone.

Powerful techniques, such as nuclear and isotopic techniques find an indispensable role in studying the carbon cycle and ocean acidification. The ocean's capacity to store carbon, the influence of ocean acidification on marine organisms, past and future changes due to ocean acidity can be widely explored using these techniques (Zahed *et al.*, 2021). Hence the elimination of carbon dioxide is the need of the hour to mitigate the threat of climatic fluctuations and ocean acidification. Biotechnological applications such as carbon sequestration-based technologies play a vital role, wherein biofuel can provide a sustainable solution for carbon sequestration, which is a non-toxic and cost-effective approach. Most interestingly, novel genomic approaches such as recombinant DNA technology and gene-splicing technology produce genetically modified organisms to make renewable biofuel and biomaterial production (Grama *et al.*, 2022). Microorganism-based biosurfactants, biopolymers and biochars emerge as sustainable future technologies to mitigate ocean acidification (Zahed *et al.*, 2021).

The epigenetic response potential of fishes (Pittman *et al.*, 2013) and marine invertebrates (Sanford and Kelly, 2011) suggests some level of adaptive capacity to climate change, but there are significant knowledge gaps. The capacity of marine organisms to adapt to increased acidity is largely unknown, as few studies have considered acclimation times of more than a few months (Reid *et al.*, 2019). Aquaculture-based populations are less likely to be impacted due to the potential to manipulate culture conditions (Richards *et al.*, 2015). Selective breeding programs for desirable traits are already common in aquaculture, and this may provide additional options for climate change adaptation. Generational improvements have the potential to occur rapidly for some traits. Some of the most globally produced culture species are sea bass, cobia and pompano. There is also compelling anecdotal evidence that environmental stressor tolerance traits may "evolve" under culture conditions. Ellis *et al.* (2017) reported that finfish in recirculated aquaculture systems (RAS) are routinely raised in high-CO_2 environments, often in excess of end-of-century predictions, sometimes >10000 μatm, with no apparent ill effects. While this apparent discrepancy from many studies on wild species could be for several reasons, it is suggested one possibility may be that aquaculture species selection and breeding under intensive culture environments may have conferred greater tolerance to high CO_2. Oyster breeding programs have experienced a decline in survival and yield due to changes in oceanic conditions (de Melo *et al.*, 2016). Further, salinity-tolerant strains of shark catfish *Pangasianodon hypophthalmus* are now being developed to adapt to saline water intrusion from floods in the Mekong Delta, Vietnam (Nguyen *et al.*, 2018). Adaptation through domestication of a variety of species tolerant to climatic variations may also reduce dependency on wild-caught seeds (Oyebola and Olatunde, 2019).

Compared to capture fisheries, which have risen at an average annual rate of 1.2%, and terrestrial farmed meat-production systems, which have increased at an average annual rate of 2.8% since 1970, aquaculture has had tremendous growth of 8.9% per year as a sector for food production (Niku *et al.*, 2014; Kumar *et al.*, 2022). The aquaculture industry has grown rapidly, which has increased environmental effects, including the production of large amounts of polluted effluent that contains uneaten feed, feces and aquatic discharge that contains nutrients as well as different organic and inorganic compounds like ammonium, phosphorus, dissolved organic carbon and organic matter. According to the FAO, the financial losses from a disease epidemic in the aquaculture sector are expected to be over USD 9 billion annually or almost 15% of the value of global fish and shellfish output. Traditional approaches to stress reduction, such as the use of antibiotics and disinfectants, have had limited success and also have extra negative effects on human health, including allergies and toxicity as well as changes to the natural microbiota of the human gut.

To reduce stressful situations in aquatic habitats and boost crustacean immune reactivity, natural products or substances derived from plants are required. The enhanced heat shock protein response, which plays a significant role in stabilizing the protein structure and function and producing protective immunity against abiotic and biotic stresses, is an interesting behavioral and physiological adaptive mechanism that crustaceans have developed to avoid the problems caused by environmental change. However, if the unfavorable environmental condition persists, it outpaces the body's natural defense mechanisms and impairs immunity, growth, survival, metabolism and reproduction. To preserve aquatic animals' homeostasis and prevent integrative and/or multiple stressors, management options, such as the use of heat shock protein-inducing (Hspi) chemicals, may offer a promising practical and sustainable method (Kumar *et al.*, 2022). For the avoidance of stressful situations in crustacean aquaculture, natural compounds from medicinal plants and marine seaweeds are viewed as promising alternatives.

To completely utilize these traits or enhance traits at a rate adequate to meet the rate of climate change, genomic approaches may be required. When it is not feasible to incorporate climate change-related characteristics into a broodstock program, keeping as much genetic diversity in the breeding should be a goal. The preservation of genetic variation may guarantee the survival of rare alleles linked with resilience to a potential illness or greater survival to environmental stressors (Gurney-Smith *et al.*, 2017). Advances in commercial-scale cryopreservation enable "gene banking", or the retention of genetics from one parent (male) regardless of continued inclusion in a broodstock program, and viable larval preservation protocols have been and are being developed for some marine shellfish, depending on species (Adams *et al.*, 2004). As a result, gene or germplasm storage strategies have been promoted as biological protection for future aquaculture breeding and stock selection requirements (Hulata, 2001). Currently, chromosomal remodeling and sex modification methods are used to generate polyploidy and uniparental chromosome transmission, as well as to produce fish offspring with the potential to accomplish pubertal ablation, mutant survival enhancement, and cloning rapidly (Lakra and Das 1998; Baroiller *et al.*, 1999). Furthermore, molecular screening methods, immunostimulants and vaccinations are becoming increasingly popular around the globe to increase disease resilience in aquatic species (Karunasagar and Karunasagar, 1999). Biotechnological tools are highly essential across the world, particularly in mariculture, to mitigate the effects of climate change on aquatic species. The 70-kilodalton (kDa) heat shock protein (HSP70) chaperones have been used as potential biomarkers for environmental protection and the role they may play in cellular defenses against climate change effects to predict a possible future geographic distribution of a wide range of marine species within the context of global warming, particularly thermal stress, salinity change and acidification (Yusof *et al.*, 2022).

6.6 CONCLUSION

Climate change is widely recognized as the most significant modern threat to the environment and humanity. According to the Intergovernmental Panel on Climate Change (IPCC), 3.3 billion people globally are susceptible to climate change. Microbes are said to play important roles in climate change. They produce and consume carbon dioxide, methane and nitrous oxide, the three major gases responsible for 98% of the increased warming [i.e. photo- or chemoautotrophic growth (cyanobacteria, algae, nitrifiers), methanotrophy (methane oxidizers) and nitrous oxide reduction (denitrifiers)].

It is observed that climate change may modify the spatial distribution pattern and negatively impact the abundance of commercially important tunas in the Indian Ocean. Feeding and spawning habitat may shift towards higher latitudes and/or deeper water (if not limited by oxygen and light conditions). Fisheries and national economics will have to adapt to climate change impacts. Some coastal countries located at higher latitudes may benefit from the spatial redistribution of tunas. The combined effects of climate and socio-economic changes may lead to a decline in fish abundance and eventually a migration of fish into deeper water, however, recent technological advancements like heat shock proteins (Hsps) act as the agent of being an immune booster and increasing disease resistance will present a significant advancement in reducing stressful conditions in the aquaculture system. To reduce these negative effects, which could have disastrous consequences, we need to create a mitigation program that uses biotechnological tools to develop a fish type that can survive in adverse environmental conditions.

ACKNOWLEDGMENTS

The authors are thankful to the Ministry of Earth Sciences, Govt. of India, for extending their support for the present study. The authors are grateful to Dr. G.A. Ramadass, Director, NIOT, Chennai for his continual encouragement, support and guidance. We would also like to thank the scientific and supporting staff of NIOT's MBT in Chennai, India, for their timely help.

REFERENCES

Abatenh, E., Gizaw, B., Tsegaye, Z., & Tefera, G. (2018). Microbial function on climate change – a review. *Environment Pollution and Climate Change*, 2(1), 1–6.

Adams, S. L., Smith, J. F., Roberts, R. D., Janke, A. R., Kaspar, H. F., Tervit, H. R., ...& King, N. G. (2004). Cryopreservation of sperm of the Pacific oyster (*Crassostrea gigas*): development of a practical method for commercial spat production. *Aquaculture*, 242(1–4), 271–282.

Adhikari, S., Keshav, C. A., Barlaya, G., Rathod, R., Mandal, R. N., Ikmail, S., ...& Sundaray, J. K. (2018). Adaptation and mitigation strategies of climate change impact in freshwater aquaculture in some states of India. *Journal of Fisheries Sciences*, 12(1), 16–21.

Bahri, T., Barange, M., & Moustahfid, H. (2018). Chapter 1: climate change and aquatic systems. In: *Impacts of Climate Change on Fisheries and Aquaculture, Synthesis of Current Knowledge, Adaptation and Mitigation Options*, Eds. M. Barange, T. Bahri, M. C. M. Beveridge, K. L. Cochrane, S. Funge-Smith, & F. Poulaine. FAO: Rome, pp. 1–17.

Baker, S., Hoover, E., & Sturmer, L. (2013). *The Role of Salinity in Hard Clam Aquaculture*. Gainesville, FL: University of Florida.

Barange, M., Merino, G., Blanchard, J. L., Scholtens, J., Harle, J., Allison, E. H., ...& Jennings, S. (2014). Impacts of climate change on marine ecosystem production in societies dependent on fisheries. *Nature Climate Change*, 4(3), 211–216.

Barati, B., Zeng, K., Baeyens, J., Wang, S., Addy, M., Gan, S. Y., & Abomohra, A. E. F. (2021). Recent progress in genetically modified microalgae for enhanced carbon dioxide sequestration. *Biomass and Bioenergy*, 145, 105927.

Baroiller, J. F., Guiguen, Y., & Fostier, A. (1999). Endocrine and environmental aspects of sex differentiation in fish. *Cellular and Molecular Life Sciences CMLS*, 55, 910–931.

Bopp, L., Resplandy, L., Orr, J. C., Doney, S. C., Dunne, J. P., Gehlen, M., ...& Vichi, M. (2013). Multiple stressors of ocean ecosystems in the 21st century: projections with CMIP5 models. *Biogeosciences*, 10(10), 6225–6245.

Borgesa, A. V., & Gypensb, N. (2010). Carbonate chemistry in the coastal zone responds more strongly to eutrophication than ocean acidification. *Limnology and Oceanography*, 55(1), 346–353.

Britz, P. J., Hecht, T., & Mangold, S. (1997). Effect of temperature on growth, feed consumption and nutritional indices of *Haliotis midae* fed a formulated diet. *Aquaculture*, 152(1–4), 191–203.

Brodie, J., Williamson, C. J., Smale, D. A., Kamenos, N. A., Mieszkowska, N., Santos, R., ...& Hall-Spencer, J. M. (2014). The future of the northeast Atlantic benthic flora in a high CO_2 world. *Ecology and Evolution*, 4(13), 2787–2798.

Cinner, J. E., Adger, W. N., Allison, E. H., Barnes M. L., Brown, K., Cohen, P. J.,...& Morrison, T. H. (2018). Building adaptive capacity to climate change in tropical coastal communities. *Nature Climate Change*, 8, 117–123.

Cochrane, K., De Young, C., Soto, D., & Bahri, T. (2009). Climate change implications for fisheries and aquaculture. *FAO Fisheries and Aquaculture Technical Paper*, 530, 212.

de Melo, C. M. R., Durland, E., & Langdon, C. (2016). Improvements in desirable traits of the Pacific oyster, *Crassostrea gigas*, as a result of five generations of selection on the West Coast, USA. *Aquaculture*, 460, 105–115.

Domenici, P., Steffensen, J. F., & Marras, S. (2017). The effect of hypoxia on fish schooling. *Philosophical Transactions of the Royal Society B: Biological Sciences*, 372(1727), 20160236.

Doney, S. C., Ruckelshaus, M., Emmett Duffy, J., Barry, J. P., Chan, F., English, C. A., ...& Talley, L. D. (2012). Climate change impacts on marine ecosystems. *Annual Review of Marine Science*, 4, 11–37.

Dopson, M., González-Rosales, C., Holmes, D. S., & Mykytczuk, N. (2023). Eurypsychrophilic acidophiles: from (meta) genomes to low-temperature biotechnologies. *Frontiers in Microbiology*, 14, 1149903.

Dove, M. C., & Sammut, J. (2007). Impacts of estuarine acidification on survival and growth of Sydney rock oysters *Saccostrea glomerata* (Gould 1850). *Journal of Shellfish Research*, 26(2), 519–527.

Dueri, S., Bopp, L., & Maury, O. (2014). Projecting the impacts of climate change on skipjack tuna abundance and spatial distribution. *Global Change Biology*, 20(3), 742–753.

Ellis, R. P., Urbina, M. A., & Wilson, R. W. (2017). Lessons from two high CO_2 worlds – future oceans and intensive aquaculture. *Global Change Biology*, 23(6), 2141–2148.

FAO (2016). *The State of World Fisheries and Aquaculture 2016. Contributing to Food Security and Nutrition for All*. Rome, p. 200.

Fry, F., & Hart, J. S. (1948). The relation of temperature to oxygen consumption in the goldfish. *The Biological Bulletin*, 94(1), 66–77.

Gazeau, F., Alliouane, S., Bock, C., Bramanti, L., López Correa, M., Gentile, M., ...& Ziveri, P. (2014). Impact of ocean acidification and warming on the Mediterranean mussel (*Mytilus galloprovincialis*). *Frontiers in Marine Science*, 1, 62.

Gazeau, F., Parker, L. M., Comeau, S., Gattuso, J. P., O'Connor, W. A., Martin, S., ...& Ross, P. M. (2013). Impacts of ocean acidification on marine shelled molluscs. *Marine Biology*, 160, 2207–2245.

Gleizer, S., Ben-Nissan, R., Bar-On, Y. M., Antonovsky, N., Noor, E., Zohar, Y., ...& Milo, R. (2019). Conversion of *Escherichia coli* to generate all biomass carbon from CO_2. *Cell*, 179(6), 1255–1263.

Gobler, C. J., DePasquale, E. L., Griffith, A. W., & Baumann, H. (2014). Hypoxia and acidification have additive and synergistic negative effects on the growth, survival, and metamorphosis of early life stage bivalves. *PloS One*, 9(1), e83648.

Grama, S. B., Liu, Z., & Li, J. (2022). Emerging trends in genetic engineering of microalgae for commercial applications. *Marine Drugs, 20*(5), 285.

Gray, J. S. (1997). Marine biodiversity: patterns, threats and conservation needs. *Biodiversity & Conservation, 6*(1), 153–175.

Green, M. A., Waldbusser, G. G., Reilly, S. L., Emerson, K., & O'Donnell, S. (2009). Death by dissolution: sediment saturation state as a mortality factor for juvenile bivalves. *Limnology and Oceanography, 54*(4), 1037–1047.

Gruber, N. 2011. Warming up, turning sour, losing breath: ocean biogeochemistry under global change. *Philosophical Transactions: Mathematical, Physical and Engineering Sciences, 369*, 1980–1996.

Gubbins, M., Bricknell, I. & Service, M. (2013). Impacts of climate change on aquaculture. *Marine Climate Change Impacts Partnership: Science Review*, 318–327. www.mccip. org.uk/sites/default/files/2021-

Gurney-Smith, H. J., Wade, A. J., & Abbott, C. L. (2017). Species composition and genetic diversity of farmed mussels in British Columbia, Canada. *Aquaculture, 466*, 33–40.

Hagstrom, G. I., & Levin, S. A. (2017). Marine ecosystems as complex adaptive systems: emergent patterns, critical transitions, and public goods. *Ecosystems, 20*, 458–476.

Hamdan, R., Kari, F., Othman, A., & Samsi, S. M. (2012). Climate change, socio-economic and production linkages in East Malaysia aquaculture sector. *2012 International Conference on Future Environment and Energy IPCBEE* (Vol. 28). Singapore: IACSIT Press.

Hargreaves, J. A. (2014). Editor's note: climate change is here now. *Aquaculture Magazine, 45*(3).

Hay, C. C., Morrow, E., Kopp, R. E., & Mitrovica, J. X. (2015). Probabilistic reanalysis of twentieth-century sea-level rise. *Nature, 517*(7535), 481–484.

Helen, O. H., Miri, T., Obileke, K., Hart, A., Anumudu, C., & Al-Sharify, Z. T. (2021). Minimizing carbon footprint via microalgae as a biological capture. *Carbon Capture Science & Technology, 1*, 100007.

Hew, C. L., Fletcher, G. L., & Davies, P. L. (1995). Transgenic salmon: tailoring the genome for food production. *Journal of Fish Biology, 47*, 1–19.

Hilmi, N., Allemand, D., Dupont, S., Safa, A., Haraldsson, G., Nunes, P. A., ...& Cooley, S. R. (2013). Towards improved socio-economic assessments of ocean acidification's impacts. *Marine Biology, 160*, 1773–1787.

Hobman, E. V., Mankad, A., Carter, L., & Ruttley, C. (2022). Genetically engineered heat-resistant coral: an initial analysis of public opinion. *PloS One, 17*(1), e0252739.

Hulata, G. (2001). Genetic manipulations in aquaculture: a review of stock improvement by classical and modern technologies. *Genetica, 111*, 155–173.

IPCC (2013). Climate change 2013: the physical science basis. In: *Contribution of Working Group I to the Fifth Assessment Report of the Intergovernmental Panel on Climate Change*, Eds. T. F. Stocker, D. Qin, G.-K. Plattner, M. Tignor, S. K. Allen, J. Boschung, A. Nauels, Y. Xia, V. Bex, & P. M. Midgley. Cambridge University Press: Cambridge, UK and New York, NY, USA, p. 1535.

IPCC (2018). Global warming of 1.5°C. In: *An IPCC Special Report on the Impacts of Global Warming of 1.5°C Above Pre-industrial Levels and Related Global Greenhouse Gas Emission Pathways, in the Context of Strengthening the Global Response to the Threat of Climate Change, Sustainable Development, and Efforts to Eradicate Poverty*, Eds. V. Masson-Delmotte, P. Zhai, H.-O. Pörtner, D. Roberts, J. Skea, P. R. Shukla, A. Pirani, W. Moufouma-Okia, C. Péan, R. Pidcock, S. Connors, J. B. R. Matthews, Y. Chen, X. Zhou, M. I. Gomis, E. Lonnoy, T. Maycock, M. Tignor, & T. Waterfield. www.ipcc.ch/site/ass ets/uploads/sites/2/2022/06/SR15_Summary_Volume_HR.pdf

Jahan, A., Nipa, T. T., Islam, S. M., Uddin, M. H., Islam, M. S., & Shahjahan, M. (2019). Striped catfish (*Pangasianodon hypophthalmus*) could be suitable for coastal aquaculture. *Journal of Applied Ichthyology, 35*(4), 994–1003.

Jensen, Ø., Dempster, T., Thorstad, E. B., Uglem, I., & Fredheim, A. (2010). Escapes of fishes from Norwegian sea-cage aquaculture: causes, consequences and prevention. *Aquaculture Environment Interactions, 1*(1), 71–83.

Jha, D. K., Kirubagaran, R., Vinithkumar, N. V., Dharani, G., & Madeswaran, P. (2018). The Andaman and Nicobar Islands. In: *World Seas: An Environmental Evaluation*, Ed. C. Sheppard. UK and USA: Academic Press, pp. 185–209.

Jones R. W. (2011). *Applications of Palaeontology: Techniques and Case Studies*. Cambridge University Press. pp. 267–271. ISBN 978-1-139-49920-0.

Joshi, G., Verma, P., Meena, B., Goswami, P., Peter, D. M., Jha, D. K., ...& Dharani, G. (2023). Unraveling the potential of bacteria isolated from the equatorial region of Indian Ocean in mercury detoxification. *Frontiers in Marine Science*, 2022, 9.

Kais, S. M., & Islam, M. S. (2018). Impacts of and resilience to climate change at the bottom of the shrimp commodity chain in Bangladesh: a preliminary investigation. *Aquaculture, 493*, 406–415.

Karunasagar, I., & Karunasagar, I. (1999). Diagnosis, treatment and prevention of microbial diseases of fish and shellfish. *Current Science, 76*(3), 387–399.

Kemp, W. M., Boynton, W. R., Adolf, J. E., Boesch, D. F., Boicourt, W. C., Brush, G., ...& Stevenson, J. C. (2005). Eutrophication of Chesapeake Bay: historical trends and ecological interactions. *Marine Ecology Progress Series, 303*, 1–29.

Koblinsky, C. J., Hildebrand, P., LeVine, D., Pellerano, F., Chao, Y., Wilson, W., ...& Lagerloef, G. (2003). Sea surface salinity from space: science goals and measurement approach. *Radio Science, 38*(4), 29–31.

Krishnan, P., Roy, S. D., George, G., Srivastava, R. C., Anand, A., Murugesan, S., ...& Soundararajan, R. (2011). Elevated sea surface temperature during May 2010 induces mass bleaching of corals in the Andaman. *Current Science, 100*(1), 111–117.

Kumar, P. S., Pillai, G. N., & Manjusha, U. (2014). El Nino southern oscillation (ENSO) impact on tuna fisheries in Indian Ocean. *SpringerPlus, 3*, 1–13.

Kumar, V., Roy, S., Behera, B. K., & Das, B. K. (2022). Heat shock proteins (Hsps) in cellular homeostasis: a promising tool for health management in crustacean aquaculture. *Life, 12*(11), 1777.

Lakra, W. S., &Das, P. (1998). Genetic engineering in aquaculture. *Indian Journal of Animal Sciences (India), 68*(8) Special Issue, 873–879.

Lemasson, A. J., Hall-Spencer, J. M., Fletcher, S., Provstgaard-Morys, S., & Knights, A. M. (2018). Indications of future performance of native and non-native adult oysters under acidification and warming. *Marine Environmental Research, 142*, 178–189.

Levin, R. A., Voolstra, C. R., Agrawal, S., Steinberg, P. D., Suggett, D. J., & Van Oppen, M. J. (2017). Engineering strategies to decode and enhance the genomes of coral symbionts. *Frontiers in Microbiology*, 8: 1220. doi: 10.3389/fmicb.2017.01220

Longhurst, A. R. (2010). *Ecological Geography of the Sea*. Elsevier.

Luening, E. (2013). After the hurricane. *Aquaculture North America, 4*(2), 1.

Lukacs, J. R. (Ed.). (2013). *The People of South Asia: The Biological Anthropology of India, Pakistan, and Nepal*. Springer.

Mahajan, A. U., Sunil Kumar, C. S., Kumar, P., Chakradhar, B., & Badrinath, S. D. (1996). Environmental quality assessment of Port Blair in Andaman Islands. *Environmental Monitoring and Assessment, 41*, 203–217.

Maulu, S., Hasimuna, O. J., Haambiya, L. H., Monde, C., Musuka, C. G., Makorwa, T. H., ...& Nsekanabo, J. D. (2021). Climate change effects on aquaculture production: sustainability implications, mitigation, and adaptations. *Frontiers in Sustainable Food Systems*, 5, 609097.

Mukherjee, B. N. (2001). *Nationhood and Statehood in India: A Historical Survey* (Vol. 3). Daya Books.

Müller, M., & Kersten, S. (2003). Nutrigenomics: goals and strategies. *Nature Reviews Genetics*, 4(4), 315–322.

Ngoan, L. D. (2018). Effects of climate change in aquaculture: case study in Thua Thien Hue Province, Vietnam. *Biomedical Journal of Scientific & Technical Research*, 10(1), 7551–7552.

Nguyen, L. A., Pham, T. B., Bosma, R., Verreth, J., Leemans, R., De Silva, S., & Lansink, A. O. (2018). Impact of climate change on the technical efficiency of striped catfish, *Pangasianodon hypophthalmus*, farming in the Mekong Delta, Vietnam. *Journal of the World Aquaculture Society*, 49(3), 570–581.

Niku, K. O., Ekmekçi, F. G., Özsoy, E., Kirankaya, Ş., Kokkola, T., Emecen, G., ...& Atalay, M. (2014). Natural thermal adaptation increases heat shock protein levels and decreases oxidative stress. *Redox Biology*, 3, 25–28.

Oyebola, O. O., & Olatunde, O. M. (2019). Climate change adaptation through aquaculture: ecological considerations and regulatory requirements for tropical Africa. *Agriculture and Ecosystem Resilience in Sub Saharan Africa: Livelihood Pathways Under Changing Climate*, Eds. Y. Bamutaze, S. Kyamanywa, B. Singh, G. Nabanoga, & R. Lal. Springer, Cham, pp. 435–472. https://doi.org/10.1007/978-3-030-12974-3_20

Pankhurst, N. W., & Munday, P. L. (2011). Effects of climate change on fish reproduction and early life history stages. *Marine and Freshwater Research*, 62(9), 1015–1026.

Paukert, C. P., Lynch, A. J., & Whitney, J. E. (2016). Effects of climate change on North American inland fishes: introduction to the special issue. *Fisheries*, 41(7), 329–330.

Pittman, K., Yúfera, M., Pavlidis, M., Geffen, A. J., Koven, W., Ribeiro, L., ...& Tandler, A. (2013). Fantastically plastic: fish larvae equipped for a new world. *Reviews in Aquaculture*, 5, S224–S267.

Pörtner, H. O., & Farrell, A. P. (2008). Physiology and climate change. *Science*, 322(5902), 690–692.

Pörtner, H. O., & Peck, M. A. (2010). Climate change effects on fishes and fisheries: towards a cause-and-effect understanding. *Journal of Fish Biology*, 77(8), 1745–1779.

Priyadharsini, P., Nirmala, N., Dawn, S. S., Baskaran, A., SundarRajan, P., Gopinath, K. P., & Arun, J. (2022). Genetic improvement of microalgae for enhanced carbon dioxide sequestration and enriched biomass productivity: review on CO_2 bio-fixation pathways modifications. *Algal Research*, 66, 102810.

Rashid, I., Nagpure, N. S., Srivastava, P., Kumar, R., Pathak, A. K., Singh, M., & Kushwaha, B. (2017). HRGFish: a database of hypoxia responsive genes in fishes. *Scientific Reports*, 7(1), 1–9.

Reid, G. K., Filgueira, R., & Garber, A. (2015). Revisiting temperature effects on aquaculture in light of pending climate change. In: *Aquaculture Canada 2014 Proceedings of Contributed* Papers, Eds. J. Wade, T. Jackson, & K. Brewer-Dalton. Aquaculture Association of Canada, St. Andrews, pp. 85–91.

Reid, G. K., Gurney-Smith, H. J., Flaherty, M., Garber, A. F., Forster, I., Brewer-Dalton, K., ...& De Silva, S. (2019). Climate change and aquaculture: considering adaptation potential. *Aquaculture Environment Interactions*, 11, 603–624.

Remen, M., Oppedal, F., Imsland, A. K., Olsen, R. E., & Torgersen, T. (2013). Hypoxia tolerance thresholds for post-smolt Atlantic salmon: dependency of temperature and hypoxia acclimation. *Aquaculture*, 416, 41–47.

Richards, R. G., Davidson, A. T., Meynecke, J. O., Beattie, K., Hernaman, V., Lynam, T., & van Putten, I. E. (2015). Effects and mitigations of ocean acidification on wild and aquaculture scallop and prawn fisheries in Queensland, Australia. *Fisheries Research, 161*, 42–56.

Ries, J. B., Cohen, A. L., & McCorkle, D. C. (2009). Marine calcifiers exhibit mixed responses to CO_2-induced ocean acidification. *Geology, 37*(12), 1131–1134.

Robinson, R. A., Learmonth, J. A., Hutson, A. M., Macleod, C. D., Sparks, T. H., Leech, D. I., ...& Crick, H. Q. (2005). *Climate Change and Migratory Species*. British Trust for Ornithology, p. 414.

Rodrick, G. B. (2008). Effect of temperature, salinity, and pesticides on oyster hemocyte activity. *Florida Water Resources Journal, 86*, 4–14.

Sahu, B. K., Begum, M., Khadanga, M. K., Jha, D. K., Vinithkumar, N. V., & Kirubagaran, R. (2013). Evaluation of significant sources influencing the variation of physico-chemical parameters in Port Blair Bay, South Andaman, India by using multivariate statistics. *Marine Pollution Bulletin, 66*(1–2), 246–251.

Salisbury, J., Green, M., Hunt, C., & Campbell, J. (2008). Coastal acidification by rivers: a threat to shellfish? *Eos, Transactions American Geophysical Union, 89*(50), 513–513.

Sanford, E., & Kelly, M. W. (2011). Local adaptation in marine invertebrates. *Annual Review of Marine Science, 3*, 509–535.

Seggel, A., & De Young, C. (2016). Climate change implications for fisheries and aquaculture: summary of the findings of the Intergovernmental Panel on Climate Change Fifth Assessment Report, by Anika Seggel, Cassandra De Young and Doris Soto. FAO Fisheries and Aquaculture Circular No. 1122. Rome, Italy.

Sherwood, S., & Fu, Q. (2014). A drier future? *Science, 343*(6172), 737–739.

Siikavuopio, S. I., James, P., Lysne, H., Sæther, B. S., Samuelsen, T. A., & Mortensen, A. (2012). Effects of size and temperature on growth and feed conversion of juvenile green sea urchin (*Strongylocentrotus droebachiensis*). *Aquaculture, 354*, 27–30.

Sinha, E., Michalak, A. M., & Balaji, V. (2017). Eutrophication will increase during the 21st century as a result of precipitation changes. *Science, 357*(6349), 405–408.

Smajgl, A., Toan, T. Q., Nhan, D. K., Ward, J., Trung, N. H., Tri, L. Q., ...& Vu, P. T. (2015). Responding to rising sea levels in the Mekong Delta. *Nature Climate Change, 5*(2), 167–174.

Stewart, H. A., Aboagye, D. L., Ramee, S. W., & Allen, P. J. (2019). Effects of acute thermal stress on acid–base regulation, haematology, ion-osmoregulation and aerobic metabolism in channel catfish (*Ictalurus punctatus*). *Aquaculture Research, 50*(8), 2133–2141.

Stocker, T. F., Qin, D., Plattner, G. K., Alexander, L. V., Allen, S. K., Bindoff, N. L., ...& Xie, S. P. (2013). Technical summary. In: *Climate Change 2013: The Physical Science Basis. Contribution of Working Group I to the Fifth Assessment Report of the Intergovernmental Panel on Climate Change*, Eds. T. F. Stocker, D. Qin, G.-K. Plattner, M. Tignor, S. K. Allen, J. Boschung, A. Nauels, Y. Xia, V. Bex, & P. M. Midgley. Cambridge University Press, pp. 33–115.

Sumaila, U. R., Cheung, W. W., Lam, V. W., Pauly, D., & Herrick, S. (2011). Climate change impacts on the biophysics and economics of world fisheries. *Nature Climate Change, 1*(9), 449–456.

Thomsen, J., Gutowska, M. A., Saphörster, J., Heinemann, A., Trübenbach, K., Fietzke, J., ...& Melzner, F. (2010). Calcifying invertebrates succeed in a naturally CO_2-rich coastal habitat but are threatened by high levels of future acidification. *Biogeosciences, 7*(11), 3879–3891.

Tiedje, J. M., Bruns, M. A., Casadevall, A., Criddle, C. S., Eloe-Fadrosh, E., Karl, D. M., ...& Zhou, J. (2022). Microbes and climate change: a research prospectus for the future. *Mbio, 13*(3), e00800-22.

Tilman, D., Fargione, J., Wolff, B., D'antonio, C., Dobson, A., Howarth, R., ... & Swackhamer, D. (2001). Forecasting agriculturally driven global environmental change. *Science*, 292(5515), 281–284.

Vaishampayan, P., Probst, A., Krishnamurthi, S., Ghosh, S., Osman, S., McDowall, A., ... & Venkateswaran, K. (2010). *Bacillus horneckiae* sp. nov., isolated from a spacecraft-assembly clean room. *International Journal of Systematic and Evolutionary Microbiology*, 60(5), 1031–1037.

Waldbusser, G. G., Voigt, E. P., Bergschneider, H., Green, M. A., & Newell, R. I. (2011). Biocalcification in the eastern oyster (*Crassostrea virginica*) in relation to long-term trends in Chesapeake Bay pH. *Estuaries and Coasts*, 34, 221–231.

Weiman, S. (2015). Microbes help to drive global carbon cycling and climate change. *Microbe Magazine*, 10(6), 233–238.

Wells, M. L., Trainer, V. L., Smayda, T. J., Karlson, B. S., Trick, C. G., Kudela, R. M., ...& Cochlan, W. P. (2015). Harmful algal blooms and climate change: learning from the past and present to forecast the future. *Harmful Algae*, 49, 68–93.

Wilkinson, R. J., Longland, R., Woolcott, H., & Porter, M. J. (2010). Effect of elevated winter–spring water temperature on sexual maturation in photoperiod manipulated stocks of rainbow trout (*Oncorhynchus mykiss*). *Aquaculture*, 309(1–4), 236–244.

Yusof, N. A., Masnoddin, M., Charles, J., Thien, Y. Q., Nasib, F. N., Wong, C. M. V. L., ...& Bharudin, I. (2022). Can heat shock protein 70 (HSP70) serve as biomarkers in Antarctica for future ocean acidification, warming and salinity stress? *Polar Biology*, 45(3), 371–394.

Zahed, M. A., Movahed, E., Khodayari, A., Zanganeh, S., & Badamaki, M. (2021). Biotechnology for carbon capture and fixation: critical review and future directions. *Journal of Environmental Management*, 293, 112830.

Zhang, P., Zhao, T., Zhou, L., Han, G., Shen, Y., & Ke, C. (2019). Thermal tolerance traits of the undulated surf clam *Paphia undulata* based on heart rate and physiological energetics. *Aquaculture*, 498, 343–350.

7 Interannual and Seasonal Patterns of Soil Evapotranspiration in Response to Vegetation Dynamics as Revealed from Satellite Time Series Data

Ali P. Yunus and A.C. Narayana

7.1 INTRODUCTION

Soil evaporation (E_s) is a critical environmental variable in various energy contexts (Dong *et al.*, 2020). For instance, soil evaporation is the major contributor variable in the evapotranspiration (ET), where the interception becomes more prominent (Qubaja *et al.*, 2020). Understanding and managing soil evaporation is essential for optimizing crop growth and is an essential component in ecosystem hydrology (Ras-Yaseef *et al.*, 2010; Suárez *et al.*, 2023). It has been noted that excessive soil evaporation can lead to moisture stress in plants, affecting crop yield. For example, Zhang *et al.* (2017) analyzed the seasonal trends of evapotranspiration in a field composed of tomato plantations, suggesting that the crop yields decreased significantly when irrigation water is used. Contrarily, insufficient evaporation can result in waterlogging; accordingly, E_s become the largest flux in the evapotranspiration component in areas having sparse vegetation and limited water availability (Yinglan *et al.*, 2019). Moreover, E_s plays a crucial role in water balance dynamics, and surface energy conversion process (Li *et al.*, 2022; Miralles *et al.*, 2011). Therefore, quantifying and modeling of E_s are vital for sustainable agriculture, water resource management and in climate studies.

Soil evaporation is found to decline as the vegetation develops (Hooke *et al.*, 2012; Lin., 2010), however, quantifying the E_s dynamics in relationship to vegetation is scarce. In addition, the rate of soil evaporation is sensitive to temperature and

DOI: 10.1201/9781003485995-7

other climate-driven mechanisms. As temperatures rise, the potential for increased soil evaporation also escalates. However, this relationship is not always valid; for instance, Chattopadhyay and Hulme (1997), noted a good correlation between the declining trend in ET and the concurrent increases in relative humidity and reductions in radiation. Their study concluded that despite an increase in temperature over the Indian region during 1961–1992, there shows a decreasing trend in ET at the same period. Such a complex interplay between temperature and ET, as well as that of vegetation cover can produce complex dynamics of soil moisture in spatial and temporal scale.

Despite the significance of soil evaporation in the Earth's hydrological cycle, there is a need for a more comprehensive understanding of the spatial and temporal variability of soil evaporation at various scales, from local to global, under changing climatic conditions. Here, we aim to analyze the role of vegetation cover changes and land surface temperature dynamics in altering soil evaporation patterns in the Peninsular India, incorporating remote sensing technologies that can enhance our ability to predict and monitor soil evaporation dynamics.

7.2 STUDY AREA AND DATASETS

The study area, Peninsular India (PI) covers an area of about 5,00,000 km² (Figure 7.1). The region is strongly influenced by the southwest summer monsoon and the northeast monsoon. The maximum annual precipitation over the Peninsular basins ranges over 3000 mm in the west to 500 mm in the interior peninsula, and a mean value of about 1250 mm. The annual average temperature varies between 17 and 43 °C with a mean value about 34 °C. The elevation ranges between sea level and 2695 m. Six major river basins (Narmada, Tapti, Mahanadi, Krishna, Godavari and Cauvery) and several smaller rivers originating in the Western Ghats encompasses the study area. Tropical evergreen and semi-evergreen forests are characteristics of the southern Western Ghats, which receive more than 2000 mm annual rainfall, while interior peninsular region is covered with savanna vegetation. Previously, Madhu *et al.* (2015) studied the trends of evapotranspiration (ET) and its response to droughts over India from 1901 to 2007. They used the Hargreeves and Samani method (1982), based on the temperature and latitude longitude-dependent extra-terrestrial solar radiation for the estimation of ET, and found that the trend of ET was increasing over India with a maximum ET increase in the West Coast India.

The soil evaporation (E_s) data for the study area is downloaded through Google's earth engine platform from the data tagged PML_V2 0.1.7: Coupled Evapotranspiration and Gross Primary Product (GPP) https://developers.google.com/earth-engine/datasets/catalog/CAS_IGSNRR_PML_V2_v017#description. The Penman-Monteith-Leuning Evapotranspiration V2 (PML_V2) products include soil evaporation at 500 m and 8-day resolution during 2000–2020. The underlying principle in construction of the E_s dataset is described in Zhang *et al.*, (2019).

In order to analyze the E_s sensitivity to vegetation dynamics, we utilized the normalized difference vegetation index (NDVI) MOD13A1 V6.1 product, which is derived from MODIS Terra sensor red and NIR reflectance composite of 16 days at a resolution of 500 m. The temperature data for the study period was obtained from

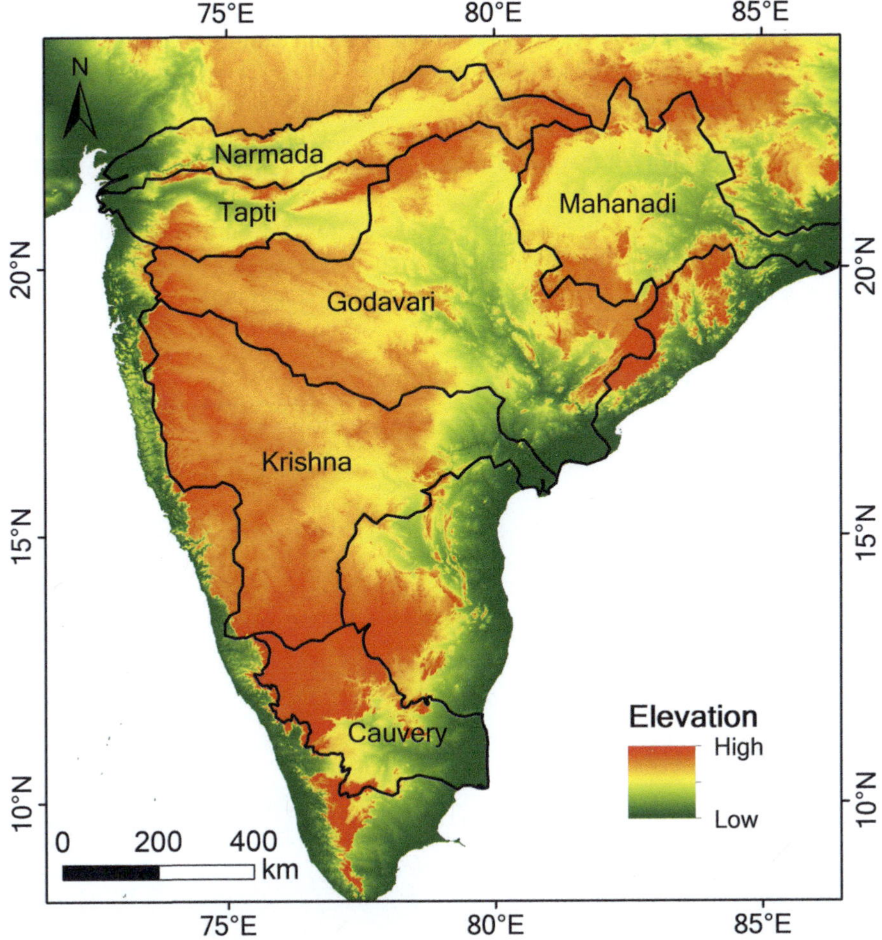

FIGURE 7.1 Study area and boundaries of the selected river basins in the Peninsular India.

the MOD11A2 V6.1 product that provides an average 8-day land surface temperature (LST).

All the images are pre-processed in the GEE platform and the statistical analysis such as trend analysis and correlation plots are carried using R. For visualization, we used ArcGIS 10.8 v.

7.3 RESULTS

7.3.1 Spatiotemporal Soil Evaporation Patterns

The spatial distribution of the annual average soil evaporation (E_s) exhibited notable variations throughout the PI region, as depicted in Figure 7.2; E_s values ranged from 0.02 to 3.52 mm, with an average of 0.91 mm. Generally, higher E_s levels were observed in the South Eastern coastal belt, while lower levels prevailed in the Western and North Eastern regions. Central Peninsular regions exhibit moderate to low E_s. The lowest E_s values were observed in the Western Ghats region, where dense vegetation is abundant, and rainfall values exceed 2000 mm annually. Regarding the E_s values in different river basins, the Cauvery basin exhibits the highest E_s (1.25 mm), followed by the Krishna (1.09 mm), Godavari (0.82 mm), Mahanadi (0.81 mm), Tapti (0.77 mm) and Narmada (0.74 mm) basins (Figure 7.2).

Overall, there was a positive trend in E_s levels across the entire Peninsular region, but at an insignificant ($p > 0.05$) level. On the other hand, significant declining trend in E_s is particularly evident in the Krishna Basin and along the coastal plains

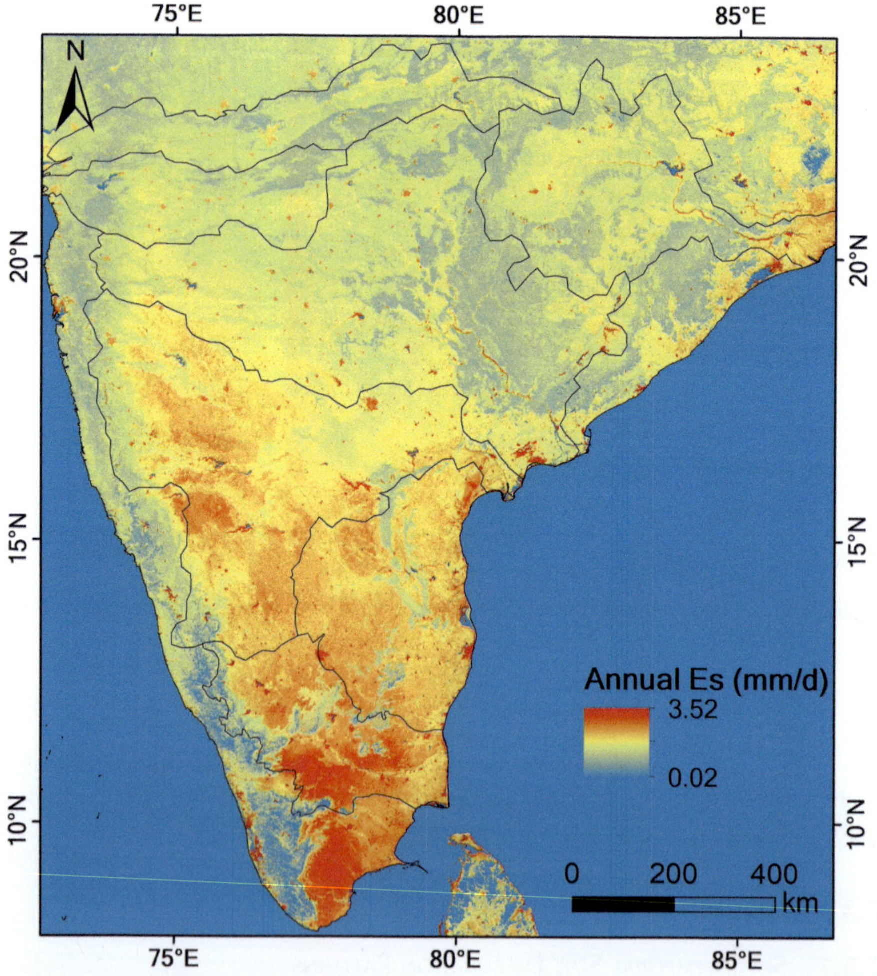

FIGURE 7.2   The spatial distribution of annual soil evaporation map (E_s: units, mm/d) during the 2000–2020 period for Peninsular India. Basin boundaries (black polygons) are overlaid on the E_s map.

of the Eastern margin (see Figures 7.3 and 7.4). The most substantial E_s decrease rates were recorded in the Krishna basin, with an average rate of –0.009865 mm/yr, followed by the Cauvery (–0.006518 mm/yr), Mahanadi (–0.005334 mm/yr), Tapti (–0.004879 mm/yr) and Godavari (–0.002177 mm/yr) basins. In contrast to other basins, the Narmada basin exhibited a positive E_s change in the last two decades, with a rate of 0.001563 mm/yr. It is important to note that the southeastern and northeastern coastal plains experienced the most significant decline in E_s, demonstrating a pronounced decreasing trend. Meanwhile, certain areas of the Western Ghats showed an increasing trend in E_s without reaching statistical significance (Figure. 7.3a and 7.3b).

7.3.2 Seasonal Soil Evaporation Patterns

For illustrative purposes, the seasonal variation of the soil evaporation (E_s) values in the Peninsular region is plotted in Figure 7.5. Seasonal patterns of distribution E_s varied widely across different seasons, may be because of changes in temperature, precipitation and vegetation seasonality. Notably, the winter period (December–January–February) exhibited the lowest E_s throughout the Peninsular region (Figure. 7.5a). During June–July–August months (south west monsoon months), E_s levels exceeded 2 mm/day in most parts, with the highest E_s values observed for Krishna basin.

During summer months, i.e., March–April–May, the Peninsular region may clearly be divided into three clusters; low values of E_s occupied in west and northwest, low to medium values in north and north east, and medium to high values in the south eastern parts. During post-monsoon phase, low values of E_s was observed in vegetated area of Western Ghats, and all along the north eastern coastal region that receives the north-east monsoon phase. Krishna and Cauvery basins, and the south eastern coastal belts records high E_s during this period.

7.3.3 Contributions of Vegetation, and Temperature to Soil Evaporation Trends

The interaction of E_s within the upper soil is determined by soil water supply, properties of vegetation and solar radiation (Huang *et al.*, 2017). These parameters are often represented by land surface temperature, precipitation and, NDVI, a proxy of vegetation dynamics, respectively. In terms of temporal changes, there was a general decline in E_s between 2000 and 2020 in the Eastern margin of PI. It is important to highlight that vegetation, as indicated by NDVI data, exhibited an overall upward trend over this timeframe (refer to Figures 7.6a and 7.6b). Figure 7.6 presents the annual average NDVI and the analysis of NDVI trends throughout the study period. It becomes apparent that, although NDVI experienced significant localized increases ($p<0.001$), but there is not a strong visual correlation between the spatial distribution of NDVI and E_s data (as depicted in Figure 7.6c).

In order to quantify the effect of vegetation greening/browning on soil evaporation characteristics, the pixel-wise scatter plot between the E_s and NDVI is shown in Figure 7.7 and Figure 7.8. From Figure 7.7, it can be seen that the higher values of NDVI corresponds to the lower E_s and vice versa. Coefficient of determination

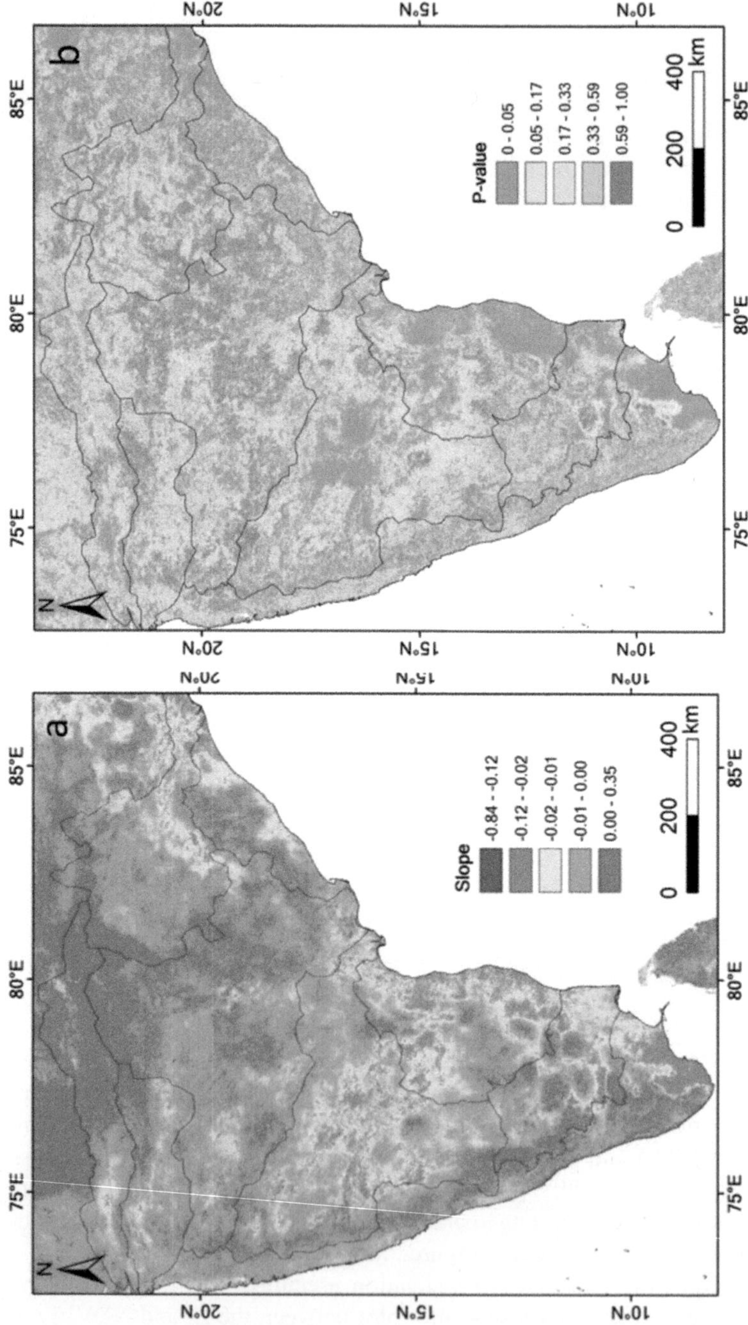

FIGURE 7.3   The spatial distribution of E_s time series trend slope values (a) and p-values (b) obtained for the year 2000–2020. Note the major changes on eastern coastal belts and in the interior of Krishna basin.

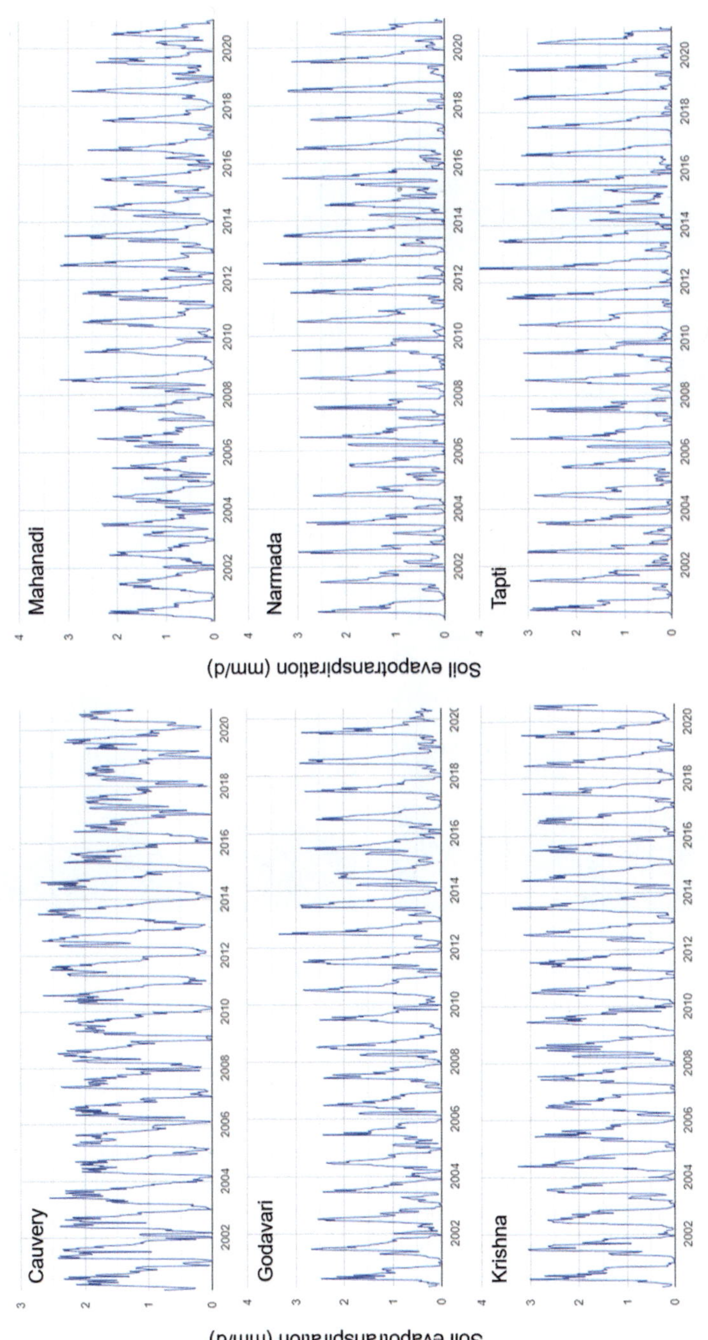

FIGURE 7.4 Time series (2000–2020) of soil evapotranspiration variation for the six major peninsular river basins.

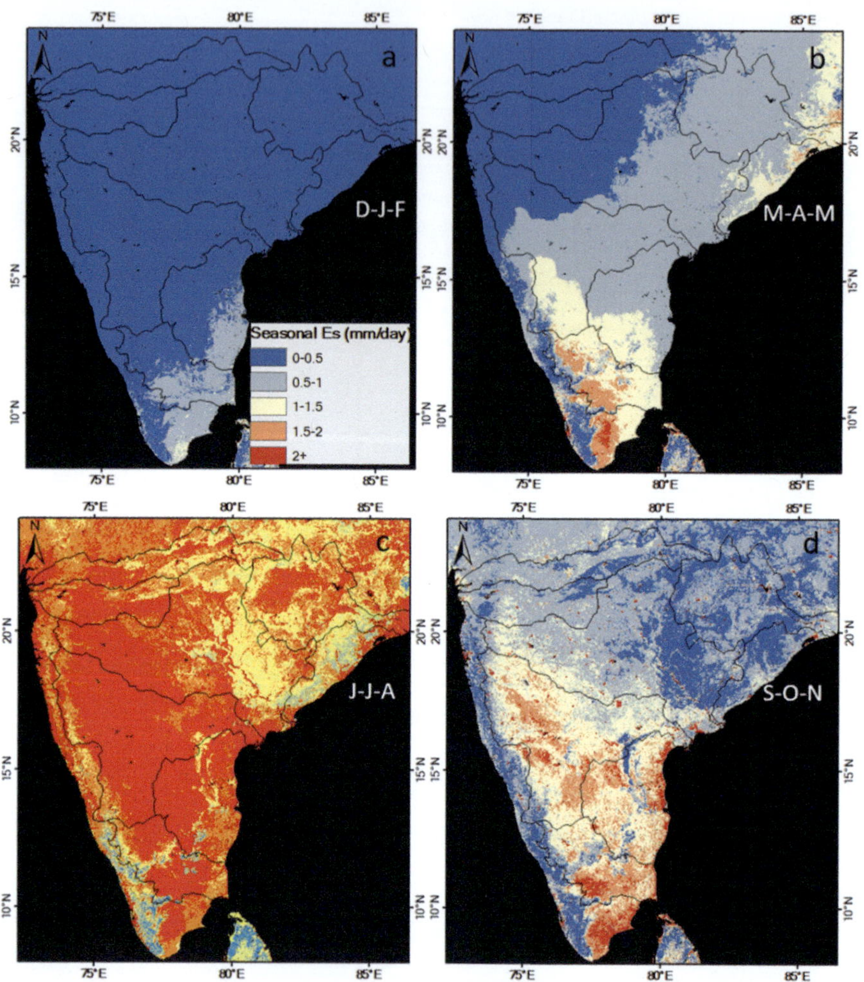

FIGURE 7.5   The spatial distribution of E_s in (a) winter (December–February), (b) summer (March–May), (c) rainy period (June–August) and (d) post-monsoon (September-–November).

was used to detect the relationship between E_s and NDVI. The determination coefficient ($R^2 = 0.29$) found between E_s and NDVI suggests that the overall trends in inter-annual E_s were marginally affected by vegetation greening in Peninsular region. On the other hand, individual basins show large varying relationship with E_s and NDVI. The highest determination coefficient ($R^2 = 0.75$; p-value <0.05) between E_s and NDVI is found in Cauvery and the lowest is found in Narmada basin ($R^2 = 0.31$; p-value <0.05). This suggests that vegetation dynamics have a significant control on declining E_s rates in certain regions.

The spatiotemporal pattern of the land surface temperature (LST) was analyzed based on the mean values of LST during 2000 to 2020 (Figure 7.9). Overall, an increasing trend in the temperature is recorded in the study area during the analysis

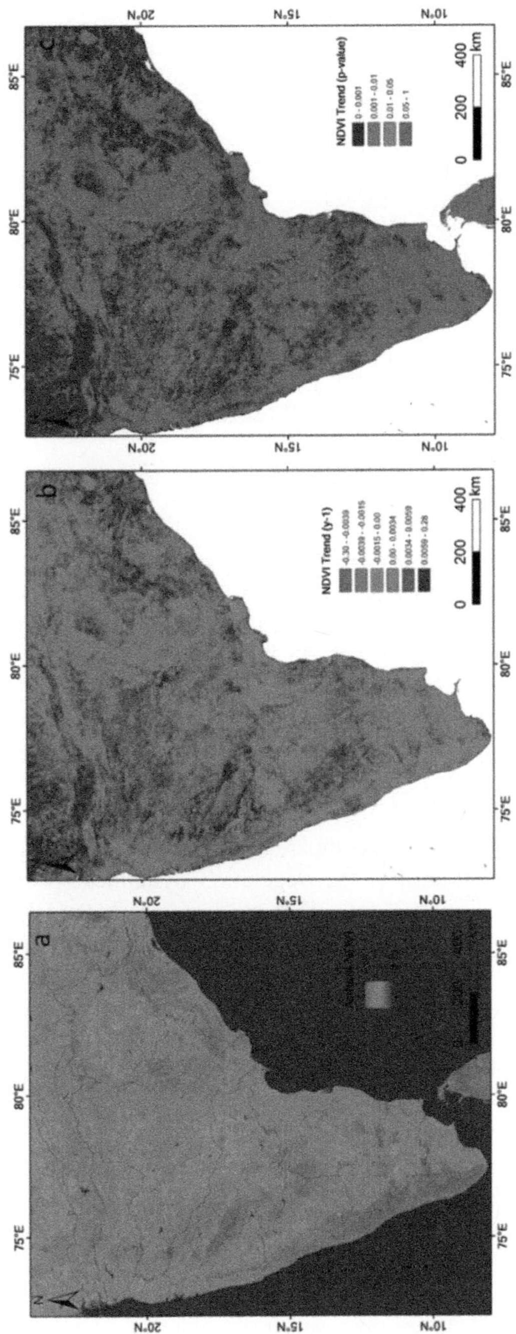

FIGURE 7.6 Spatial distribution of (a) annual mean NDVI, (b) NDVI trend and (c) *p*-value for the trends observed between 2000 and 2020.

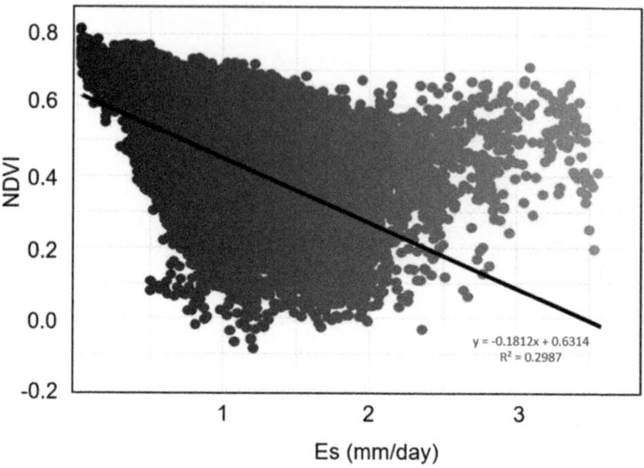

FIGURE 7.7   Pixel wise scatter plot of NDVI versus E_s.

period (Figure 7.9b). The spatial distribution of soil evaporation (E_s) is found less correlated with that of the land surface temperature (Figure 7.10). The determination coefficient between long-term mean annual E_s estimated from the models and temperature is about 0.18. However, similar to the NDVI relationship, individual basins exhibit varying correlations, in which the highest R^2 is noted for the Cauvery basin ($R^2 = 0.68$), followed by Tapti ($R^2 = 0.28$) and Krishna basin ($R^2 = 0.17$). In contrast, low values of R^2 are noted for the Godavari ($R^2 = 0.06$), Narmada ($R^2 = 0.04$) and Mahanadi ($R^2 = 0.07$) basins.

7.4 DISCUSSION AND CONCLUSIONS

In this study, we evaluated and compared soil evaporation (E_s) dynamics in the Peninsular India in relation with the vegetation dynamics (proxied from NDVI) and land surface temperature (LST). Soil evaporation derived from the Penman–Monteith–Leuning evapotranspiration model, temperature and NDVI data from the MODIS sensor for a period of 2000 to 2020 were used in the investigation. By analyzing the time series of E_s calculated by the P–M method, the annual trends of E_s coefficients were obtained for the study area.

The following conclusions may be drawn from the study:

(1) There is an overall increase in the values of soil evaporation in the Peninsular region during the last two decades, but without significant trend. The highest values of E_s are found during the June to August months and the lowest values found in the December to February months. The higher E_s values during the summer months can be attributed to the higher latent heat flux during these months (Madhu *et al.*, 2015). As Tthe spatial distribution pattern shows that the highest E_s levels are observed in the South Eastern coastal belt, and the lowest levels prevailed in the Western and North Eastern regions. This information provides valuable information for regional hydrological studies.

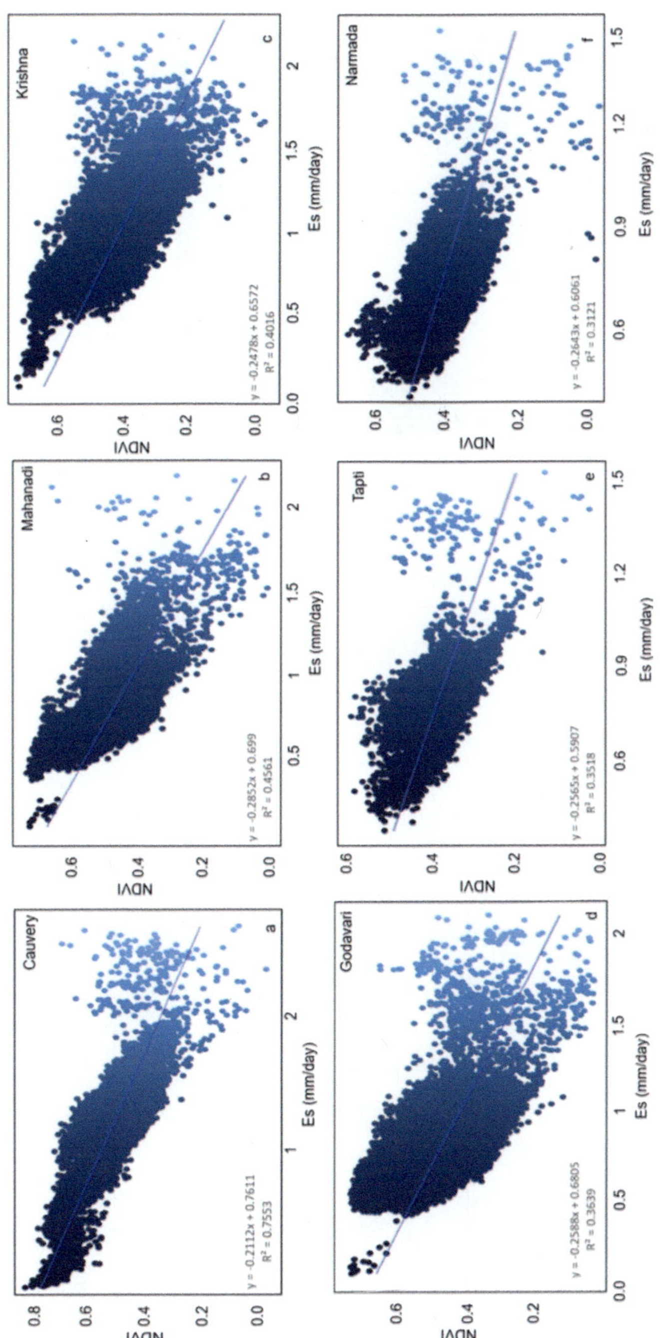

FIGURE 7.8 Pixel-wise correlation plot of NDVI versus E_s for different basins.

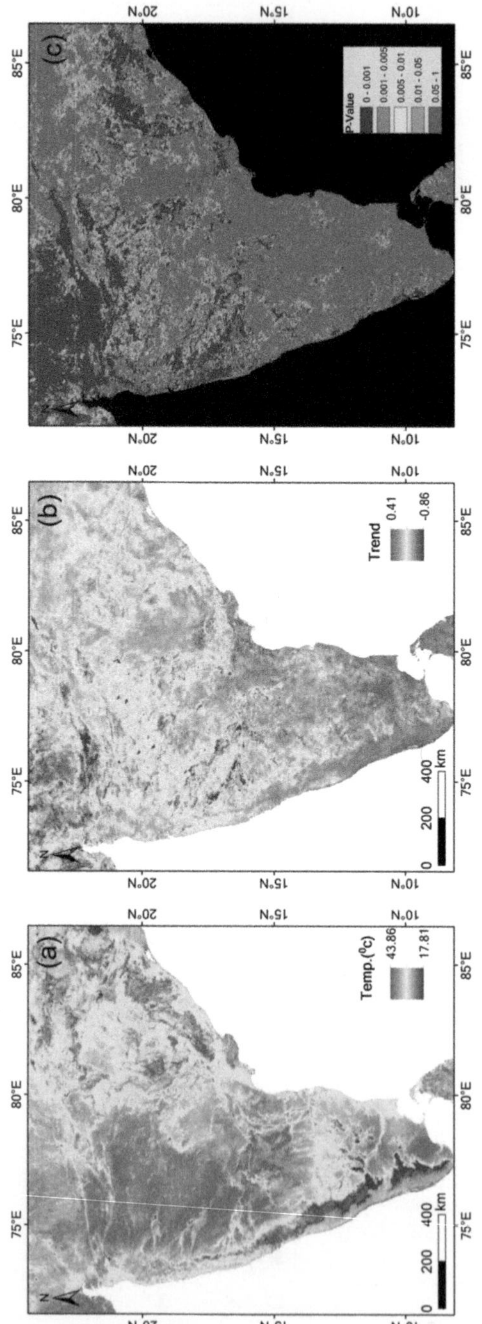

FIGURE 7.9 Spatial distribution of (a) annual temperature, (b) temperature trend and (c) *p*-values of trend signficance observed between 2000 and 2020.

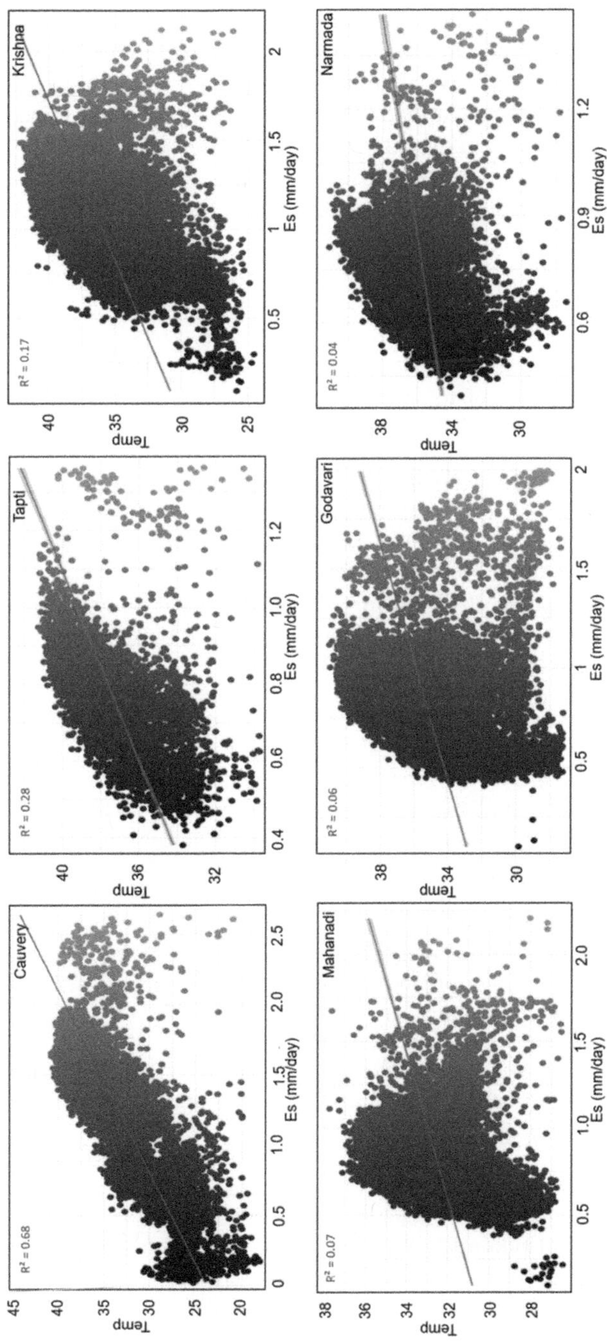

FIGURE 7.10 Pixel-wise correlation plot of temperature versus E_s for different basins.

(2) An increase in the vegetation greening trends is found in the Peninsular region during 2000 to 2020. These increasing trends in vegetation may jointly lead to a decreasing trend in soil evaporation in the region. Noted correlation is found between E_s and NDVI for majority of the studied basins.

(3) For the whole catchment, there is no significant trend in the LST except for the region enclosing Narmada basin and adjoining areas, where the LST is found significantly decreased in the past two decades. However, no clear correlation is noted between the E_s and LST for the Narmada basin. On the other hand, E_s and LST is positively well correlated in Cauvery and Krishna Basins.

REFERENCES

Chattopadhyay, N. and Hulme, M., 1997. Evaporation and potential evapotranspiration in India under conditions of recent and future climatic change. Agricultural and Forest Meteorology, 87(1), pp. 55–74.

Dong, J., Dirmeyer, P.A., Lei, F., Anderson, M.C., Holmes, T.R., Hain, C., and Crow, W.T., 2020. Soil evaporation stress determines soil moisture-evapotranspiration coupling strength in land surface modeling. Geophysical Research Letters, 47(21), p. e2020GL090391.

Hargreaves, G.H. and Samani, Z.A., 1982. Estimating potential evapotranspiration. Journal of the Irrigation and Drainage Division, 108(3), pp. 225–230.

Hooke, J. and Sandercock, P., 2012. Use of vegetation to combat desertification and land degradation: Recommendations and guidelines for spatial strategies in Mediterranean lands. Landscape and Urban Planning, 107(4), pp. 389–400.

Li, W., Franssen, H.J.H., Brunner, P., Li, Z., Wang, Z., Wang, Y., and Wang, W., 2022. The role of soil texture on diurnal and seasonal cycles of potential evaporation over saturated bare soils–Lysimeter studies. Journal of Hydrology, 613, p. 128194.

Lin, B.B., 2010. The role of agroforestry in reducing water loss through soil evaporation and crop transpiration in coffee agroecosystems. Agricultural and Forest Meteorology, 150(4), pp. 510–518.

Madhu, S., Kumar, T.L., Barbosa, H., Rao, K.K., and Bhaskar, V.V., 2015. Trend analysis of evapotranspiration and its response to droughts over India. Theoretical and Applied Climatology, 121, pp. 41–51.

Miralles, D.G., De Jeu, R.A.M., Gash, J.H., Holmes, T.R.H., and Dolman, A.J., 2011. Magnitude and variability of land evaporation and its components at the global scale. Hydrology and Earth System Sciences, 15(3), pp. 967–981.

Qubaja, R., Amer, M., Tatarinov, F., Rotenberg, E., Preisler, Y., Sprintsin, M., and Yakir, D., 2020. Partitioning evapotranspiration and its long-term evolution in a dry pine forest using measurement-based estimates of soil evaporation. Agricultural and Forest Meteorology, 281, p. 107831.

Raz-Yaseef, N., Rotenberg, E., and Yakir, D., 2010. Effects of spatial variations in soil evaporation caused by tree shading on water flux partitioning in a semi-arid pine forest. Agricultural and Forest Meteorology, 150(3), pp. 454–462.

Suárez, F., Oportus, T., Mendoza, M., Aguirre, I., Godoy, V., and Muñoz, J.F., 2023. Evaporation processes in the Silala River basin. Wiley Interdisciplinary Reviews: Water, p. e1638.

Yinglan, A., Wang, G., Liu, T., Xue, B., and Kuczera, G., 2019. Spatial variation of correlations between vertical soil water and evapotranspiration and their controlling factors in a semi-arid region. Journal of Hydrology, 574, pp. 53–63.

Zhang, H., Xiong, Y., Huang, G., Xu, X., and Huang, Q., 2017. Effects of water stress on processing tomatoes yield, quality and water use efficiency with plastic mulched drip irrigation in sandy soil of the Hetao Irrigation District. Agricultural Water Management, 179, pp. 205–214.

Zhang, Y., Kong, D., Gan, R., Chiew, F.H., McVicar, T.R., Zhang, Q., and Yang, Y., 2019. Coupled estimation of 500 m and 8-day resolution global evapotranspiration and gross primary production in 2002–2017. Remote Sensing of Environment, 222, pp. 165–182.

8 Rainfall Induced Landslide Hazards and Forecast under Changing Climatic Patterns and Associated Adaptations

Devesh Walia

8.1 INTRODUCTION

Meghalaya in particular, because of its structural complexity, rock types and climatic conditions, is characterized by well-developed mantle of waste, which are easily susceptible and highly prone to landslides, debris flow, ground subsidence and associated damages. The increasing population and rapid urbanization with changing climatic patterns have considerably enhanced the vulnerability of landslide occurrences and have increased risk to human life and property and needs adaptation. The landslide is mostly triggered by events of heavy rainfall (frequent incessant and torrential), unstable slope, anthropogenic interference and seismicity. The incidences of landslides have been on the increase in recent times due to the rapid pace of development especially of roads in the hilly regions. It is very important in the present context where expansion of roads (NH40) is being carried out over the plateau.

8.2 STUDY AREA

The study area is part of Survey of India topographic sheet no. 78 N/16 & 78 O/13, located in the northern part of Ribhoi District of Meghalaya. It is bounded by 91°48′0″ E to 91°56′0″ E longitude and 25°54′0″ N to 26°6′0″ N latitude (Figure 8.1). Most of the area exhibits a rugged topography with high mountains and intermittent linear valleys. The total area of the watershed is 118.23 sq. km. The area of study is selected on the basis of landslide zone close to road (NH 40) and the digging for the road has already undertaken decreasing the load from the toe area, and may increase the landslide incidences along NH 40, and communication and supplies to Mizoram, Tripura and Lower Assam along with Meghalaya may get effected due to landslide event. However, the presence of heavy infrastructure and high rainfall further increases the landslide vulnerability and hence the forecast and prevention are of utmost importance especially at the two selected sites (i.e., near CMJ University, Jorabat and Pahamlang). Further, combined action of hydrological and

DOI: 10.1201/9781003485995-8

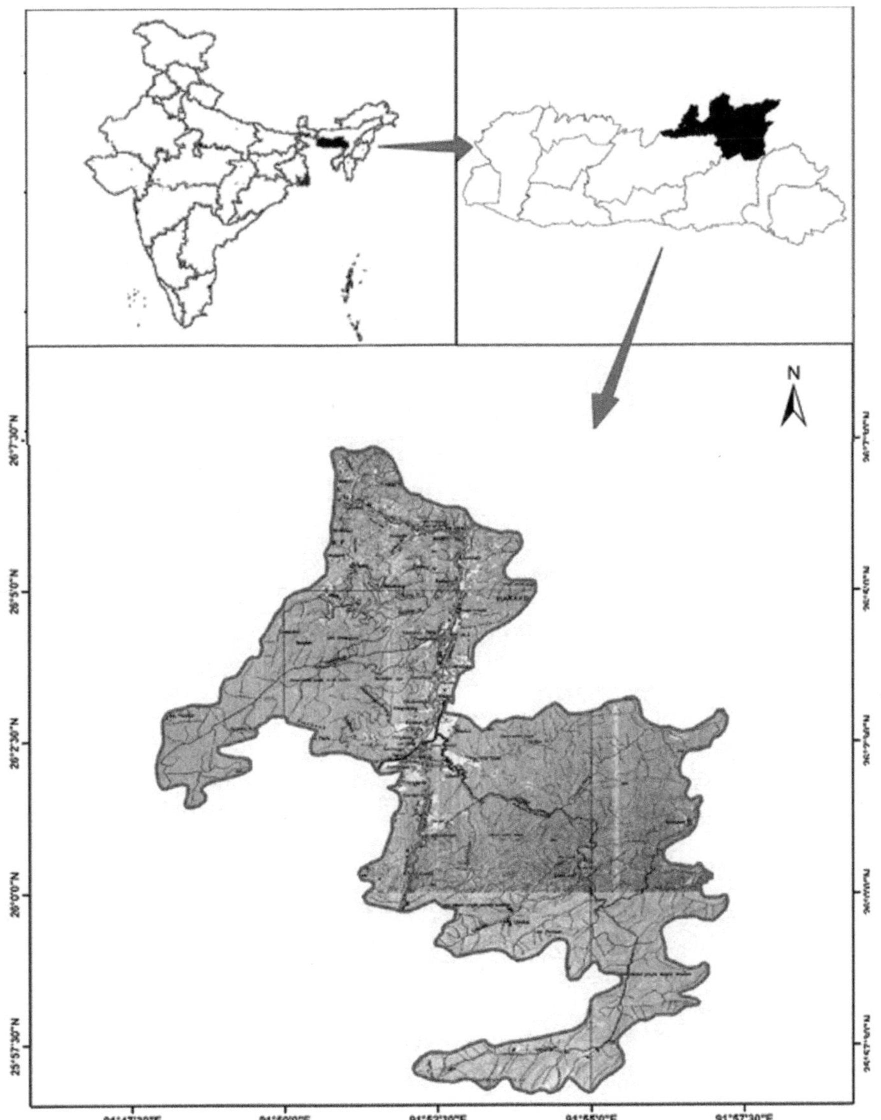

FIGURE 8.1   Location map of the study area. The study area lies in the NE region of India and lies close to the Guwahati Shillong road, which connects to Tripura and Mizoram.

land use/anthropogenic factors, in which the major triggers are water and toe removal, incidences of landslides in the State of Meghalaya have assumed great significance and such a study is of great importance and significance.

Landslide Location 1 (Near CMJ University (91°52'40.44"E, 26° 5'45.11"N): The landslide occurred in the vicinity of CMJ University, Jorabat (along NH40, Ghy-Shillong Road). University building was built on top of the slope cut, just below is

FIGURE 8.2 Landslide site near CMJ University. The site is close to the Guwahati-Shillong National Highway, which connects the state of Mizoram and Tripura.

NH 40 where movement of vehicles take place. Landslide occurred along the slope because of excessive slope cutting, the slope made up of hill cut material, weight of the building and poor drainage system. Geomorphologically the area is covered with hills with moderately dissected steep slopes with moderate to sparse vegetation (Figure 8.2).

Landslide Location 2 (Pahamlang, near Byrnihat (91°52'03.693"E, 26° 01'26.034"N): The landslide occurred at Pahamlang (17 Mile) Village along NH40. The area is covered by loose debris. Landslide occurred along the slope because of high degree of slope cutting, the slope made up of hill cut material and poor drainage system. Geomorphologically the area is covered with hills with dissected steep slopes and moderate vegetation (Figure 8.3).

8.3 PHYSIOGRAPHY OF THE AREA

The area supports a varied topography exhibiting altitudinal extremes of 60 m to 735 m above mean sea level. The structural, lithological and geomorphological features [Figure 8.4(a) and 8.4(b)] control the direction of flow of the tributaries. The available water in the watershed as a surface runoff is dependent upon rainfall, changes in soil moisture, stream characteristics such as frequency length and base flow, soil and land use, and hydrological characteristics such as groundwater gains and losses, evapotranspiration. These varying relief and rugged topography influence the amount and intensity of rainfall. The climate of the area is largely controlled by Southwest monsoon and seasonal winds. The average annual rainfall is about 2000 mm to 23000 mm

FIGURE 8.3 Landslide Pahamlang is a site around 25 km south of the first site and lies close to the road.

with summer monsoon starting from May to September and the winter monsoon from December to February. The maximum rainfall takes place during the month of June to September. The climate is most pleasant with short summer days. The northern part being close to Brahmaputra valley remains warm and humid most of the year except

TABLE 8.1
Physiography and Area Distribution

Description	Area	Percentage
Low dissected hills	19.26	16.29
Moderately dissected hills	98.97	83.71

in winter where temperature falls to 5 °C or even less. On the basis of the physio-graphic features, two distinct geomorphic units have been identified, viz., (i) low dissected hills covering about 16% of the total area, (ii) moderately dissected hills covering about 83% of the total area (Table 8.1). The physiography map of the study area is shown in Figure 8.4.

8.4   GEOLOGY

Geologically the study area is a part of the rigid massif of the Shillong plateau, a detached part of peninsular Gondwana land cratonic block (Table 8.2). The prevailing rock types in the study area is Banded gneiss, Porphyritic granite, Ferruginous clay and Sillimanite bearing quartzite (Figure 8.5). The gneisses are traversed by numerous veins of quartz and pegmatite. The outcrops of the gneissic rocks are rarely seen as they have undergone deep weathering all over the district. However, good outcrops can be seen along Guwahati-Shillong Road sections. The general foliation trend of these rocks is N 70 22° E – S 70° W but trends like E-W and N200E – S200W are also seen.

The Nongpoh pluton, intruded into the gneissic complex, exposed in the southern part of the study area towards Nongpoh. The porphyritic granite covers about 2.37 % of the study area. These rocks are gray to pink in color, coarse grained, hard and compact, comprises K-feldspar phenocrysts, quartz, feldspar, biotite and amphibole in groundmass (Sadiq *et al.*, 2014). The Quaternary fluvial sediments occur in the extreme northern part of the district bordering Assam, forming part of Brahmaputra valley, with a thickness ranging between 3 to 20 m.

8.5   FACTORS INFLUENCING THE LANDSLIDE PROBLEM IN THE STUDY AREA

The slope along a section of NH-40 possesses a mature topography with no signifi-cant signs of mass movement in the recent past. Possibly, these slopes conquered to its naturally stable slope condition through time, and vegetation cover strengthens them to retain against failure. However, heavy earth cutting for road widening causes traffic snarls due to landslides in several places. Several new landslide incidences were reported from the NH-40 highway during the monsoon period. These landslide events, once experienced, become the treasury for repeated recurrence, making the highway unsafe for commuters.

Landslides occur on soil slopes or a combination of soil and bedrock slopes in the study area. Almost 80% of the section is covered by *in situ* soil and earthy

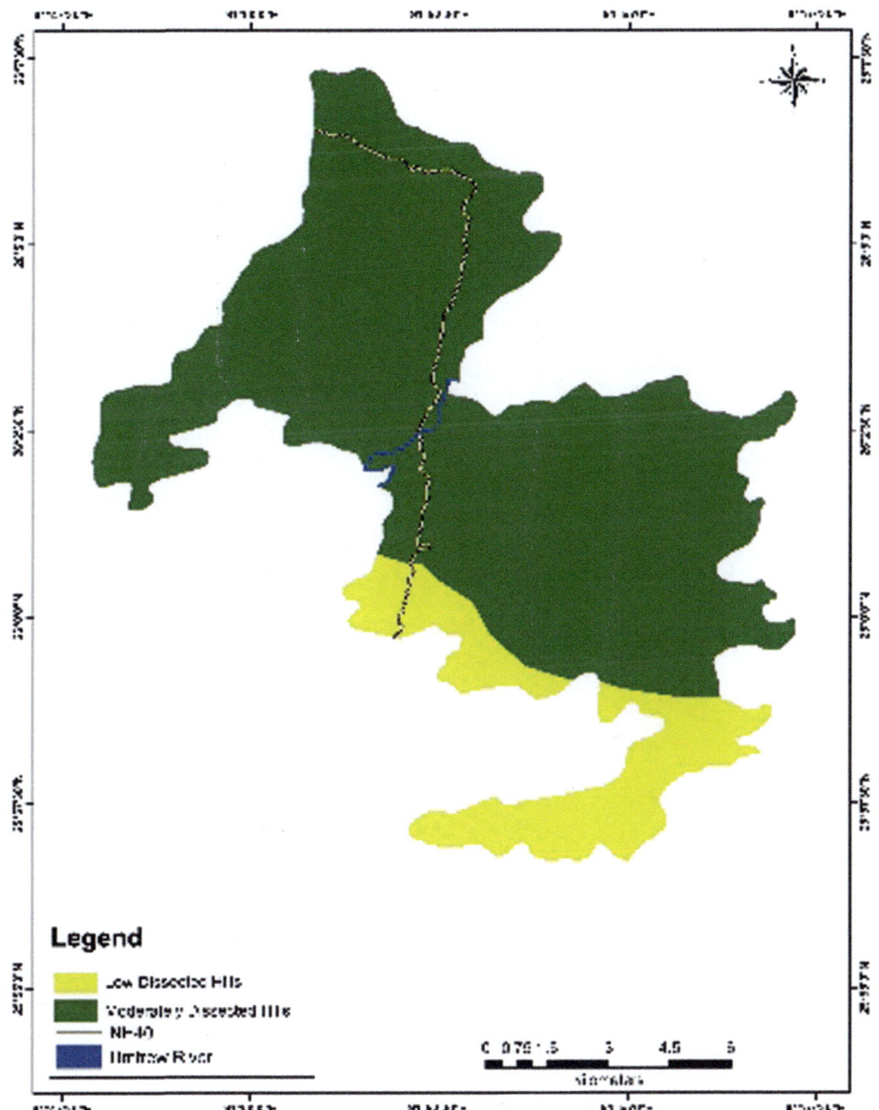

FIGURE 8.4 (a) Geomorphology and (b) drainage map of the study area depicting the rock types and the drainage patterns which is influenced by the recent tectonic features as the drainage pattern is trellis to sub-trellis.

materials, while the rest is covered by rocks or weathered rock masses. The process of weathering is very fast using climatic conditions of the region. Many factors are involved in quick forming soil process, including rain, temperature variation, chemical change, the wind, plant growth, animal activity and our human enthusiasm for development. The factors responsible are mainly human interventions that include

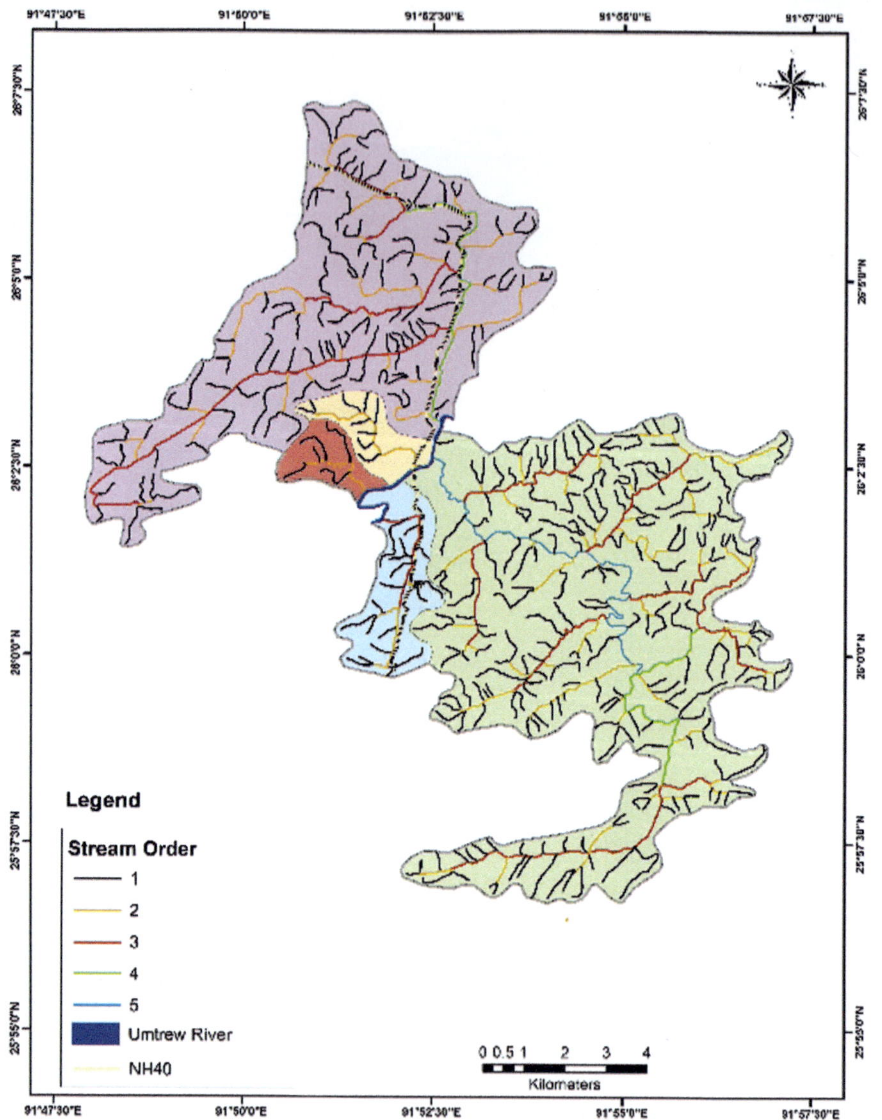

FIGURE 8.4 Continued

unscientific heavy earth cuttings leading to steeper slopes and vegetation clearance that led to slope failure triggered by saturation of soil mass, toe erosion, etc.

The hill slopes were excavated without proper geotechnical assistance; thereby, excavated slopes are steep to very steep with cliffs at places. Slope height varies from about 10 to 40 m with average height about 20 m. The exposed cut slopes are sensitive to minor changes in the factors that affect their stability and show signs of instability whenever triggered by some natural or anthropogenic factors. The excavation of hills

TABLE 8.2
Lithology and Area Distribution

Lithology	Area	Percentage
Highly oxidized ferruginous silty clay	1.97	1.62
Migmatite/banded gneiss	116.86	95.95
Sillimanite bearing quartzite	0.08	0.06
Grey to pink porphyric granite	2.89	2.37

also led to the removal of the existing vegetation of these slopes. Trees and small vegetation cover over soil slopes help to maintain the stability of the slope by absorbing large quantities of water from the soil, controlling total moisture content and its roots that tighten soil mass, and resist against surface water runoff, and diminish the likelihood of a landslide. The unguided surface runoff from excavated steep soil slopes led to a mass movement during monsoon.

Rainfall usually plays a critical role in initiating a landslide, so it is a key factor to be controlled on sites with more than a low landslide risk. The climate of the area is governed by the south-west monsoon and seasonal winds and receives heavy rainfall or long period rainfall during monsoon. The study area is one of the highest rainfall-receiving areas. The heavy rainfall or long period rainfall increases the moisture content of soil slopes, rise in the water table and consequent softening of the soil. The presence of excess moisture breaks the cohesive bond of particles and, in turn, starts sliding/flowing along with surface runoff. Also, lack of proper water drainage caused water logging and, in turn, toe erosion on the excavated slopes.

The present study deals with landslide hazards, which are occurring frequently and effecting severely in the study area. The methodology is carried out to identify the influence of geomorphologic parameters on slope stability, geotechnical properties of soil and estimation of rainfall threshold for landslide initiation so that the early warning system can be installed in the area especially considering the changing climatic patterns.

The morphometric analysis of the watershed has been carried to determine the geomorphic parameters on slope stability with SOI maps, STRM and Landsat 8 satellite data on the scale of 1:50,000.

The AWS with rain gauge is installed at NECBDC, Byrnihat to estimate the threshold levels of rain for triggering landslides and formulate strategies for minimizing the impact of landslides.

8.6 DRAINAGE MORPHOMETRY

The quantitative morphometric analysis of drainage basins is carried out in three aspects, such as linear (one dimension), areal (two dimensions) and relief (three dimensions) in order to evaluate the morphometric parameters of the study area (Table 8.3). The watershed shows trellis to sub-trellis drainage pattern indicating the structural control.

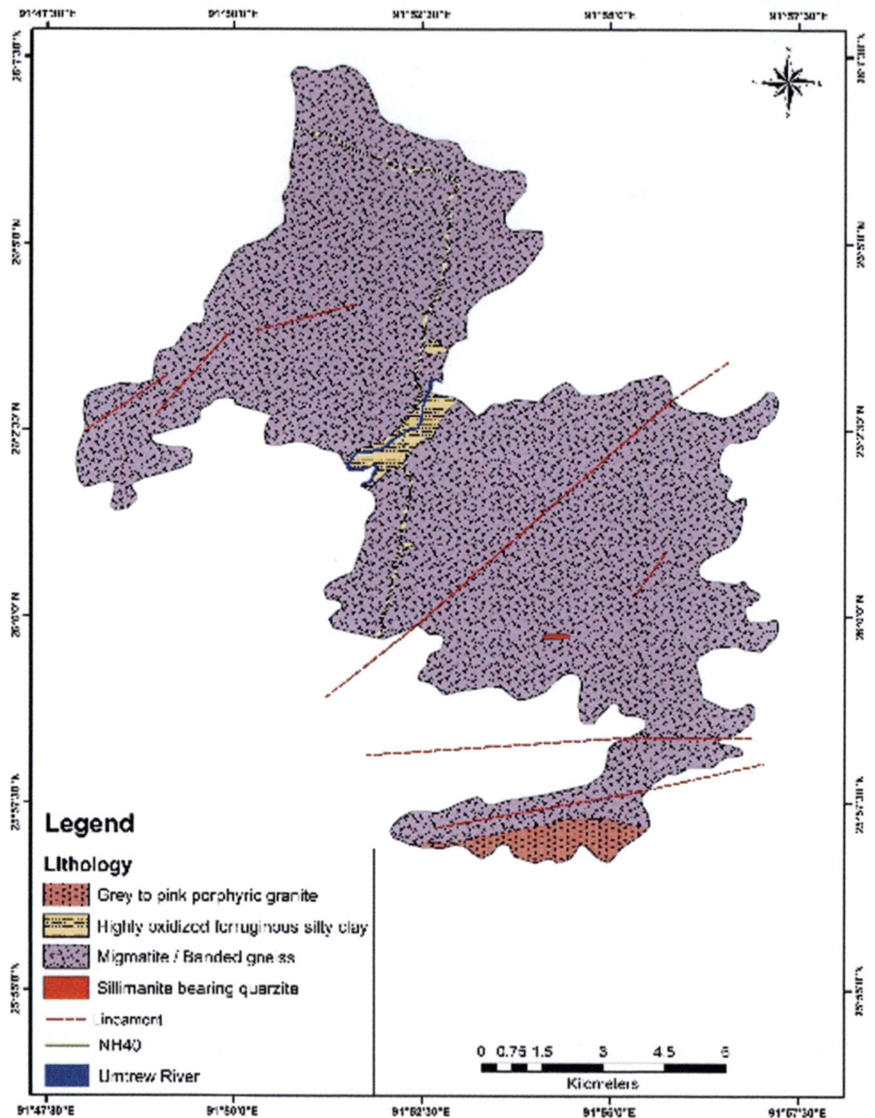

FIGURE 8.5   Geological map of the study area enumerating the rock types and the structures present in the area.

(i) **Relief ratio (R_h):** Tthe relief ratio (R_h) of maximum relief to horizontal distance along the longest dimension of the basin parallel to the principal drainage line is termed as a relief ratio (Schumm, 1956). It measures the overall steepness of a drainage basin and is an indicator of the intensity of the erosion processes operation on the slope of the basin. The relief ratio of

TABLE 8.3
Aerial Aspects of the Study Area

Sl. no.	Parameters	Watershed-1	Watershed-2	Watershed-3	Watershed-4	Watershed-5
1.	Basin area (A_u)	43.28 Sq.km	3.13 Sq.km	2.45 Sq.km	62.83 Sq.km	6.27 Sq.km
2.	Length of the basin (L_b)	11.75 km	3.23 km	1.98 km	17.16 km	4.71 km
3.	Basin perimeter (P)	41.27 km	8.70 km	7.32 km	59.45 km	14.70 km
4.	Circulatory ratio (R_c)	0.33	0.52	0.58	0.22	0.36
5.	Form factor (R_f)	0.31	0.30	0.62	0.21	0.28
6.	Drainage density (D_d)	2.75 km^{-1}	2.61 km^{-1}	2.66 km^{-1}	3.25 km^{-1}	2.64 km^{-1}
7.	Drainage frequency (D_f)	3.66	3.19	2.86	4.55	3.35
8.	Elongation ratio (E_r)	1.12	0.86	1.18	1.10	0.99
9.	Drainage texture	3.49 km^{-1}	1.15 km^{-1}	0.96 km^{-1}	4.81 km^{-1}	1.42 km^{-1}
10.	Bifurcation ratio (R_b)	5.07	9	6	4.11	4.3
11.	Drainage basin asymmetry (AF)	78.93	46.32	30.20	60.03	43.54
12.	Transverse topographic symmetry factor (T)	0.39	0.36	0.35	0.30	0.25

TABLE 8.4
Slope and Area Distribution (after Anbalagan, 1992 & BIS, 1998)

Slope Angle	Category	Area (Sq. Km.)	Percentage
< 15°	Very gentle slope	35.47	30.01
16°–25°	Gentle slope	55.77	47.20
26°–35°	Moderately steep slope	23.87	20.20
36°–45°	Steep slope	2.95	2.50
> 45°	Escarpment / cliff slope	0.11	0.09

the watershed is 0.02 (Table 8.4). A low value of relief ratio is mainly due to the resistant basement rocks of the basin and low degree of slope.

(ii) **Ruggedness number (R_n):** ruggedness number (R_n) is the product of drainage density and basin relief (Strahler, 1968). The watershed displays the ruggedness number of 0.24 and indicates that the area is extremely rugged with moderate relief and low drainage density.

(iii) **Drainage basin asymmetry (AF):** asymmetry factor or drainage basin asymmetry is an effective parameter for visualizing active tilt block tectonics in areas of both high and low seismic activity by visualizing neotectonic structures. The asymmetry factor (AF) can detect tectonic tilting on drainage basin scales and is sensitive to tilting perpendicular to the direction of the trunk. This factor allows determination of general tilt of the basin landscape irrespective of whether the tilt is local or regional (Hare and Gardener, 1985). When the AF is greater than 50; the main channel has shifted towards the downstream left side of the drainage basin and if AF is less than 50; the channel has shifted towards the downstream right side of the drainage basin (Hare and Gardener, 1985). The watershed of the study area has the AF range from 30.20–78.93, which indicates that the main channel has shifted suggesting the tectonic activity in the area.

 (iv) **Transverse topographic symmetry (T):** the transverse topographic symmetry factor, which is 0.33, also indicates that the basin is in the process of tilting. This shows medium dispersion of values indicating that the stream has migrated from the mid-line.

8.6.1   Basin Elongation Ratio (Re)

Elongation ratio is the ratio of the diameter of a circle of the same area as the drainage basin and the maximum length of the basin (Schumm, 1965). Basin elongation ratio < 0.5 is characteristic of tectonically active basins, values ranging from 0.5 to 0.75 reflect slightly active basins, and values > 0.75 reflect inactive basin settings (Bull and McFadden, 2020). The elongation ratio of the study area is 0.24, which indicates that Umpri watershed is tectonically active as Re < 0.5.

8.6.2   Drainage Basin Asymmetry Factor (AF)

The drainage basin asymmetry is an effective parameter in studying active tilt block tectonics in areas of both high and low seismic activity by studying neotectonic structures. This method is used to elucidate the formation of drainage basin and the state of tectonic control in its development (Cox, 1994; Hare & Gardner, 1985). It is determined by the following formula:

$$AF = 100(A_r/A_t)$$

where "A_r" is the area of the basin to the right bank of the stream facing downstream and "A_t" is the total area of the drainage basin.

   If AF is 50, it suggests a stable setting and there is no tilt in the basin and if AF is more or less 50, it may suggest a tilt and it indicates that the basin is tectonically active and still uplifting (Keller and Pinter, 2002). The Umpri watershed has AF equal to 44.55, which indicates that the main channel has shifted right suggesting tectonic activity in the area.

8.6.3   Transverse Topographic Symmetry Factor (T)

The transverse topographic symmetry factor estimates the amount of asymmetry of a river within a basin and how this asymmetry varies in length. For each segment, the ratio of the distance is taken from the basin mid-line to the active meander-belt midline (D_a) and to the basin divide. This index is calculated with the formula:

$$T = D_a/D_d,$$

where, D_a = the distance from the mid-line of the drainage basin to the midline of the active meander belt and D_d the distance from the basin mid-line to the basin divide.

Perfectly symmetric basin has value of transverse topographic symmetry (T) as zero, as the asymmetry increases T also increases and approaches the value of one indicating areas of potential tilting influence on drainage pattern. Transverse topographic symmetry factor is calculated for different segments of stream channels and indicates preferred migration of streams perpendicular to the drainage basin axis (Keller and Pinter, 2002).

The Umpri watershed has an average transverse topographic symmetry factor of 0.32, which indicates that the basin is in the process of tilting (Figure 8.6). This medium dispersion of value suggests that the stream has migrated from the mid-line. However, the amount of shifting is less.

8.6.4 HYPSOMETRY INTEGRAL

Hypsometric integral is a dimensionless parameter proposed by Strahler in 1952. It is one of the most useful parameters that describes and analyzes the distribution of elevations in an area. It also helps in explaining the erosion that had taken place in the watershed during the geological time scale due to hydrologic processes and land degradation factors (Bishop *et al.*, 2003). Hypsometric integral (HI) is calculated by the following formula: HI = mean elevation – minimum elevation / maximum elevation – minimum elevation.

Keller and Pinter (2002) described the values of HI on the basis of the erosional status, which is also an indicator of the tectonic condition of the basin as HI ≤ 0.3 (old stage/less active), it means watershed is fully stabilized. 0.3 – 0.6 (mature stage/ active) indicates the watershed is susceptible to erosion or tectonically active. HI ≥ 0.6 (young stage/highly active) indicates the watershed is highly susceptible to erosion. The mean value of HI of the Umpri watershed is 0.6, which is tectonically active and is susceptible to erosion (Figure 8.7).

8.6.5 MOUNTAIN FRONT SINUOSITY (SMF)

Mountain front sinuosity is extensively used geomorphic parameters to identify areas on the basis of the tectonic activity. It is defined as the total length of the mountain front as measured along the prominent break in slope along the foot of a mountain and the straight line length of the mountain front. This index is calculated with the formula;

$$S_{mf} = L_{mf}/L_s$$

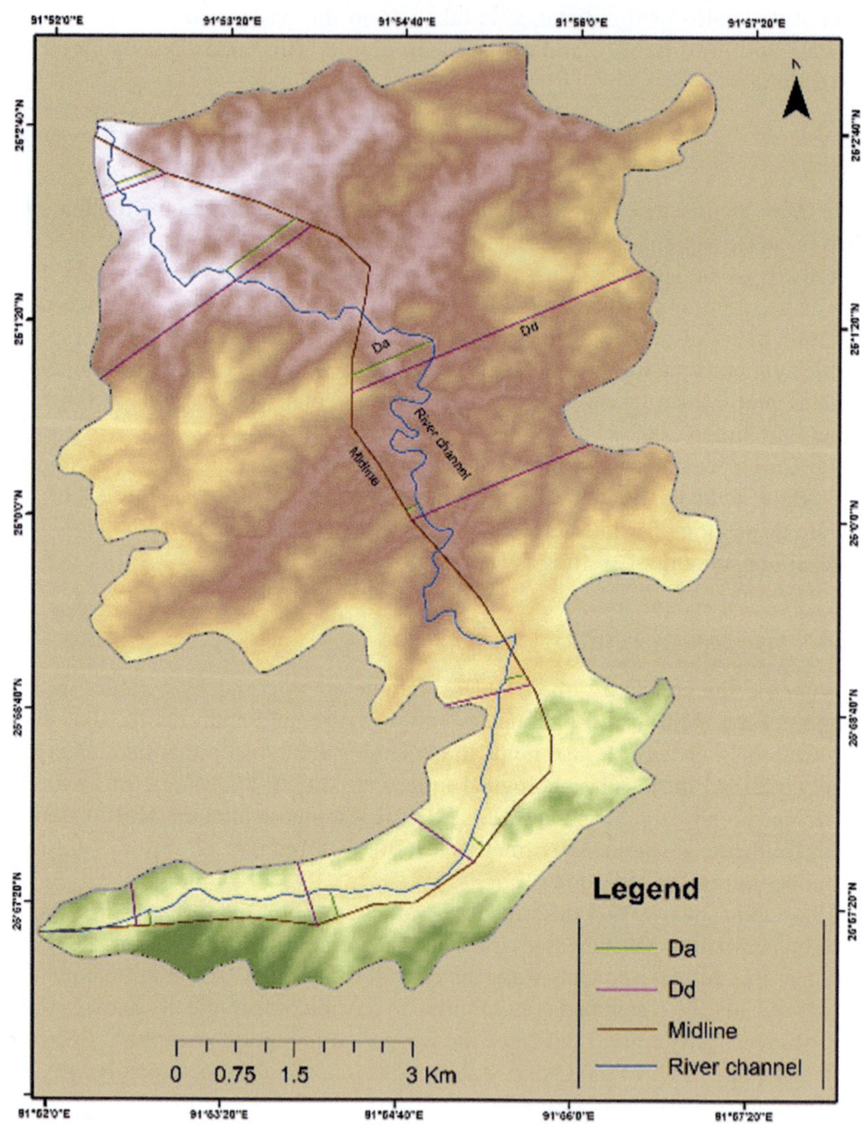

FIGURE 8.6   Transverse topographic map of the Umpri watershed.

where

S_{mf} = mountain front sinuosity index.

L_{mf} = total length of the mountain front.

L_s = straight line length of the mountain front.

The values of S_{mf} less than 1.4 indicate tectonically active areas, S_{mf} values between 1.4–3.0 indicate slightly active areas and S_{mf} values greater than 3.0 indicate inactive

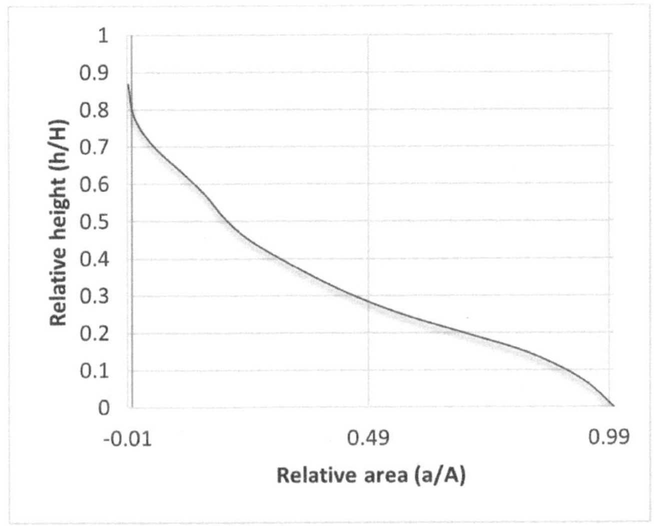

FIGURE 8.7 Graph showing the hypsometric integral of Umpri watershed.

setting (Bull and McFadden, 1977). The mean value of mountain front sinuosity index of the study area is 1.11, which suggests that the Umpri watershed is tectonically active (Figure 8.8).

8.6.6 Channel Sinuosity (*S*)

The channel sinuosity is used to understand the role of tectonism in the basin. According to Muller (1968), it is the ratio between the stream length (S_l) to the valley length (V_l). It is calculated by the following formula: $S = S_l/V_l$

where
  S = channel sinuosity.
  S_l = stream length.
  V_l = valley length.

The index value of 1.0 indicates a straight river course, while values between 1.0 to 1.5 indicates a sinuous river shape and values greater than 1.5 represents a meandering course (Leopold *et al.*, 1964). The channel sinuosity of the Umpri watershed is 1.19 indicating a sinuous course and appears to be tectonically active.

8.6.7 STREAM-LENGTH GRADIENT INDEX (SL)

The stream-length gradient index (SL) is calculated along a river and is used to evaluate the relationship between potential tectonic activity, rock resistance, topography and length of the stream (Hack, 1973; Keller and Pinter, 2002). It is a very useful parameter to evaluate if change in stream slope is due to rock resistance or tectonic deformation

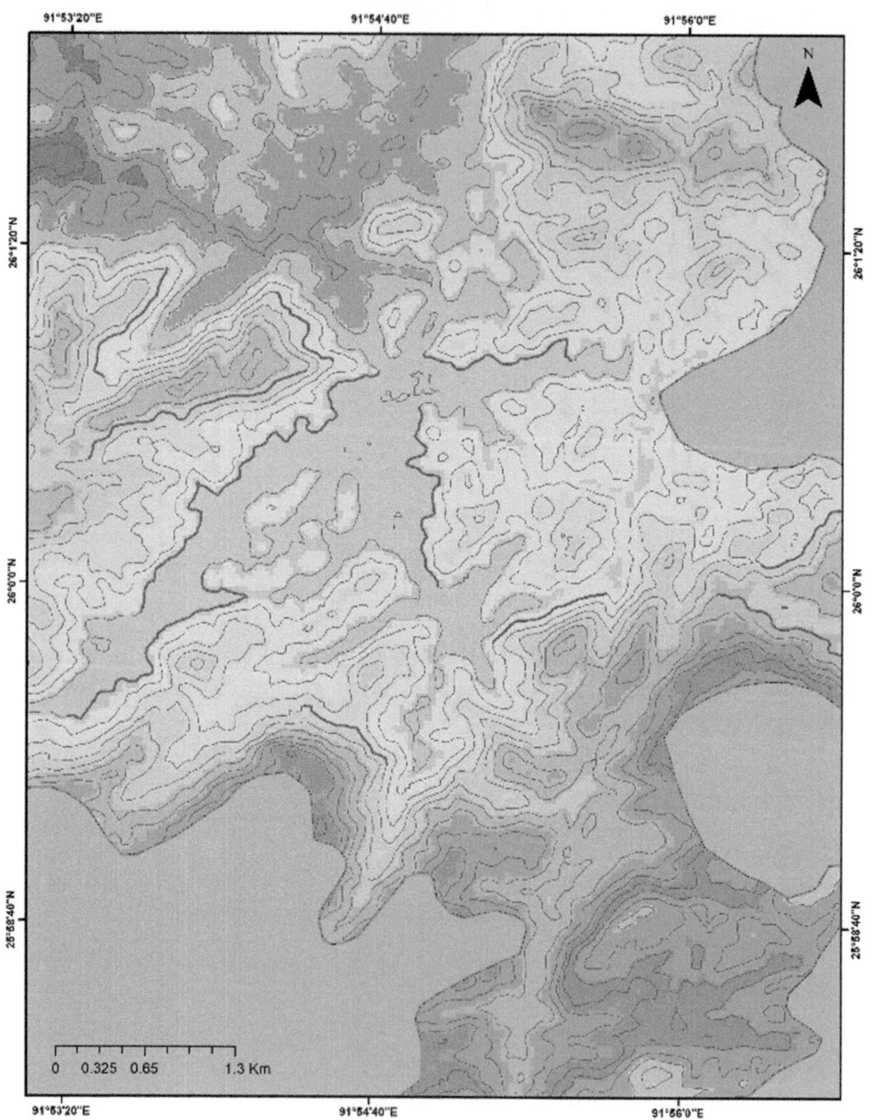

FIGURE 8.8   Mountain front sinuosity map of the Umpri watershed.

in particular, if it has a vertical component (Keller and Pinter, 2002). The stream-length gradient index (SL index) was proposed by Hack (1973) and is calculated as

$$SL = (\Delta H / \Delta L) \, / \, L$$

where
SL = stream length gradient.
$\Delta H / \Delta L$ = channel slope or gradient of particular reach.

ΔH = change in elevation of the reach.
ΔL = length of the reach.
L = total length from mid-point of the reach of interest upstream to the highest point on the channel.

The high SL values are reported where rivers cross hard rocks reflecting high tectonic activity while the low SL index values indicate relatively low tectonic activity and suggest less resistant and softer underlying rock types (Keller and Pinter, 1996). In the study area, values of the SL index range from 7.6 m to 529.61 (Figure 8.9). The variations of SL index values across the drainage basin might be due to the lithological variations or influenced by tectonic activity.

8.6.8 The Ratio of Valley Floor Width to Valley Feight (Vf)

Valley floor width to valley height ratio (V_f) is an important parameter to measure whether the river is actively downcutting and incision in uplifted areas (Bull and McFadden, 1977). It is expressed as

$$V_f = 2\ V_{fw}/\ (E_{ld} - E_{sc}) + (E_{rd} - E_{sc})$$

where V_{fw} is the width of the valley floor, E_{sc} is the elevation of the valley floor, E_{ld} and E_{rd} are the elevations of the left and right valley divides.

Low values of V_f (<0.5) are associated with higher rates of uplift and incision and the flat-floored, wide valleys with high V_f (>0.8) indicate attainment of the base-level of erosion and tectonic quiescence (Keller and Pinter, 1996). In the present study (Figure 8.10), the values of V_f range from 0.04 to 0.56 suggesting the tectonic movements and uplift. The valley floor morphology seems to have been controlled by the upliftment on the faulted segments and also the folded and deformed behavior of the banded gneiss or migmatite rocks of the Meghalaya Gneissic Complex.

8.7 SLOPE PATTERN

Slope analysis is an important parameter in geomorphic studies. It affects land qualities such as runoff, erosion hazard, moisture balance and landslide. An understanding of slope distribution is essential as the slope map provides data for planning, settlement, mechanization of agriculture, deforestation, planning of engineering structures, morpho conservation practices, etc. (Sreedevi et al., 2005). The nature of the slope in the area is highlighted by the fact that more than 20% of its geographical area falls in the moderate to steep slope category. This combined with gentle slopes and very gentle slopes together constitutes nearly 80% of the study area. Considering slope as one of the important factors for landslide study, the area would be left with only about 30% very gently sloping land surface (Figure 8.11a).

8.7.1 Relative Relief

Relative relief map represents the difference between the maximum and minimum heights within an individual facet. It shows the major breaks in the slopes of the

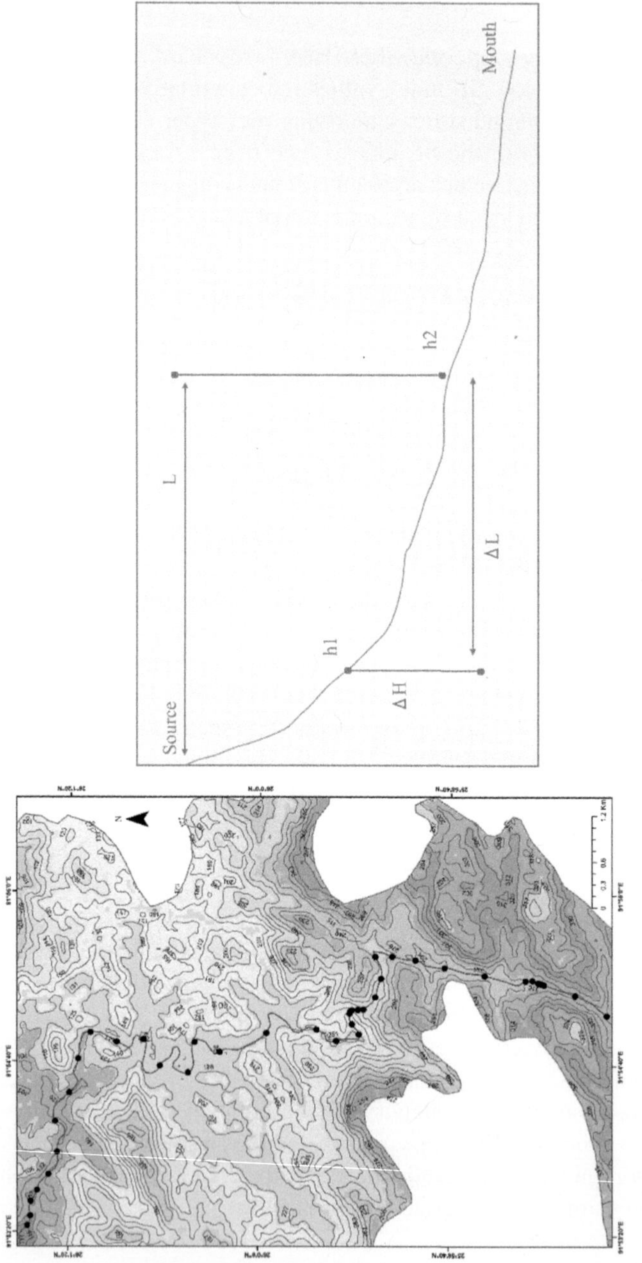

FIGURE 8.9   Schematic representation of stream-length gradient index.

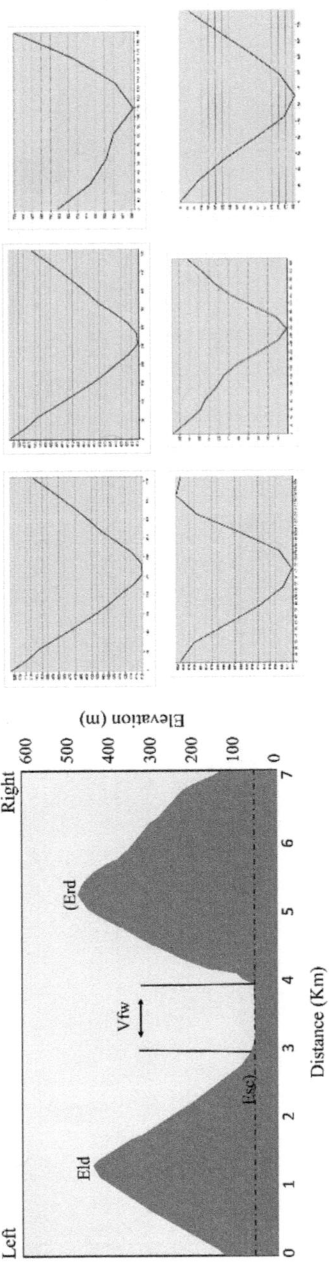

FIGURE 8.10 Schematic representation of Ratio of valley floor width to valley height.

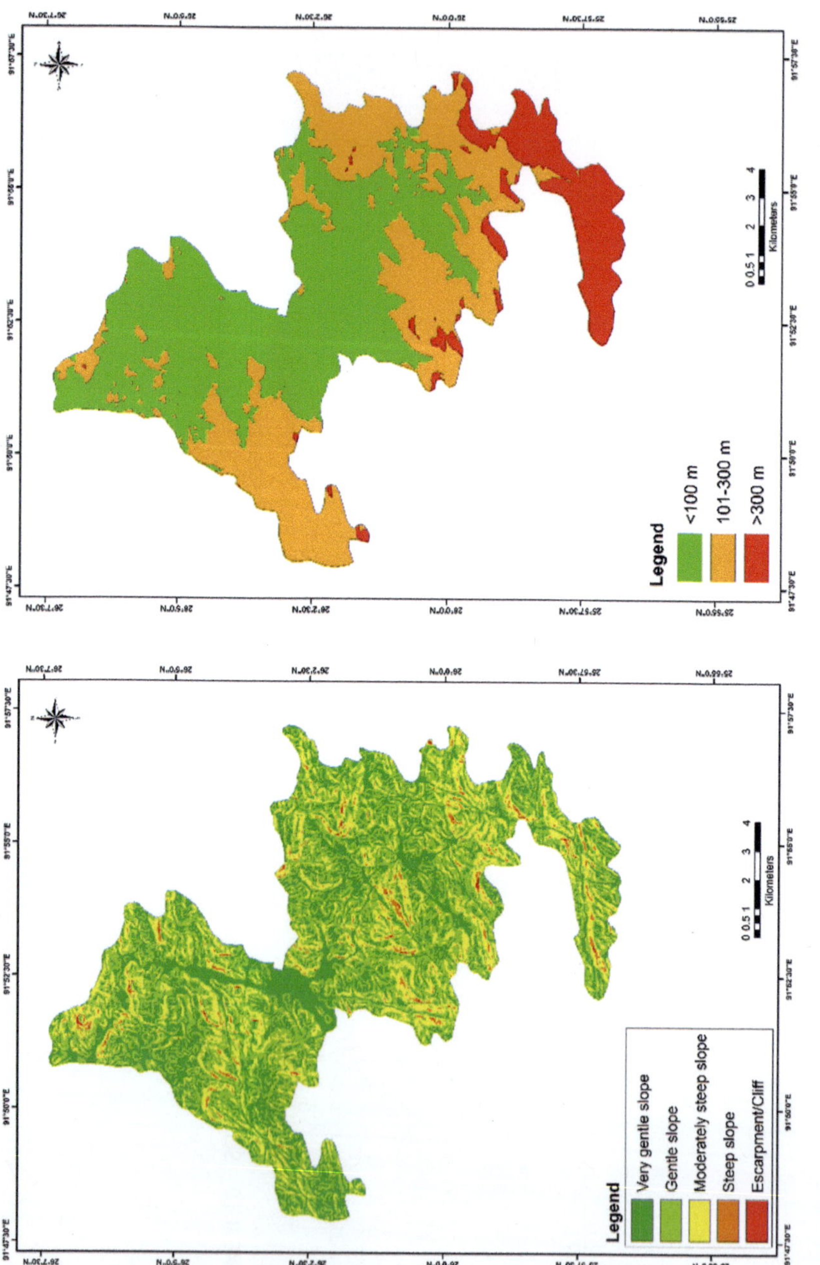

FIGURE 8.11   (a) Slope pattern and (b) Relative relief of the study area.

TABLE 8.5
Relative Relief and Area Distribution (after Anbalagan, 1992 & BIS, 1998)

Relative Relief	Category	Area (Sq. Km)	Percentage
< 100 m	Low relief	58.24	49.26
101–300 m	Medium relief	45.26	38.28
>300 m	High Relief	14.72	12.45

area and plays a significant role in landform development, drainage development, surface and sub-surface water flow, permeability and erosional properties of the terrain. The general relief of the study area is excessive, except in the northern part. The elevation varies from 60 to 735 m above mean sea level with general aspect towards the south. The southern part which has comparatively high elevation, but is less to moderately excessive whereas the northern part is comparatively less in height. About 50% of the area fall under medium to high relief (Table 8.5 and Figure 8.11b).

8.8   GEOTECHNICAL INVESTIGATION

Landslides are exclusively produced by water or moisture content in the slope-forming materials. This is the reason why landslides are more common during the rainy season. However, many other parameters defining the overall physical as well as mechanical properties of the soil and rock, directly and indirectly cause landslides. So, a detailed analysis and study of soil and rock mechanical properties of slide affected as well as potential areas will form one of the main bases of the present investigation. In the study area, soil mechanical properties – water/moister content, porosity, particle size distribution and classification of shear strength determinations – are one of the main aspects of the research activity. The samples have been collected for determining of shear strength and other parameters of soil and rock from the two selected landslide area, viz., near CMJ University and Pahamlang along NH40. The soils were collected vertically, and a series of laboratory tests were carried out to ascertain the geotechnical properties of these soils and rock (Table 8.6).

The geotechnical characterization of the slope-forming geomaterials is extremely important for the analysis of slope models (Sharma *et al.*, 2016; Singh *et al.*, 2017). The soil samples from two sites, i.e., near CMJ University and Pamalang along NH 40, which developed over pre-Cambrian gneisses were collected. The characteristics of the soils were investigated in the laboratory to study the physical and geotechnical properties of soil at NIT, Meghalaya. Following the American Society for Testing and Materials, laboratory investigations of the soils have been conducted, namely specific gravity (G), grain size analysis, Atterberg's limits [i.e., liquid limit (LL), plastic limit (PL)], free swell index, standard proctor compaction characteristics [i.e., optimum moisture content (OMC) and maximum dry density (MDD)] and shear strength characteristics (i.e., cohesion and angle of internal friction, see Table 8.7).

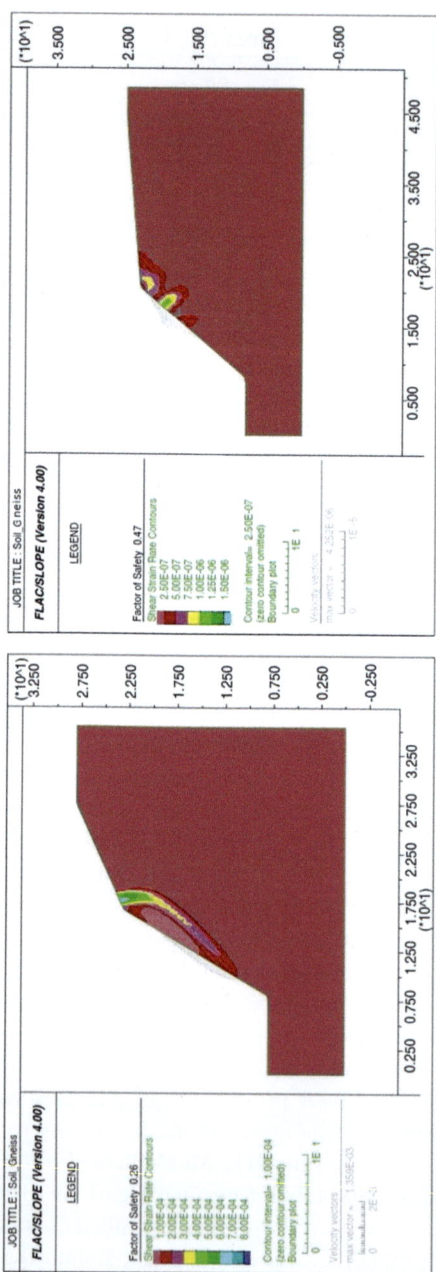

FIGURE 8.12 FLAC/Slope models (a) Pahamlang and (b) near CMJ University signifies maximum shear strain rate, velocity vectors and characteristic failure zones.

TABLE 8.6
Physical Properties of Soil

Location	Specific Gravity	Shrinkage limit (SL %)	Liquid limit (LL %)	Plastic limit (PL %)	Plasticity index (PI %)	Free swell index (%)	Grain Size Distribution Gravel (%)	Sand (%)	Silt (%)	Clay (%)
				Consistency Limit			Grain Size Distribution			
CMJ University	2.62	17.80	25.59	48.63	22.98	7.25	7.07	25.00	38.39	29.30
Pahamlang	2.08	17.28	25.07	48.11	22.46	6.73	6.55	24.48	37.87	28.78

TABLE 8.7
Geotechnical Properties

Location	Natural moisture content (%)	Optimum moisture content (%)	Maximum dry density (g/cm³)	Maximum wet density (g/cm³)	Unconfined compressive strength (kPa)	Cohesion (kPa)	Angle of internal friction (°)
CMJ University	24.94	29.44	1.51	1.94	54.50	19.16	25.94
Pahamlang	24.42	28.92	0.99	1.42	53.98	18.64	25.42

8.8.1 DETERMINATION OF FACTOR OF SAFETY

The use of numerical techniques in the stability analysis is amplified with the rapid development of computing efficiency. The factor of safety (FoS) for slope may be computed by reducing the shear strength of rock or soil in stages until the slope fails. A slope may fail if the shear strength of slope mass is less than the active, effective shear stresses. FoS is a dimensionless value that provides important information about the overall strength of a slope and is defined as the ratio of prevailing shear strength to the shear stress required for equilibrium [Eq. (8.1)]. In normal practice, FoS values greater than unity indicate the stable slope, whereas values lower than unity indicate unstable slope conditions. The FoS is calculated as

$$C_r = \frac{C}{\text{SRF}} \tag{8.1}$$

$$\text{Tan}(\theta) = \frac{\tan(\theta)}{\text{SRF}} \tag{8.2}$$

where SRF is the strength reduction factor, and C and (θ) are original shear strength parameters (cohesion and friction angle).

This method is based on the shear strength reduction technique, and so, to attain the correct SRF value, it is vital to trace the value of FoS that allows slope to fail. The FoS is calculated by gradually reducing the shear strength of slope mass to fetch the slope to a state of equilibrium. Thus, a series of models are obtained through using trial values of factor (F_{trial}) to reduce the cohesion (C) and internal friction angle (ϕ) until the mathematical solution became non-converging. This non-converging SRF termed critical SRF is equivalent to the FoS of the slope (Singh et al., 2017). In fast lagrangian analysis of continua (FLAC), a bracketing approach similar to that proposed by Dawson et al. (1999) is used for the analysis.

The two soil slope profiles with landslide events were analyzed for stability using the strength properties of soil determined at OMC. The behavior of soil slope cuttings

TABLE 8.8
Geometry of Soil Slopes

Slope height (m) and slope angle (^{0})

Location	Cut slope
CMJ University	47 m, 70^0
Pahamlang	54 m, 82^0

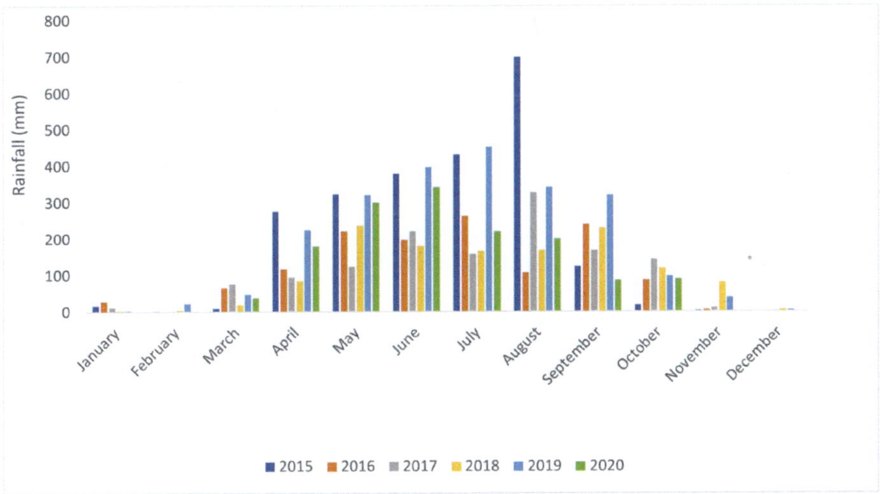

FIGURE 8.13 Monthly rainfall (mm) from 2015–2020.

and overlying natural slope profiles was studied. The sloping geometry of both cut slopes and overlying natural slope of these two soil profiles is tabulated in Table 8.8. These two soil profiles were analyzed by FDM using two-dimensional FLAC/Slope software. The FoS of these two slopes that varies from 0.93 and 1.2 indicates the completely unstable soil slopes at OMC. Failure occurs at a certain height from the base as indicated by contours of shear strain rate. The velocity vectors indicate movement of soil mass downwards and away from the slope (Figure 8.13a). The soil mass of both slopes was oversaturated either by heavy or prolonged rainfall or increased level of the water table during the period as specified by higher NMC (by 10%) than OMC in the soil mass. Failure surface propagates at the top of the cut slope with downward and outward movement (Figure 8.13b).

Further, the field observations and laboratory results indicate that the soil exposed contains a significant amount of silt and clay size particles (more than 60%). Consistency limit test involves liquid limit, plastic limit and shrinkage limit tests and is presented in Table 8.1. Soil classification according to the plasticity index (ASTM,

2010) indicates that soils are highly plastic (PI > 17). Soil samples have higher natural moisture content (NMC) than optimum moisture content (OMC), which is due to the presence of more void spaces and water-absorbing clay minerals. This proves that the excess quantity of moisture drives the initiation of landslides to cause soil and debris slides/flows especially near CMJ University where the probability of landslide occurrence has increased in the recent past, posing a threat to the existing infrastructure including the road connecting the State of Tripura, Mizoran, Meghalaya and Barak Valley.

8.9   MITIGATION MEASURES

The preventive measures to stabilize slopes at Pahamlang of about a 4 m height concrete wall using gabion mesh filled with rock pieces at the toe of slopes and cemented drainages. However, the slope near CMJ University needs to be stabilized because of the weight of the University building at the crown, the slope will fail as driving forces will exceed the resisting forces.

Since most of the landslides along this road section are related to soil slopes and triggered by rainfall mainly during monsoon, it is better to take long-term stability measure for most of this slide's area. The field investigation indicates that the area can be stabilized with certain precautions and preventive measures. At sites experiencing recurring landslides, slopes may be re-excavated for flattening of slopes of higher heights along with benching on slopes. At these sites, geonets or bionet can be installed for the protection from surface water runoff, erosion and jute matting to promote vegetation growth on soil slopes.

8.10   RAINFALL THRESHOLD FOR LANDSLIDE INITIATION

Rainfall is considered as an important factor responsible for landslide triggering, especially in areas characterized by heavy precipitation during monsoon. The predisposing factors of a landslide are intrinsic of the local specific conditions including slope angle, soil thickness, slope exposure, slope curvature, land use, hydraulic conditions, historical landslides, geomorphological units, lithology, tracks and man-made cuts. Landslides caused due to rainfall are due to the build-up of pore pressure into the ground. The adoption of rainfall thresholds helps in developing an alarm system, which would help local people to evacuate in case of emergencies and notify the civil authorities about such incidents. The correlation between rainfall and land sliding is well known, but no attempt has yet been made to assess the precise amount and duration of rainfall that triggers mass movement. Rainfall thresholds can be obtained either by physical model or by empirical model.

The main objective of the study is to derive local rainfall thresholds for landslide occurrences along NH40 using daily rainfall and landslide data for the year 2015–2020, also to describe the monsoon rainfall as low intensity and long duration with interruptions with intermittent heavy bursts, and to check whether the presence of such rainfall pattern emphasizes the importance of antecedent rainfall.

TABLE 8.9
Impact of Landslides in Ribhoi District from 2014–2019 (source District Disaster Management Plan, Ribhoi District, 2019, Volume-I)

Year of Occurrence	No. of Villages Affected	Population Affected	Impact on Life	Damage to Houses (Partly & Fully)	Damage to Infrastructure (Partly & Fully)	Damages to Crops	Live-stock
2014–2015	1	Nil	1	Nil	Nil	Nil	Nil
2015–2016	13	38	Nil	21	Nil	yes	Nil
2016–2017	11	16	Nil	6	Nil	8.6 ha	Nil
2017–2018	4	55	Nil	2	Nil	6.3 ha	Nil
2018–2019	15	30	Nil	30	Nil	Nil	Nil

Another objective is to plot the daily rainfalls on the day of failure for all the landslide events against the antecedent rainfalls to interpret the bias of landslide occurrences. The methodology used consists of two key components: (i) collection of rainfall data, which caused slides along with the date and intensity; (ii) using empirical methods to determine the relationship between rainfall and landslide occurrence.

Seven historical landslide events that occurred during 2015 to 2020, which led to loss of life and property, including loss of cattle and agriculture land in and around the study area were considered to determine the rainfall threshold development. The events were identified from the technical reports of Disaster Management Authority, Ribhoi District, previous literatures and local newspaper reports. The rainfall data for the landslide events occurring at the study area was collected from various government agencies and the Automatic Weather Station with rain gauge at CBCT, Byrnihat maintained by North-Eastern Hill University, Shillong (Figure 8.11). The year of thelandslide events and theirimpact on socio-economic conditions of the Ribhoi district are listed in Table 8.9.

8.10.1 Development of Empirical Rainfall Threshold

In general, many landslides occur in the study area during the monsoon. The landslides in the study area can be classified as debris flow type, which frequently causes significant damages to roads and other infrastructures. The rainfall threshold developed by an empirical model based on daily and antecedent rainfall requires parameters, such as landslide history, antecedent rainfall and daily rainfall on the event day. When using antecedent rainfall measurements to predict landslide occurrence, a key difficulty is the definition of the period over which to accumulate the precipitation. Review of the literature reveals a significant scatter in the considered periods. This large variability may be attributed to different factors, including (i) diverse lithological, morphological, vegetation and soil conditions, (ii) different climatic regimes and meteorological circumstances leading to slope instability, (iii) and heterogeneity

TABLE 8.10
Landslide Events with Corresponding Daily and Antecedent Rainfall

Event date	Average daily rainfall (on event date)	3 day antecedent rainfall
17-06-2017	219	137
06-07-2017	157	99
18-06-2018	180	87
02-05-2019	236	121
23-08-2019	341	215
22-09-2020	216	102
27-09-2020	306	117

and incompleteness in the rainfall and landslide data used to determine the thresholds. The temporal probability for landslide initiation of shallow debris flow type landslides in the cut slopes of road and rail sections between Mettupalayam and Coonoor (19 km stretch) of the Nilgiris district using antecedent and the daily rainfall-based empirical threshold model was estimated. The method estimates the rainfall threshold based on antecedent and daily rainfall with respect to past landslide events. Hence, the rainfall threshold for the study area is developed based on antecedent and daily rainfall. The data may be used in creating warning and suitable mitigation measures.

In the study, 3 day antecedent rainfall has been considered for the estimation of rainfall threshold. To find the empirical threshold equation threshold (R_T), a scatter plot was prepared based on daily rainfall and corresponding 3 day antecedent rainfall with respect to the landslide events (Table 8.10). An envelope curve was drawn and represented by linear mathematical equation. The equation provides the rainfall threshold for landslide occurrences.

The rainfall threshold equation is represented by the linear mathematical equation as given below.

$$R_T = 84.078 + 1.246\ R3ad$$

where R_T = rainfall threshold in mm, $R3ad$ = 3 day antecedent rainfall in mm.

The threshold analysis based on daily and 3 day antecedent rainfall in conjunction with past landslide events revealed that, either very high magnitude of daily rainfall or a very high amount of antecedent rainfall is required to trigger landslide in the study area (Figures 8.14). The threshold indicates that, when the daily rainfall crosses 150 mm, there is a possibility of landslide occurrence even if there is no antecedent rainfall. When the 3 day antecedent rainfall (R_{3ad}) exceeds 84.078 mm, even a continuous normal rainfall is capable of triggering a landslide in the region. The recent landslides along NH 40, Jowai road is the best evidence of major landslide occurrence when both daily and 3 day antecedent rainfall exceed the threshold limit. The slope of the study area received a maximum of 215 mm of 3 day antecedent rainfall and 341 mm of daily rainfall on the day of landslide occurrence in the year 2009.

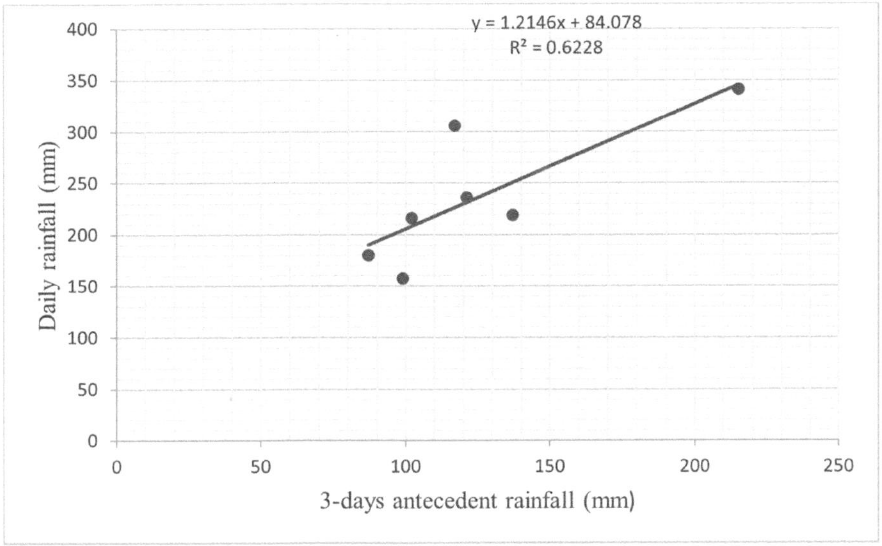

FIGURE 8.14   Rainfall Threshold Envelope Curve for the study area.

8.11   CONCLUSIONS

The study area is one of the seismically active regions and displays high seismic activity, tilting and settlement of the river channel. Based on investigations, it is observed that the watershed shows trellis to sub-trellis drainage patterns, indicating high structural control on the drainage development. Low drainage density, low frequency and coarse drainage texture indicates the presence of permeable geologic material with high infiltration having low groundwater recharging characters and vegetated slopes. The high bifurcation ratio indicates structural control of drainage representing neotectonics activity of the study area. The circulatory ratio and form factor depict diversified slope, low discharge of runoff, high permeability of the subsoil and structurally controlled drainage system. The asymmetry factor also proves that watershed is under tilting and hence vulnerable to geohazard. Thus, the linear aspects, aerial aspects and the relief aspects reveal that the watershed is in the process of evolution as the basin is in the process of tilting.

Since the region is in the process of tilting, the soil slopes are prone to failure during rainy season due to heavy or prolonged rain that led to a rapid increase in soil moisture ratio and water logging at the toe of slopes. The FoS, maximum velocity vectors and their respective developed contours of shear strain rate have been calculated for these vulnerable slopes by FLAC/Slope software. During the monsoon period, rainwater flows from these newly excavated steeper slopes with no vegetation cover that also saturates the top layers that led to behave the silty soil as cohesion-less which, in turn, flows/slides with flowing water. These soil slopes require proper support from mechanical structures like retaining walls with a proper drainage system. At recurring landslide sites, slopes may be re-excavated for flattening along with benching of the slopes. It can help to stabilize landslide-prone slopes. At these sites, geonets or bionet can be installed for the protection from surface water runoff, erosion and jute matting to promote vegetation growth on soil slopes. After installation, remedial measures must be inspected routinely and maintained as the lack of maintenance may cause renewed landslide movements.

The study indicates that landslides in the region would occur when the daily rainfall crosses 150 mm and 3 day antecedent rainfall (R_{3ad}) exceeding 84.078 mm. These thresholds can be used as a predictive tool, but due care needs to be taken when used as an early warning system. When the actual rainfall reaches towards the threshold limit, it is recommended to give an alert for the people in that region to safely move from the risk area. Rainfall threshold correlation will be useful in order to save lives and property loss by issuing advance alerts. With changing climatic patterns, the landslide activity may increase many fold and hence there is a need for early warning so as to prevent disastrous landsliding.

REFERENCES

Bull, W. B., & McFadden, L. D., (2020), Tectonic geomorphology north and south of the Garlock fault, California. In D. O. Doehring (Ed.), *Geomorphology in Arid Regions*. Routledge, Abingdon, pp. 115–138.

Bishop, M. P., Shroder Jr, J. F., & Colby, J. D., (2003), Remote sensing and geomorphometry for studying relief production in high mountains. *Geomorphology*, 55(1-4), pp. 345–361.

Bull, W. B., & McFadden, L. D., (1977), Tectonic geomorphology north and south of the Garlock fault, California. In D. O. Doehring (Ed.), *Geomorphology in Arid Regions*, Proceedings of the 8th Annual Geomorphology Symposium: Binghamton, New York, State University of New York Publications in Geomorphology, pP. 115–138.

Dawson, E. M., Roth, W. H., & Drescher, A., (1999), Slope stability analysis by strength reduction. *Geotechnique*, 49(6), pp. 835–840.

Hack, J. T., (1973), Stream-profile analysis and stream-gradient index. *J Res. US Geol. Survey*, 1(4), pp. 421–429.

Hare, P.W., & Gardner, T.W., (1985), Geomorphic indicators of vertical neotectonism along converging plate margins, Nicoya Peninsula, Costa Rica. In M. Morisawa & J. T. Hach (Eds.), *Tectonic Geomorphology*. Allen and Unwin, Australia, pp. 75–104.

Keller, E. A., & Pinter, N., (1996), *Active Tectonics* (Vol. 338). Prentice Hall, Upper Saddle River, NJ.

Keller, Edward A., & Pinter, N., (2002), *Geomorphic indices of active tectonics, active tectonics earthquake, uplift and landscape* (2nd ed.), Prentice Hall, New Jersey, p. 121.

Leopold, L. B., Wolman, M. G., Miller, J. P., & Wohl, E. E., (2020), *Fluvial Processes in Geomorphology*. Courier Dover Publications, New York.

Mueller, J. E., (1968), An introduction to the hydraulic and topographic sinuosity indexes. *Annals Assoc Am. Geo.*, 58(2), pp. 371–385.

Schumm, S. A., (1956), Evolution of drainage systems in badlands at Perth Amboy, New Jersey. *Bull. Geol. Soc. Am.*, 36, pp. 597–646.

Schumm, S. A. (1965). Quaternary paleohydrology. In H. E. Wright & D. G. Frey (Eds.), *The Quaternary of the U.S.* Princeton University Press, Princeton, pp. 783–794.

Sreedevi, P. D., Subrahmanyam, K., & Ahmed, S., (2005), The significance of morphometric analysis for obtaining groundwater potential zones in a structurally controlled terrain. *Environmental Geology*, 47, pp. 412–420.

Strahler, A. N., (1956), Quantitative geomorphology of drainage basins and channel networks. In Chow, V. (Ed.), *Handbook of Applied Hydrology,* McGraw-Hill Book Company, New York, pp. 4.39–4.76.
Strahler, A. N., (1968), Quantitative geomorphology. In Fairbridge, R.W. (Ed.), *The Encyclopedia of Geomorphology.* Reinhold, New York, pp. 898–911.

9 Hotspots of Land Use Land Cover Change of Upper Meghna River Basin, India

Amritee Bora, Devesh Walia, B.S. Mipun and Vishwa Ranjan Sinha

9.1 INTRODUCTION

Human activities have modified the environment and ecosystem to a considerable extent (Goldewijk, 2001). Land use land cover change (LULCC) is a very basic as well as the most widely studied anthropogenic phenomenon that correlates geoscience, environmental science as well as social sciences. A change in land use is the major underlying cause of global environmental change (Sala, 2000; Duraisomy *et al.*, 2018). It has been explored that during the last centuries, the intensity and scale of these modifications have increased significantly (Watson *et al.*, 1996; Goldewijk, 2001). About three-quarters of the Earth's land surface has been altered by humans within the last millennium (Arneth *et al.*, 2010; Winkler *et al.*, 2021). According to the FAO (Food and Agriculture Organization, UN) 2001 report, during the period of the 1990s the world forest covers were converted to other land uses at 0.38 per cent per year. About 17% of the Earth's land surface has changed at least once between six (major) land categories from 1060 to 2019, summing up all of the individual change events (including areas of multiple changes), the total land change extent is 43 million sq. kilometres (Winklear *et al.*, 2021).

These changes are more dominant and noticeable in developing countries, which are undergoing rapid urbanization. Thus, LULCC is the most common phenomenon to be studied in different regions of the world including South-East Asia (Vadrevu and Ohara, 2020). Monitoring land use/cover change (LUCC) has become an important theme of research as these changes influence the global fluid system; the atmosphere, world climate and sea level (Asubamteng, 2007). LUCC is also a core subject co-supported by the International Geosphere-Biosphere Program (IGBP) and International Human Dimensions Program (IHDP) (Li *et al.*, 2017).

This chapter attempts to identify the hotspots of land use/cover change during the last two decades of the upper Meghna River basin and evaluate the annual land use change rate within a time frame of 5 years.

DOI: 10.1201/9781003485995-9

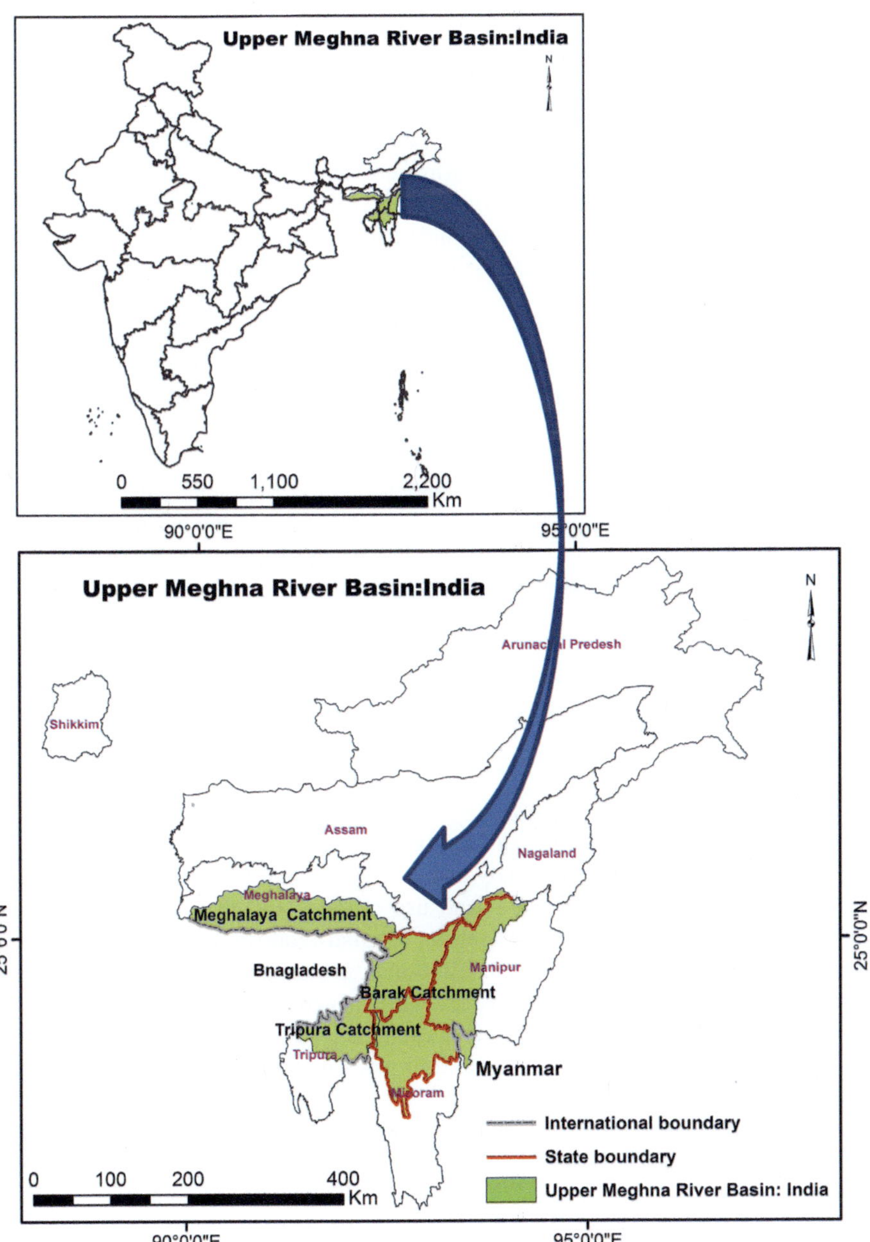

FIGURE 9.1 Upper Meghna River Basin, India.

9.2   STUDY AREA

India and Bangladesh share 54 transboundary sub-basins and all of them are part of the drainage system of the Ganga Brahmaputra Meghna (GBM) Basin. Among those, 29 transboundary sub-basins come under the Meghna River basin. Meghna River was formed due to the joining of the river Surma and river Kushiyara; together it is known as Barak River, in India. Thus, the river often is called the Barak-Meghna River. It originates in the hilly region of eastern India and constitutes the *Upper Meghna River basin*. At Chandrapur (Bangladesh) it joins with the Padma River and is known as *Lower Meghna* till its mouth.

Figure 9.1 shows the Indian part of the upper Meghna River Basin. According to the IUCN 2018 report, despite its transboundary nature, the biodiversity and livelihood significances of the Meghna River Basin are less discussed, when it comes to discourse on transboundary water governance between India and Bangladesh. As per records, the total area covered by the Meghna basin is 82,000 sq. km of which 47,000 sq. km (57%) of the area comes under the Indo-Myanmar part. However, the present study reveals that by combining India and Myanmar the basin covers an area of approximately about 43,961.57 sq. km. The basin includes six North East Indian states (Assam, Meghalaya, Manipur, Tripura, Nagaland and Mizoram) and the north-east corner of Myanmar as well. For convenience of study, the Meghna River basin is sub-divided into three catchments, i.e., Barak catchment, Tripura catchment and Meghalaya catchment. Geographically the basin lies between 26° 00′ N and 23° 05′ N latitude and 90° 00′ E and 94° 05′ E longitude.

9.3   METHODOLOGY

Identification of hotspots using satellite data is one of the effective methodological approaches in terms of LULCC study and is able to identify and highlight the region that falls under rapid and frequent land use changes and the land cover becomes vulnerable and is at high risk of losing its original biodiversity. Over the last three decades, the availability of global satellite data products has aided in the development of spatially explicit global and regional land cover databases at various spatial resolutions (Grekausis *et al.*, 2015). However, each satellite data with a certain spatial resolution has its own limitation in terms of image interpretation and mapping. The hotspot identification for the present study is based on the frequency of land use land cover changes in six different time periods such as 2005–2010, 2005–2015, 2005–2019, 2010–2015, 2010–2019 and 2015–2019 (Table 9.1). The study has used Landsat data, with 30 m of spatial resolution for the hotspot assessment.

Each image was projected to Universal Transverse Mercator projection, with spheroid and datum as WGS 84, Zone 46 North projection parameters. Visual image interpretation technique has been used for the initial assessment of LULCC with a minimum mapping unit (MMU) of 0.8 hectares. In order to quantify the hotspots, the G_i^* statistic was computed, which is a measure of the degree of spatial clustering of a local sample and its difference from the expected value. It is calculated as the sum of the difference between values in the local sample and the mean and is standardized as a "z" score. For a statistically significant positive z score, the larger the z score

TABLE 9.1
List of Selected Landsat Images of the Study Area, India

Sl. No	Year	Satellite	Sensor	Path	Row	Image Acquisition Date
1	**2005**	LT5	TM	135	42	10-Feb-05
2		LT5	TM	135	43	1-Mar-06
3		LT5	TM	135	44	16-Jan-05
4		LT5	TM	136	42	8-Mar-06
5		LT5	TM	136	43	13-Apr-05
6		LT5	TM	136	44	17-Mar-06
7		LT5	TM	137	42	1-Mar-06
8		LT5	TM	137	43	16-Feb-05
9		LT5	TM	138	42	27-Feb-05
1	**2010**	LT7	ETM+	135	42	8-Feb-10 & 30-Jan-15
2	**&**	LT7	ETM+	135	43	27-Feb-09 & 23-Mar-15
3	**2015**	LT7	ETM+	135	44	30-Jan-10 & 11-Apr-15
4		LT7	ETM+	136	42	6-Mar-10 & 21-Feb-14
5		LT7	ETM+	136	43	21-Jan-10 & 9 Mar-15
6		LT7	ETM+	136	44	23-Mar-09 & 17-Mar-15
7		LT7	ETM+	137	42	7-Mar-09 & 11-Mar-15
8		LT7	ETM+	137	43	17-Apr-08 & 17-Mar-15
9		LT7	ETM+	138	42	27-Feb-11 & 24-Mar-15
1	**2019**	LS8	OLI/TIRS	135	42	16-Jan-19
2		LS8	OLI/TIRS	135	43	2-Mar-18
3		LS8	OLI/TIRS	135	44	17-Mar-19
4		LS8	OLI/TIRS	136	42	9-Mar-18
5		LS8	OLI/TIRS	136	43	3-Mar-19
6		LS8	OLI/TIRS	136	44	9-Feb-18
7		LS8	OLI/TIRS	137	42	17-Mar-19
8		LS8	OLI/TIRS	137	43	3-Feb-19
9		LS8	OLI/TIRS	138	42	9-Mar-18

indicates more intense clustering of high values and is helpful for hotspot identification. For statistically significant negative z scores, the smaller the z score is, the more intense the clustering of low values (cold spot). The G_i^* statistic is a two-tailed statistical experiment, where positive values of G_i^* represent clusters that are, on average, greater than the mean (hotspots), the negative values represent clusters that are less than the mean (cold spots) (Getis and Ord, 1992, 1995). It represents the frequency of detection as a hotspot relative to the number of inputs covered in the area. The following equation is used for the calculation of Getis-Ord G_i;

$$G_i^* = \frac{\sum_{j=1}^{n} w_{i,j} x_j - \bar{X} \sum_{j=1}^{n} w_{i,j}}{S\sqrt{\dfrac{n\sum_{j=1}^{n} w_{i,j}^2 - \left(\sum_{j=1}^{n} w_{i,j}\right)^2}{n-1}}} \tag{9.1}$$

$$\bar{X} = \frac{\sum_{j=1}^{n} x_j}{n}$$

$$S = \sqrt{\frac{\sum_{j=1}^{n} x_j^2}{n} - \left(\bar{X}\right)^2}$$

where x_j is the attribute value for feature j, $W_{i,j}$ is the spatial weight between feature i and j, and n is equal to the total number of features. The G_i^* statistic is a z score.

An annual land use change rate (ALUCR), of susceptible and frequently changing land use patterns are calculated for a better understanding of the change ratio. ALUCR (Tian et al., 2014; Guite and Bora, 2018) is calculated in per cent per year

$$\text{ALUCR}_{a,t} = \frac{\left(\text{LU}_{a,t} - \text{LU}_{a,t-1}\right)/\text{LU}_{a,t-1}}{N_t - N_{t-1}} \times 100, \tag{9.2}$$

where t is the current year and time $t-1$ is the former year $\text{LU}_{a,t}$ and $\text{LU}_{a,t-1}$ are the total land area of one land use class in square kilometres at two different times, N is the total number of years (time frame) from time t (current year) to time $t-1$ (former year).

9.4   RESULT AND ANALYSIS

The land use land cover classes of the basin area are classified as aquaculture, cropland, forest cover, grassland, orchard and other plantation, other land, settlement, wetland and river.

In total, 11,631 grids (2 km × 2 km) have been created using fishnet to cover the Upper Meghna River Basin for hotspot analysis. The study shows 1705 grids are indicating hotspots, covering an area of about 6,58,861 ha (Figure 9.2). The grids fall within 99% level confidence and have a 0.01 significant level, where p-value = 0.004998 and Z score ± 2.576. Further, the grids falling within 95% confidence have a 0.05 significant level, where p-value = 0.024998 and Z score ± 1.960. Both the grids have statistically significant hotspots. When we superimpose the

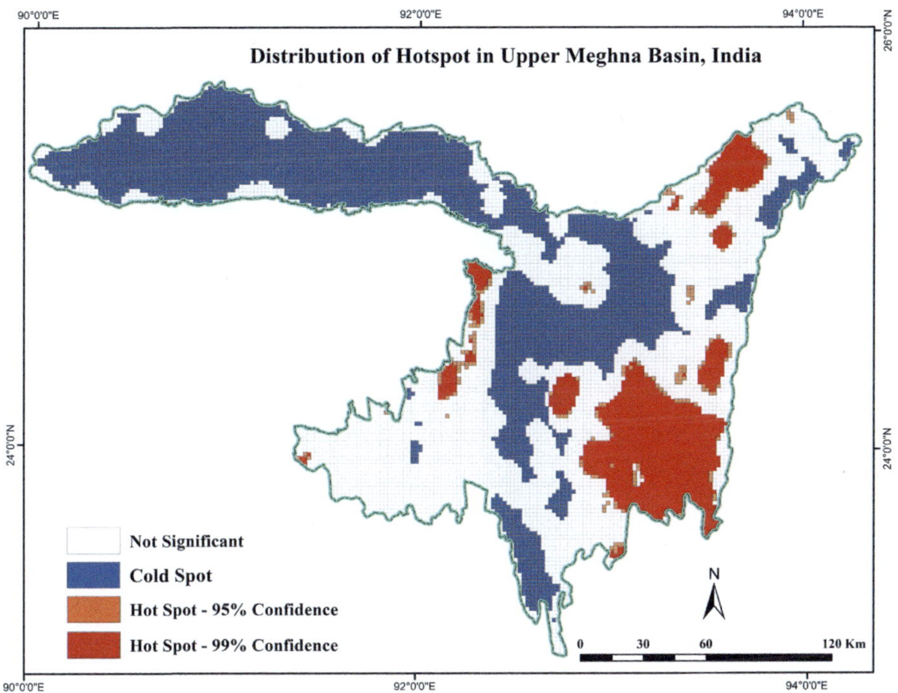

FIGURE 9.2 Distribution of hotspots in Upper Meghna River Basin, India.

locations of the villages on the grids then 826 hamlets/villages are within the identified hotspot areas (Figure 9.3). Altogether, 20 hotspot patches are identified within the study area (Table 9.2). The administrative locations of hotspots fall in 15 districts of India comprising 34 administrative blocks and 826 villages.

The hotspot patches are mostly concentrated in the Barak and Tripura catchments of the basin. Whereas in Meghalaya catchment the presence of hotspots is almost negligible. In the Barak catchment, the most frequent changes are identified regarding forests, orchards and other plantation and cropland. While calculating the annual land use change rate of cropland it is noticed that between the years 2005 to 2010, it was 2.36% per year. From 2010 to 2015 it became 2.43% per year and from the year 2015 to 2019 it is 5.36% per year. On the other hand, in terms of forest cover the annual land use change rate between the years 2005 to 2010 was 2.87% per year. From the year 2010 to 2015, the estimated rate of change was 0.34% per year and from the year 2015 to 2019 it was 0.73% per year. For orchards and other plantations it is estimated that during the years 2005 to 2010, it was 1.25% per year, from 2010 to 2015 it was 0.27% per year and from 2010 to 2019 it was 0.25% per year. The statistical analysis of the land use land cover change indicates that the cropland in Barak catchment is showing a decreasing trend, during the study period, whereas the forest cover in Barak catchment shows fluctuation as has increased between the years 2005 to 2010 by 35,852 ha, and then it decreased during 2010 to 2015 by 36,896 ha and then it increased between the years 2015 to 2019 by 63,313 ha. The orchard and other

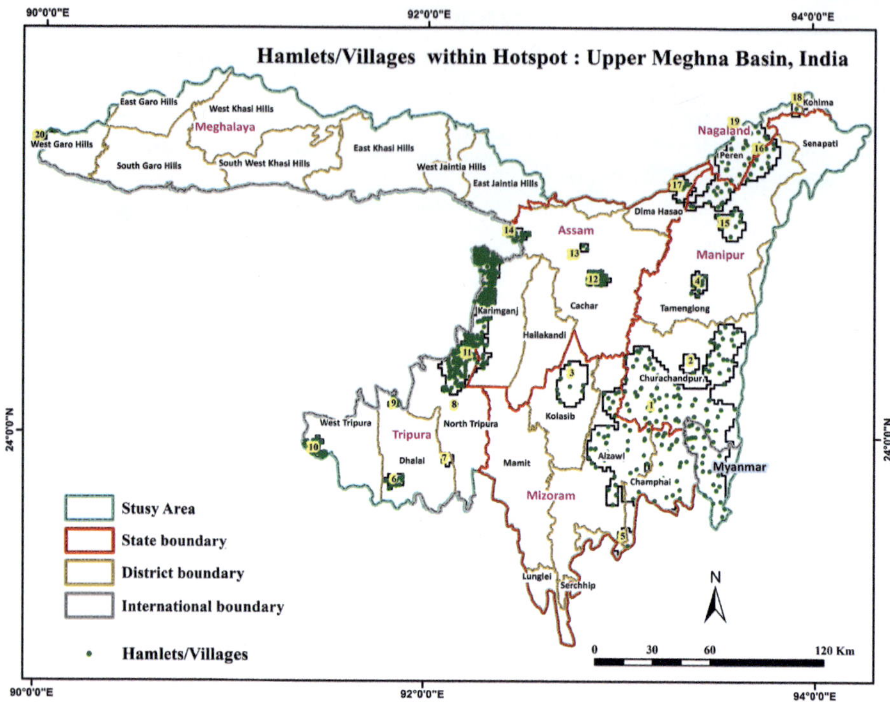

FIGURE 9.3   Hotspots in Upper Meghna River Basin, India.

TABLE 9.2
Confidence Level of Hotspots and Area Coverage in Hectors: Upper Meghna River Basin, India

Sl. No.	Total Hotspots (Grid)	Area (ha)	Confidence Level
1	1487	576175	99%
2	218	82686	95%
Total	**1705**	**658861 hectors**	

plantation land use class between the years 2005 to 2010 has increased by 3082 ha, between the years 2010 to 2015 decreased by 712 ha and between the years 2015 to 2019 has increased by 513 ha. It has been observed that these changes in forest cover and cropland are due to the prevailing practice of "Jhum" or shifting cultivation in hilly states like Manipur and Mizoram. Further, we know that "Jhum" or shifting cultivation is a typical agricultural practice that connects cropland and forest land, where both the land use classes show a pattern of indirect proportionality. Districts such as Churachandpur of Manipur and Champhai and Aizawl of Mizoram show a strong presence of shifting agriculture. In 2005 and 2019 there are some fresh and new land patches of "Jhum" that have been identified. Whereas in 2010 and 2015 rejuvenation of some old Jhum land patches can be seen.

In Tripura catchment, a similar scenario has been noticed in terms of cropland and forest cover. These frequent changes regarding forest and cropland are identified, indicating the practices of "Jhum" or shifting cultivation. In terms of change of forest cover the annual land use change rate for the year 2005 to 2010 was 0.14% per year. From 2010 to 2015 it was 0.48% per year and from the year 2015 to 2019 it was 0.68% per year. In terms of crop land annual land use change rate for the years 2005 to 2010 was 0.78% per year. From 2010 to 2015 it was 0.79% per year and from the year 2015 to 2019 the estimated rate of change was 0.15% per year. However, in the Tripura catchment, the most notable change has been identified in terms of settlement. The annual land use change rate of settlement for the years 2005 to 2010 was 0.13% per year. From 2010 to 2015 it was 7.77% per year and from 2015 to 2019 it was 7.25% per year. There is a significant expansion of settlement/built-up spotted during the periods 2010 to 2015 and 2015 to 2019 in the Karimganj district of Assam and North Tripura district of Tripura.

In the Meghalaya catchment, frequent fluctuation cannot be seen. The catchment is very much stable regarding all the classified classes. However, in terms of forest cover the annual land use change rate between the years 2005 to 2010 was .003% per year. From the year 2010 to 2015 the estimated rate of change was 0.009% per year and from the year 2015 to 2019 it was 0.01% per year. In terms of crop land annual land use change rate between the years 2005 to 2010 was 0.02% per year. From 2010 to 2015 it was 0.007% per year and from the year 2015 to 2019 it was 0.06% per year (Table 9.3). On the other hand, the annual land use change rate of settlement for the years 2005 to 2010 was 0.01% per year. From 2010 to 2015 it was 0.21% per year and from 2015 to 2019 it was 0.07% per year (Table 9.4). A scenario similar to the Barak catchment is also demarcated in the Meghalaya catchment as the cropland follows a decreasing trend whereas the forest cover fluctuates and has increased by 3316 ha between the years 2005 to 2010, decreasing by 11,286 ha between the years 2010 to 2015 and then again decreased by 12,582 ha between the years 2015 to 2019. However, the land use class orchard and other plantations is very much stagnant within this catchment. On the other hand, mainly in the plain topography of the mainland in the Tripura catchment, there is an increasing trend regarding cropland whereas forest cover shows a constant decreasing trend during the study period.

Figure 9.4 represents the land use change in Meghna Basin India, for six different periods. The LULC changes occur primarily due to indigenous practices of the forest dwellers and the expansion of settlement with population.

9.5 CONCLUSION

Quantifying the nature and pattern of land use land cover change (LULCC) is the most important aspect of land cover study. The present study reveals the dynamic nature of land use land cover in the Upper Meghna River basin, where all classes of land use land cover do not follow a single pattern of change and make the area more vulnerable. The most frequent changes that have been identified are the forest cover and cropland, which often primarily represents an inverse proportional relation

TABLE 9.3

Area Statistics of Change of Land Use Classes: Barak Catchment, Upper Meghna Basin, India

Land use Land cover Classes	2005 Area in ha	2010 Area in ha	Change 05_10 Area in ha	2015 Area in ha	Change 10_15 Area in ha	2019 Area in ha	Change 15_19 Area in ha
Aquaculture	1170	1205	+35	1205	0	1183	-22
Cropland	351141	310272	-40869	347324	-37052	273990	-73334
Forestland	2145765	2181617	+35852	2144721	-36896	2208034	+63313
Grassland	0	0	0	258	258	0	-258
Orchard and other plantation	49128	52210	+3082	51499	-712	52012	+513
Other land	239	256	+17	256	0	244	-12
Settlement	50748	50761	+13	51247	+486	57587	+6340
Wetland	428	2297	+1870	2370	+72	6123	+3753
River	8780	8780	0	8519	-261	8227	-292
Grand Total	**2607399**	**2607399**		**2607399**		**2607399**	

Note:　Here + represents an increase & - represents a decrease

TABLE 9.4

Area Statistics of Change of Land Use Classes: Tripura catchment, Upper Meghna Basin, India

Land use Land cover classes	2005 Area in ha	2010 Area in ha	Change 05_10 Area in ha	2015 Area in ha	Change 10_15 Area in ha	2019 Area in ha	Change 15_19 Area in ha
Aquaculture	6	6	0	6	0	6	0
Cropland	74135	74193	+58	74219	+26	74391	+71
Forestland	983770	983596	-174	983106	-490	982704	-402
Grassland	12954	12805	-149	12805	0	12801	-5
Extraction or Mining site	221	329	+108	421	+92	509	+88
Other Land	32892	32860	-32	32853	-7	32852	-1
Settlements	39505	39707	+202	40115	+409	40231	+116
Wetlands	22139	22126	-13	22096	-30	22128	+32
River	4030	4030	0	4030	0	4030	0
Grand Total	**1169652**	**1169652**		**1169652**		**1169652**	

Note:　Here + represents an increase & - represents a decrease

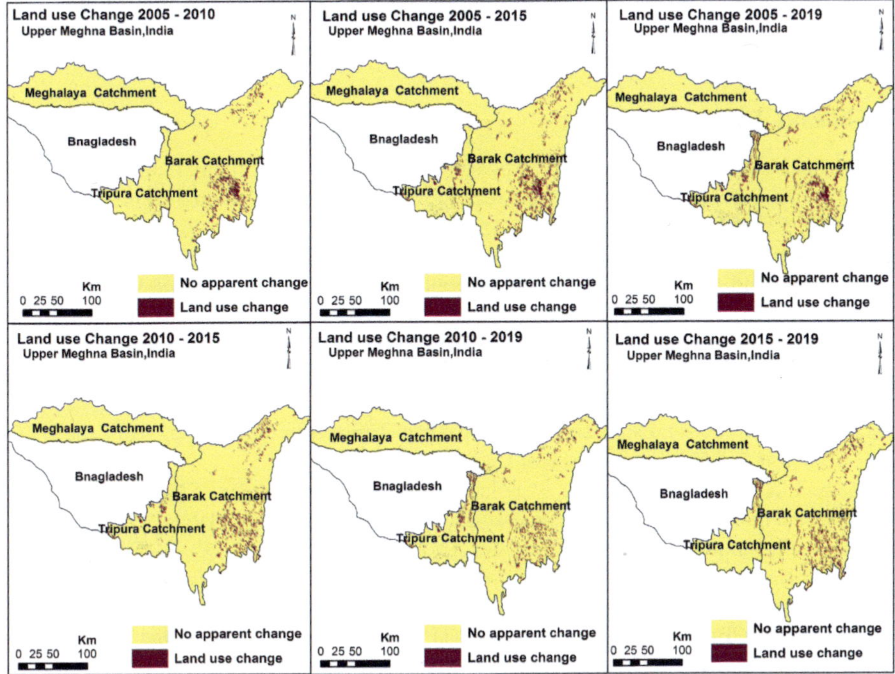

FIGURE 9.4   Land use/cover change, Upper Meghna Basin, India.

between them, and the expansion of cropland always severely affects the forest cover. Regarding global data, a global net loss of forest area is 0.8 million km^2 per year whereas expansion of global agriculture (i.e., cropland and pasture/rangeland) is 0.9 to 1.0 million km^2 per year from 1960 to 2019 (Winkler *et al.*, 2021). Such a change is due to the modifications in hill topography, rainfall patterns and changes in the climatic parameters have further aggravated in recent times.

In terms of settlement, it has been observed that it is expanding within all three catchments, and the Tripura catchment shows the highest expansion of settlement. Cropland has a decreasing trend within both the Barak and Meghalaya catchments, there is a high probability that the dynamics between forest and cropland are not only inverse but can occur with high fluctuation and not following a single trend such variations may be mostly due to the changing climatic patterns and the various factors associated with climate change leads to the modifications in the land use patterns in the upper Meghna River basin. The basin being the most vulnerable needs the attention of the people and communities to safeguard the life and livelihood of the forest-dependent communities.

ACKNOWLEDGEMENT

The present study is a part of the IUCN's funded project "Meghan River Basin Land Use Change Analysis and Atlas Development" jointly accomplished by India and Bangladesh in 2020. The authors acknowledge IUCN for providing necessary guidance and support throughout the study period.

REFERENCES

Arneth, A., Harrison, S. P., Zaehle, S., Tsigaridis, K., Menon, S., Bartlein, P. J., ... & Vesala, T. (2010). Terrestrial biogeochemical feedbacks in the climate system. *Nature Geoscience*, 3(8), 525–532.

Duraisamy, V., Bendapudi, R., & Jadhav, A. (2018). Identifying hotspots in land use land cover change and the drivers in a semi-arid region of India. *Environmental Monitoring and Assessment*, 190(9), 1–21.

FAO (2001). *Global Forest Resources Assessment 2000, Main Report.* FAO, Rome.

Getis, A., & Ord, J. K. (1992). The analysis of spatial association using distance statistics. *Geographical Analysis*, 24(3), 189–206.

Goldewijk, K. K. (2001). Estimating global land use change over the past 300 years: the HYDE database. *Global Biogeochemical Cycles*, 15(2), 417–433.

Guite, L. T. S., & Bora, A. (2018). Population growth and dynamics of land use/land cover pattern in Shillong urban agglomeration. *Hill Geographer*, XXXIV(1), 45–53.

Luyssaert, S., Jammet, M., Stoy, P. C., Estel, S., Pongratz, J., Ceschia, E., ... & Dolman, A. J. (2014). Land management and land-cover change have impacts of similar magnitude on surface temperature. *Nature Climate Change*, 4(5), 389–393.

Meyer, W. B., & Turner II, B. L. (1992). Human population growth and global land use/cover change. *Annual Review of Ecology and Systematics*, 23, 39–61.

Ord, J. K., & Getis, A. (1995). Local spatial autocorrelation statistics: distributional issues and an application. *Geographical Analysis*, 27(4), 286–306.

Sala, O. E. (2000). Global biodiversity scenarios for the year 2100. *Science*, 287(5459), 1770–1774.

Vadrevu, K. P., & Ohara, T. (2020). Focus on land use cover changes and environmental impacts in South/Southeast Asia. *Environmental Research Letters*, 15(10), 100201.

Watson, R. T., Zinyowera, M. C., & Moss, R. H. (1996). *Climate Change 1995. Impacts, Adaptations and Mitigation of Climate Change: Scientific-Technical Analyses.* IPCC.

Winkler, K., Fuchs, R., Rounsevell, M., & Herold, M. (2021). Global land use changes are four times greater than previously estimated. *Nature Communications*, 12(1), 1–10.

10 Mapping Mountain Glacier Facies Using High-Resolution Optical Satellite Data

Possibilities and Challenges

Sagar F. Wankhede, Shridhar D. Jawak,
Alvarinho J. Luis and Keshava Balakrishna

10.1 INTRODUCTION

Remote sensing (RS) of the cryosphere is a broad umbrella term comprising of several wavelength spectrums, platforms, modes of acquisition, cryospheric phenomena, and application/objective of the study. Among the available modalities of RS, satellite imagery is a reliable and cost-effective resource for monitoring and studying components of the cryosphere. Optical RS which utilizes the visible to near and shortwave infrared spectrums is most efficient due to the fine resolution available in today's satellite sensors. From moderate resolution sensors such as Landsat (30 m), high resolution sensors such as Sentinel-2A (10 m), and very high resolution (VHR) sensors such as WorldView-2 (WV-2) (2 m), optical RS provides a large repository for selection of application-specific imagery. In addition to spatial resolution, spectral resolution plays a key role in determining mappable phenomena. This is most notable when properties such as reflectance, albedo, and thermal information are considered in the information extraction process. In the examination of cryospheric phenomena, monitoring of glaciers is a focal point of concern. Glacial mass loss is a socio-economic, biogeochemical and ecological hazard (Hock *et al.*, 2017). Hence, precise monitoring of glacier phenomena is necessary for resource and disaster management. With many of the world's glaciers located in near hostile terrain and increasing logistical costs of field explorations, RS is an efficient method of monitoring glacier properties. In the context of optical RS, mapping the extent of snow and ice variations is an effective mechanism for understanding the dynamicity of a glacier.

Upon precipitation, snow undergoes accumulation, compaction, melt, refreezing to firn, and further transitioning to ice. As this mass moves along the gradient of the land, the ice picks up debris from the rock bed underneath (Benn and Evans, 2014). This debris is transferred to the surface of the glacier in the ablation zone, where ice melts and is discharged from the glacier body. In addition, rock falls from the surrounding valley adding to the supraglacial debris. Furthermore, dust and particulate matter from natural and anthropogenic sources is transported via long-range atmospheric channels and deposited on the glacier (Casey and Kääb, 2012). These stages of evolution are present throughout the body of the glacier and constitute the accumulation and ablation zones. Accumulation zone is the region of the glacier where accumulation is greater than melt while the ablation zone is characterized by greater melt. These zones are differentiated using the equilibrium line altitude (ELA), where the net accumulation and ablation is zero. This implies that tracking of accumulation and ablation zones can lead to understanding the health and recession patterns of the glacier. An increase in the ablation zone will lead to a reduction in accumulation, and a recurrent expression of this distribution leads to glacier shrinkage. These patterns of snow and ice/accumulation and ablation are called glacier facies. At the end of summer, when seasonal snow has melted, and glacier facies are visible, fine resolution optical RS can be utilized for precisely mapping the extent of the available facies. Glacier facies are characterizable using multispectral (MS) RS based upon their distinctive spectral properties. The presence of moisture, snow/ice particle size, and intermixing of dust and debris effects the reflectance of the facies (Warren, 1982; Dozier *et al.*, 1981). Multispectral/optical imagery have been extensively tested for mapping glacier facies (Jawak *et al.*, 2023; Jawak *et al.*, 2022a, 2022b, and 2022c; Mitkari *et al.*, 2022; Haq *et al.*, 2021; Luis and Singh, 2020; Zhang *et al.*, 2019; Yousuf *et al.*, 2019; Ali *et al.*, 2017; Paul *et al.*, 2016; Bhardwaj *et al.*, 2015; Robson *et al.*, 2015; Pope and Rees, 2014; Racoviteanu and Williams, 2012; De Angelis *et al.*, 2007; Gupta *et al.*, 2005)

10.1.1 THE CURRENT EXPERIMENT

In Jawak *et al.* (2018, 2019a, and 2019b) we developed three spectral indices (SIR1, SIR2, and SIR3) for mapping facies in an object-based domain. However, the three indices were only incorporated in one rule set, without any tests on threshold variability and its impact on the final map. Later in Jawak *et al.* (2022b and 2022c) we assessed the performance of OBIA. Our results suggested that OBIA is less sensitive to modulations in image processing parameters if the rules used to map facies comprise of both spatial and spectral information of the segmented objects. However, every rule set in the latest attempt was generated individually, thus, not directly comparing the modulations of specific indices. Moreover, as the previous investigation was performed to assess the impact of a variety of image processing techniques for mapping facies, threshold manipulation of key indices was not undertaken. Nevertheless, in this experiment, we revisit the customized spectral index ratios developed in Jawak *et al.* (2019a) for evaluating the specific threshold variations and the resultant necessary changes in rule set parameters to map the same facies. These results are then compared with facies maps from pixel-based image analysis (PBIA) for assessment of accuracy. This experiment is in continuation of our previous efforts to understand the mechanisms of image processing and information extraction for increasing the knowledge of VHR RS of glacier facies.

10.2   STUDY AREA AND GEOSPATIAL DATA

The Chandra-Bhaga basin is a sub-basin of the Chenab River, in the Lahaul and Spiti district of the Indian state of Himachal Pradesh (Vatsal *et al.*, 2022). The Glacier Area Mapping for Discharge from the Asian Mountains (GAMDAM) inventory reports that 25% of the Chandra-Bhaga basin is glaciated (Sakai, 2019). This basin hosts India's Himalayan research base, Himansh. The glaciers selected in this study lie between 32°50′ to 32°45′N. Samudra Tapu is the most well-known glacier in this basin. Other glaciers are named as G1, G2, G3, G4, G5, G6 and G7 (Figure 10.1). The primary data are LV2A WV-2 imagery from DigitalGlobe, Inc., Westminster, CO, USA. It was acquired on 16 October 2014 (WorldView-2 © 2014 Maxar Technologies). The multispectral (MS) product consists of Coastal (0.40–0.45 μm), Blue (0.45–0.51 μm), Green (0.51–0.58 μm), Yellow (0.565–0.625 μm), Red (0.63–0.69 μm), Red Edge (0.705–0.745 μm), NIR-1 (0.770–0.895 μm) and NIR-2 (0.86–1.04 μm). The MS product has a 2 m spatial resolution, while the panchromatic (PAN) product has a 0.5 m resolution (Jawak et al., 2018). The elevation data comprised of 30 m Advanced

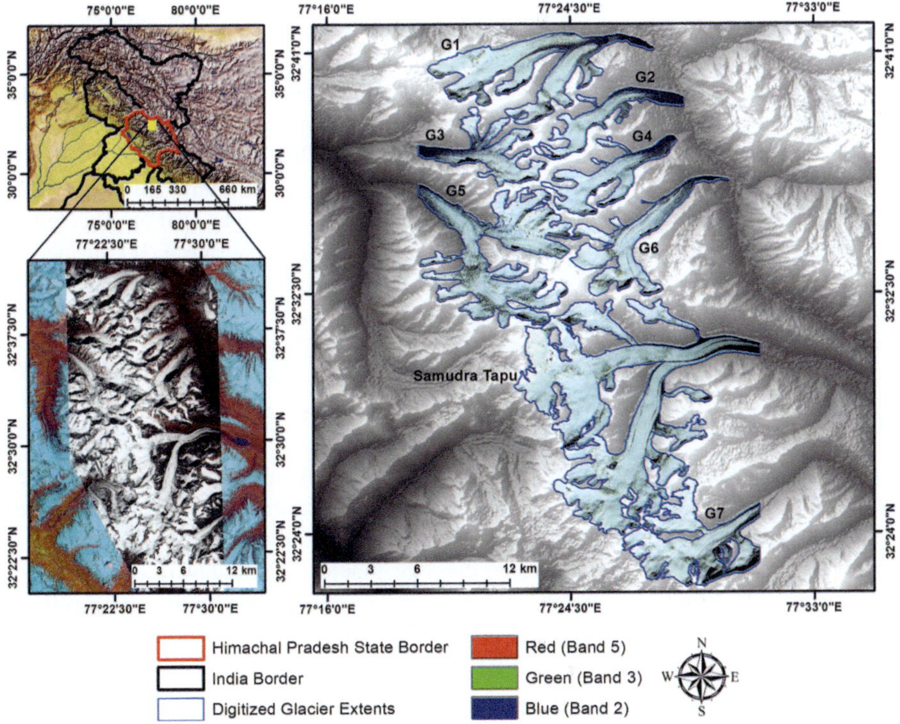

FIGURE 10.1   Geographical location of the study area. Inset maps outlining Himachal Pradesh, and India were prepared using Natural Earth (Free vector and raster map data @ naturalearthdata.com). The VHR WV-2 (WorldView-2 © 2014 Maxar Technologies) image is overlaid on Sentinel 2A (10 m) imagery (Copernicus Sentinel data 2021, processed by ESA). The extracted glacier subsets are imposed on a hill shaded ASTER GDEM v2.

Spaceborne Thermal Emission and Reflection Radiometer (ASTER) and Global Digital Elevation Model (GDEM) v2.

The WV-2 image (Figure 10.1) displays a large distribution of snow and ice facies. Recent spatiotemporal analysis of snow cover area, precipitation and temperature suggests that this region in the Himalayas is experiencing full snow cover even at temperatures close to 13°C (*Rathore et al.*, 2018). This is primarily attributed to the higher elevation of the Chandra-Bhaga basin along with increased snowfall reported for the month of data acquisition (Rathore *et al.*, 2018). Moreover, total snow cover area is observed to rise from September (Sahu and Gupta, 2020). Glacier facies function as a surface indicator of mass balance. Accurate mapping of facies in this imagery will outline the potential of VHR imagery and GEOBIA for discerning available facies.

10.3 METHODOLOGY

10.3.1 Radiometric Calibration and Extracting Glacier Subsets

The WV-2 VHR image was radiometrically calibrated by first converting digital numbers (DN) to top of atmosphere (TOA) radiance. Subsequently, TOA radiance was converted into surface reflectance using the Fast-Line-of-Sight Atmospheric Analysis of Spectral Hypercubes (FLAASH) atmospheric correction module. The FLAASH module was calibrated using image metadata, the atmospheric model and aerosol description (Atmospheric Correction Guide, 2017; Kaufman *et al.*, 1997) to derive reflectance using the MODerate resolution atmospheric TRANsmission (MODTRAN) radiative transfer code (Radiative Transfer Code, 2017). The reflectance image was overlayed on an ASTER GDEM v3 to generate a 3D surface for facilitating accurate digitization of glacial extents. The digitized extents were used to extract individual glacier subsets for further processing. Figure 10.2 outlines the broad methodology of this study.

10.3.2 Identification of Glacier Facies

Surface properties of glacier facies were assessed visually and spectrally to determine their occurrences on the individual glaciers. Table 10.1 highlights the visual and spectral assessments used to identify facies. Analysis of spatial and spectral variations are necessary for identifying each surface class of glacier facies (Jawak *et al.*, 2019a). Facies were identified using the reduction in reflectance based on moisture, refreeze and intermixing of debris. This was used to identify pixels for providing training data for the pixel-based image analysis (PBIA).

10.3.3 Pixel-Based Image Analysis (PBIA)

PBIA is performed here as a comparative analysis against the OBIA because supervised classification is known to deliver acceptable accuracies for glacier facies maps (Luis and Singh, 2020; De Angelis *et al.*, 2007). PBIA is a two-step approach, which consists of (a) assigning of training data, and (b) input of training data to

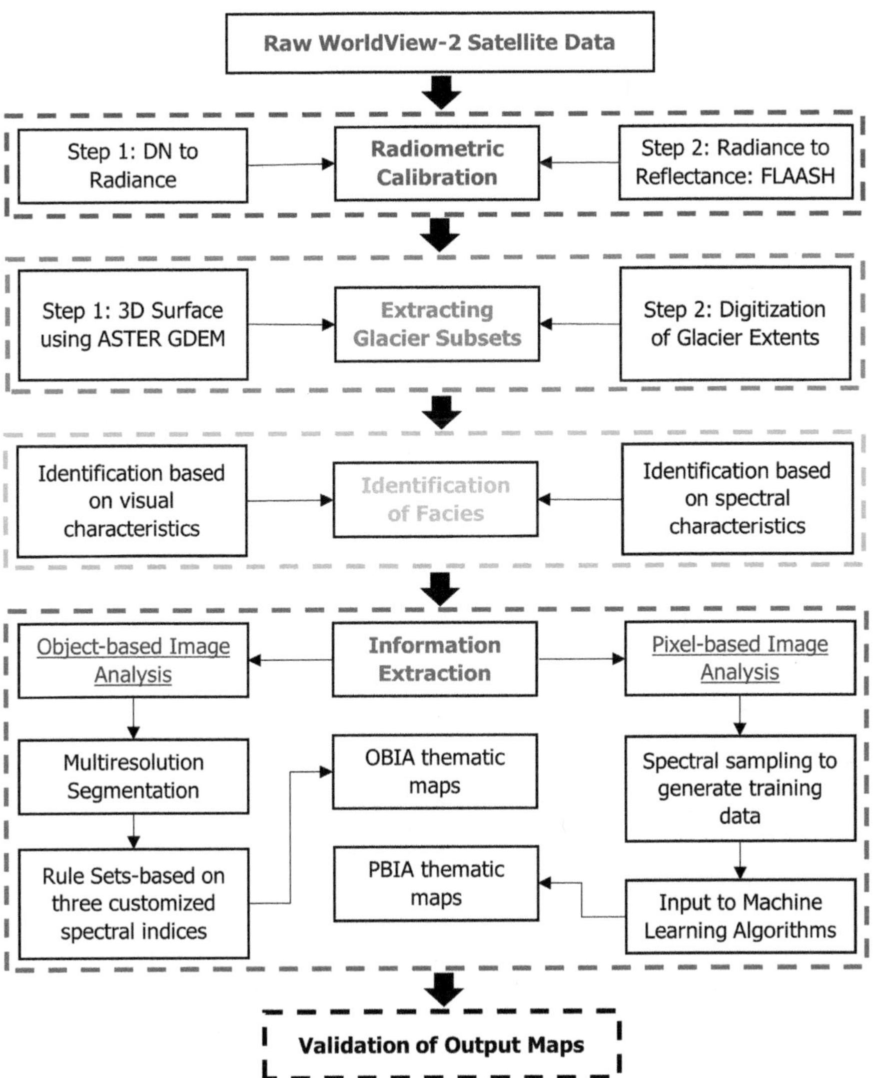

FIGURE 10.2 Methodology adopted in the current study. DN, digital numbers; FLAASH, Fast-Line-of-Sight Atmospheric Analysis of Spectral Hypercubes; ASTER, Advanced Spaceborne Thermal Emission and Reflection Radiometer; GDEM, Global Digital Elevation Model; OBIA, object-based image analysis; PBIA, pixel-based image analysis.

supervised ML algorithms. Training data was assigned by analyzing the spatial and spectral properties elaborated in section 10.3.2. The ML algorithms evaluated here are Mahalanobis distance (MHD), maximum likelihood (MXL), minimum Distance (MD), spectral angle mapper (SAM), and winner takes all (WTA). WTA is a voting method that classifies pixels based on the majority classification compiled from the

TABLE 10.1

Visual and Spectral Properties of Glacier Facies Identified in the Current Study

Facies	Visual Properties	Reflectance Properties
Snow	Smooth and continuous texture. Brightest pixels, highest elevation, and continuous smooth texture.	
Ice	Rougher texture than snow. Lower in elevation in snow. Visible striations due to its flow.	
Crevasses	Characteristic troughs and dishevelled appearance. Dark and linear features.	
Ice Mixed Debris	Lower brightness than ice. Lower in the ablation zone. Visible sporadic distribution of intermixed ice in the debris.	
Debris	No visible ice and lowest brightness. Lowest region of the ablation zone.	

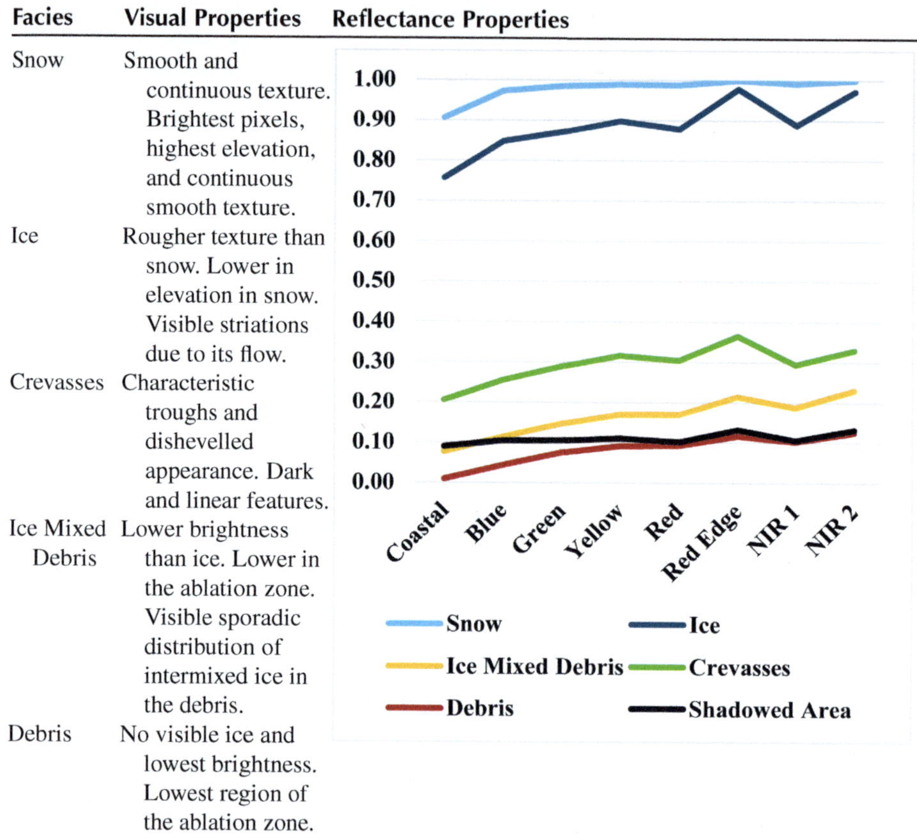

Source: The Spatial and Spectral Properties of Facies are adapted from Jawak *et al.* (2019a and 2022c).

other ML methods (www.nv5geospatialsoftware.com/learn/whitepapers/whitepaper-detail/workflow-tools-in-envi#:~:text=ENVI%20utilizes%20the%20Reed%2DXiaoli,or%20rejection%20of%20false%20positives). The ML algorithms of the current study were selected based on their performance in landcover and supraglacial classifications (Mahmon *et al.*, 2015; Jawak and Luis, 2013). Post-classification processes were not applied to avoid unintentional bias from manual assignment of parameters for each algorithm and individual classes.

10.3.4 OBJECT-BASED IMAGE ANALYSIS (OBIA)

OBIA is a knowledge and logical analysis of segmented image pixels for the assignment of thematic attributes or extraction of specific features. In this study, OBIA was implemented using (a) multi-resolution segmentation, and (b) creation of rule

TABLE 10.2
Customized Spectral Index Ratios for Mapping Glacier Facies

Index	Mathematical Expression
SIR 1	(Yellow)/((NIR1+NIR2)/2)
SIR 2	(Red Edge)/((NIR1+NIR2)/2)
SIR 3	(Blue)/((NIR1+NIR2)/2)

sets using customized spectral indices. Multi-resolution segmentation was selected to develop homogenous image objects as it permits the operator to assign custom values on parameters such as layer weights, compactness, shape and scale (Definiens Developer 7, 2007; Baatz and Schape, 2010). The layer weights were one for the blue, yellow and red bands. Two for the coastal, green and red edge bands. NIR1 and NIR2 were assigned layer weights of three and four. NIR bands were given higher weights because they are most sensitive to the occurrence of moisture. Scale, shape and compactness were parameterized using values of 200, 0.4 and 0.8, respectively. Following segmentation, rule sets were created using a combination of contextual parameters, and thresholds of customized spectral indices. The indices of the current study were developed initially in Jawak *et al.* (2019a) and then examined further in Jawak *et al.* (2019b). Table 10.2 outlines the customized spectral index ratios (SIRs) used in the current study.

In both previous studies, all three indices were assessed using only one rule set. In the current study, we evaluated the indices using four different rule sets. Each rule set incorporated different thresholds for every facies. The thresholds were common for each glacier. Table 10.3 highlights the thresholds assigned for each index for all rule sets assessed in the current study. Rule sets are named as OBIA 1, OBIA 2, OBIA 3 and OBIA 4. Shadowed snow was manually digitized to avoid complex misclassification from the spectral confusion in the resultant shadow objects.

10.3.5 EVALUATING THE ACCURACY

Field campaign was not possible as this image was acquired on 16 October 2014, which is the start of early winter. Field data was not retrieved due to the logistical challenges involved in accessing the Chandra-Bhaga basin during this period. Nevertheless, following Keshri *et al.* (2009), an equalized random sampling approach was used to assign 300 points for generating error matrices for accuracy assessment of the final maps. Products of the error matrices consist of error of commission (EC), error of omission (EO), producer's accuracy (PA), user's accuracy (UA), overall accuracy (OA), and kappa statistic (KS) (Congalton and Green, 2009).

10.4 RESULTS AND DISCUSSION

10.4.1 RESULTS

Figures 10.3, 10.4, 10.5 and 10.6 highlight the EC, EO, PA and UA of the facies maps in the current study. Accuracy measures are presented for each glacier to highlight variations between results.

TABLE 10.3
Thresholds of Each Index for the Rule Sets used in the Current Study to Characterize Glacier Facies

Algorithm	Indices and Parameters	Snow	Ice	Debris	IMD	Crevasses
OBIA 1	*Index-Wise Threshold*	SIR2 < 0.70 SIR2 < 0.74 SIR2 < 0.75	SIR1 < 0.63 SIR1<0.70and SIR2 < 0.79 SIR1 < 0.68 and SIR2 < 0.80	SIR3 < 0.00 SIR3 < 0.10	SIR3 < 0.35 SIR3 ≤ 0.45	SIR1 > 0.70 SIR1 > 0.65 SIR1 > 0.65 SIR1 < 0.67 and SIR2 < 0.80
	Contextual Parameters	Relative Border to Snow > 0.3	---	---	---	Area < 400 pixels
OBIA 2	*Index-Wise Threshold*	SIR2 < 0.68 SIR2 < 0.71 SIR2 < 0.76	SIR1 < 0.61 SIR1<0.68and SIR2 < 0.77 SIR1 < 0.66 and SIR2 < 0.78	SIR3 < 0.00 SIR3 < 0.20	SIR3 < 0.32 SIR3 ≤ 0.42	SIR1 > 0.68 SIR1 > 0.63 SIR1 > 0.63 SIR1 < 0.65 and SIR2 < 0.78
	Contextual Parameters	Relative Border to Snow > 0.15	---	---	---	Area ≤ 300 pixels
OBIA 3	*Index-Wise Threshold*	SIR2 < 0.67 SIR2 < 0.70 SIR2 < 0.74	SIR1 < 0.60 SIR1<0.67and SIR2 < 0.76 SIR1 < 0.65 and SIR2 < 0.76	SIR3 < 0.10 SIR3 < 0.20	SIR3 < 0.31 SIR3 ≤ 0.40	SIR1 > 0.68 SIR1 > 0.63 SIR1 > 0.63 SIR1 < 0.65 and SIR2 < 0.78
	Contextual Parameters	Relative Border to Snow > 0.20	Relative Border to Ice > 0.20	Relative Border to Debris > 0.20	Relative Border to IMD > 0.20	Area ≤ 300 pixels and Relative Border to IMD > 0.20
OBIA 4	*Index-Wise Threshold*	SIR2 < 0.65 SIR2 < 0.68 SIR2 < 0.72	SIR1 < 0.58 SIR1<0.65and SIR2 < 0.74 SIR1 < 0.63 and SIR2 < 0.74	SIR3 < 0.12 SIR3 < 0.22	SIR3 < 0.30 SIR3 ≤ 0.38	SIR1 > 0.66 SIR1 > 0.61 SIR1 > 0.61 SIR1 < 0.63 and SIR2 < 0.76
	Contextual Parameters	Relative Border to Snow > 0.30	Relative Border to Ice > 0.30	Relative Border to Debris > 0.30	Relative Border to IMD > 0.30	Area ≤ 300 pixels and Relative Border to Crevasses > 0.30

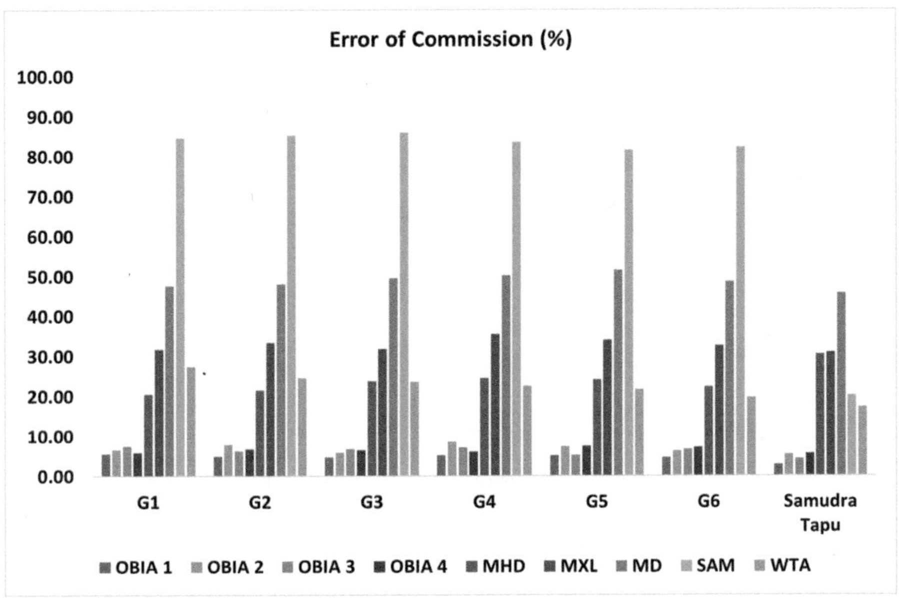

FIGURE 10.3 Column graph of error of commission for each mapping method and individual glaciers.

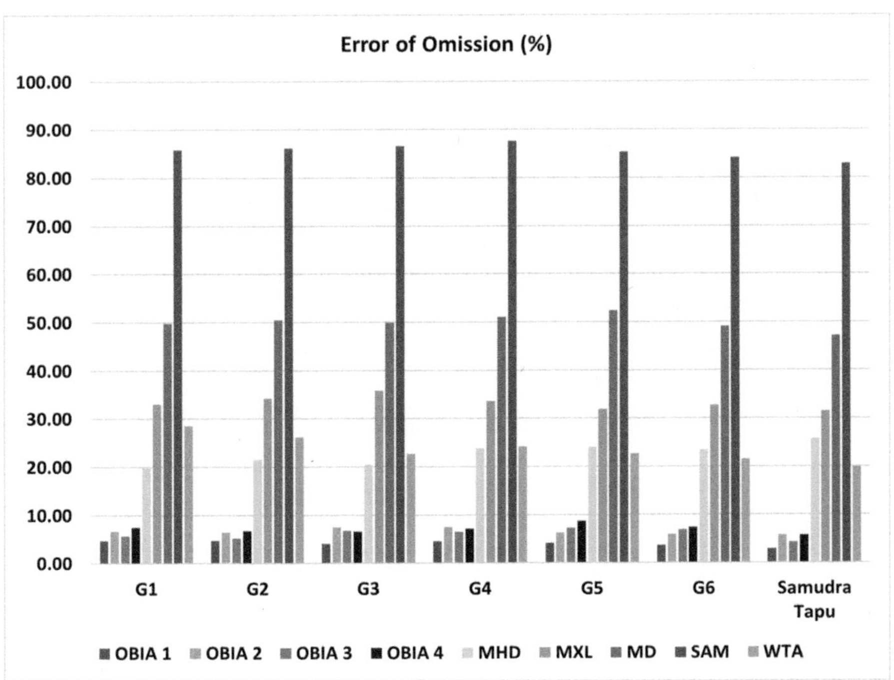

FIGURE 10.4 Column graph of error of omission for each mapping method and individual glaciers.

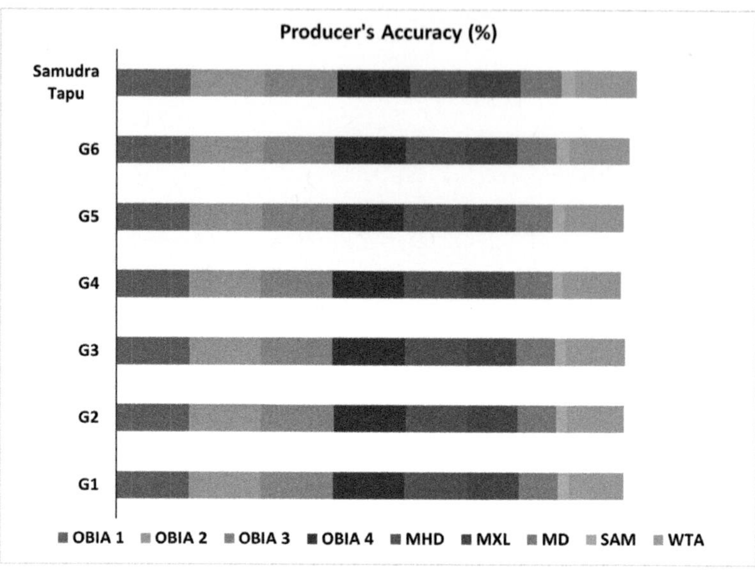

FIGURE 10.5 Stacked bar graph of producer's accuracy for each mapping method and individual glaciers.

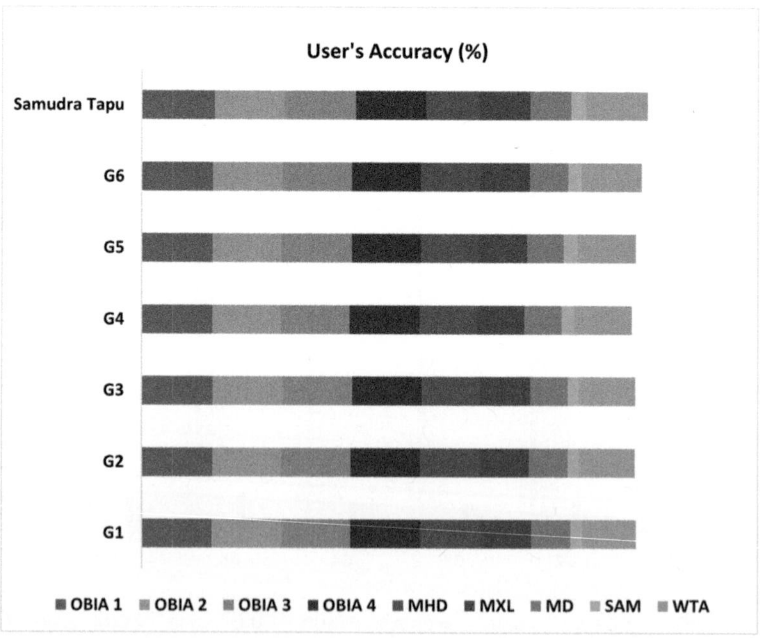

FIGURE 10.6 Stacked bar graph of the user's accuracy for each mapping method and individual glaciers.

The current results suggest that all four OBIA rule sets achieved superior results than the five supervised ML methods. Averaging the EC for all glaciers, OBIA 1 achieved the lowest overestimation (4.74%) whereas SAM achieved the maximum overestimation (74.82%). Between the individual OBIA and ML methods, OBIA 2 delivered the maximum overestimation (6.89%), and WTA yielded the lowest EC (22.40%). When the EO is averaged for all glaciers, OBIA 1 delivered the lowest underestimation (4.07%), and SAM provided the maximum underestimation (85.46). Among OBIA methods, OBIA 4 achieved the maximum underestimation (7.09%), whereas among ML methods MHD yielded the lowest underestimation (22.68%). Based on the EO and calculations of the UA (Figures 10.4 and 10.6) yield the following order of reliability: OBIA 1 > OBIA 3 > OBIA 4 > OBIA 2 > WTA > MHD > MXL > MD > SAM. Figures 10.7 and 10.8 display the OA and KS of the methods in current study.

From the OA and KS calculations (Figures 10.7 and 10.8), it is evident that PBIA ML methods deliver maximum deviations. OBIA 1 delivered the maximum average KS of 0.94, whereas OBIA 2 and OBIA 4 yielded lowest KS of 0.92, respectively. Among PBIA, MHD and MXL yielded the highest KS of 0.74 and SAM delivered the lowest average OA of 0.06. Based on the average KS, the order of performance among all the methods is OBIA 1 > OBIA 3 > OBIA 2 = OBIA 4 > MHD = WTA > MXL > MD > SAM. Figure 10.9 displays the thematic maps of

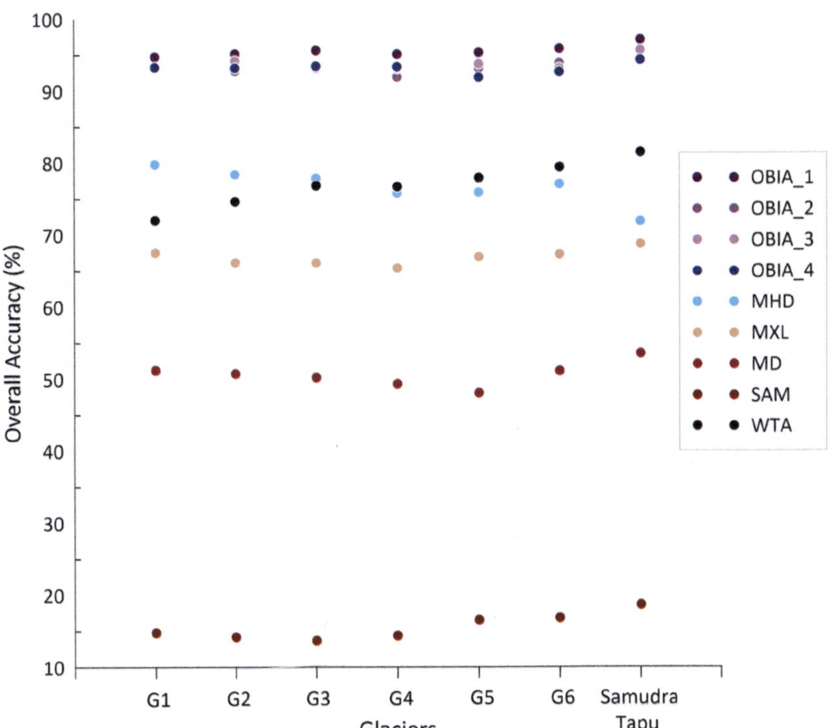

FIGURE 10.7   Scatter plot of the overall accuracy for each glacier and mapping method.

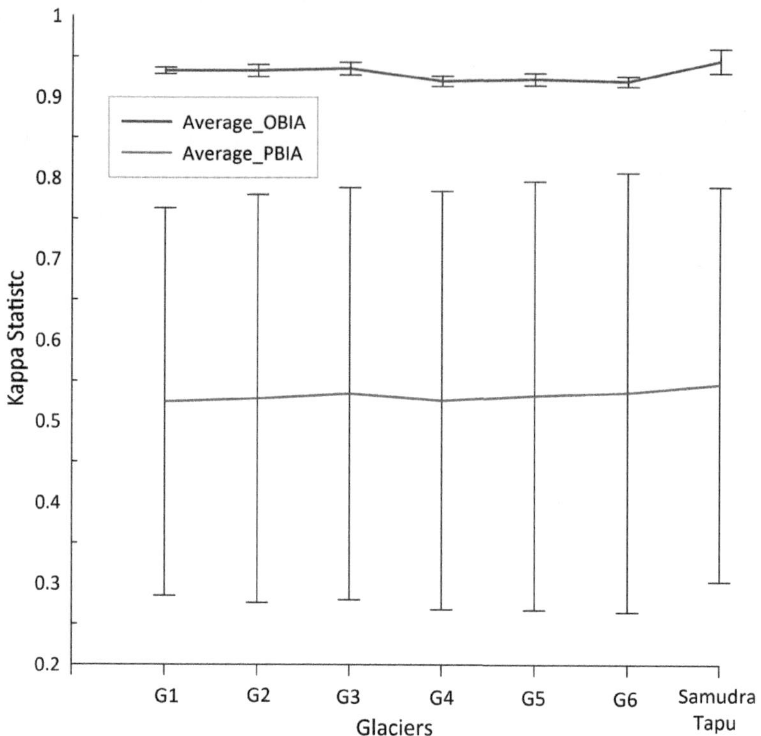

FIGURE 10.8 Kappa statistic calculated as averages for all OBIA and PBIA methods with error margins for the deviations per method.

OBIA, and Figure 10.10 highlights the classification result of MXL for the Samudra Tapu glacier.

10.4.2 Discussion

In this study, we find that OBIA outperforms all PBIA algorithms. This may be because OBIA utilizes a logical and sequential breakdown of segmented objects to assign thematic classes. This breakdown is in the form of rule sets. We utilized the same SIRs for mapping facies in four different rule set forms. Threshold changes in each rule set for the associated facies was a reduction in the index value of the magnitude from 0.03 to 0.01, but this did not necessarily translate to a corresponding reduction in the thresholds of contextual parameters. For example, OBIA 2 differed from OBIA 1 in that all SIR thresholds for each facies were reduced. Following suit, the "relative border to object class" and "area in pixels" contextual parameters were also reduced while developing the rule set (Table 10.3). However, SIR thresholds in OBIA 3 were reduced from OBIA 2, but the contextual parameter thresholds were increased. OBIA 4 also observed the same trend. In this case, the relative border to the specific thematic class parameter was added for each facies, indicating that reduction in index

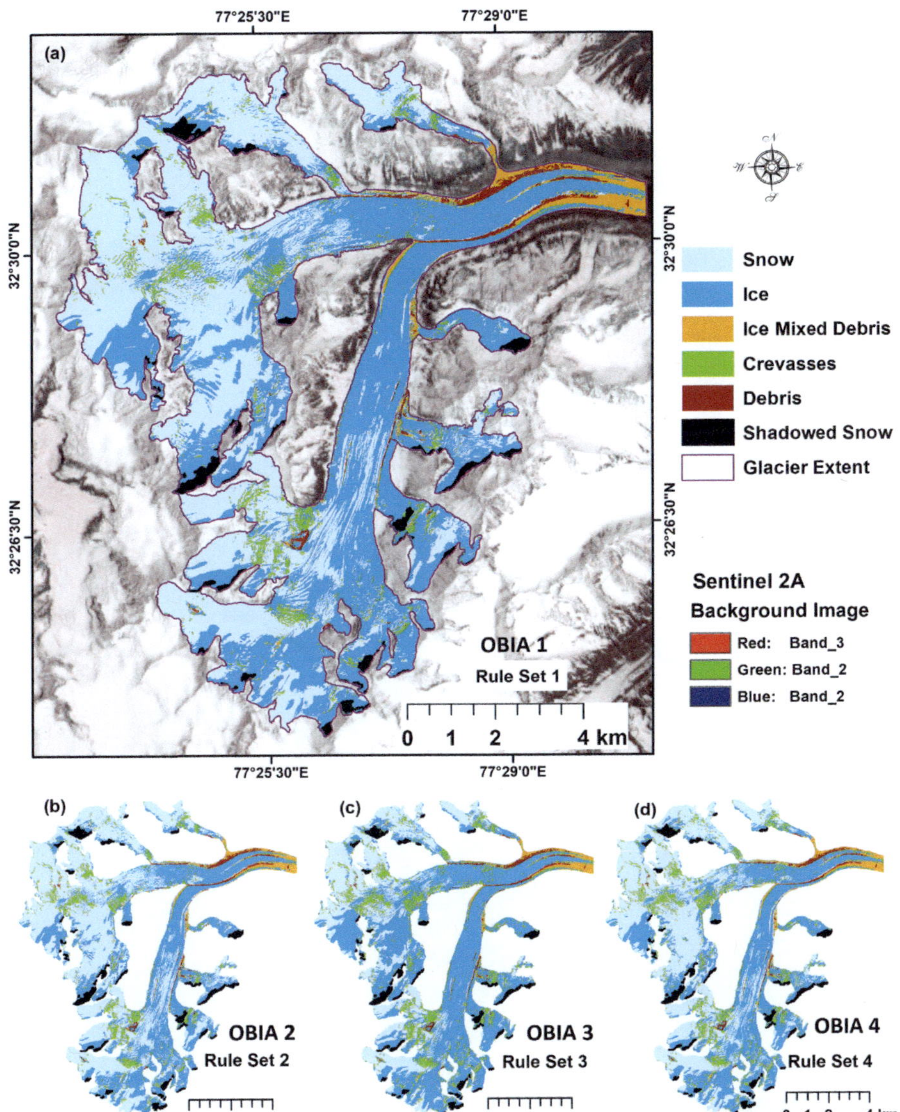

FIGURE 10.9 OBIA rule set classification thematic maps for the Samudra Tapu glacier. The classification map of rule set 1 is imposed upon Sentinel 2A (10 m) imagery (Himalaya: Copernicus Sentinel data 2021, processed by ESA).

threshold value necessitates a corresponding increase in contextual parametrization to supplement effective OBIA. This may arise because reduction in threshold will first categorize limited objects, following which relative border parameters can be used to integrate nearby objects from the same facies into the thematic class. A similar approach is extended to the area parameter, which effectively integrates pixels of a

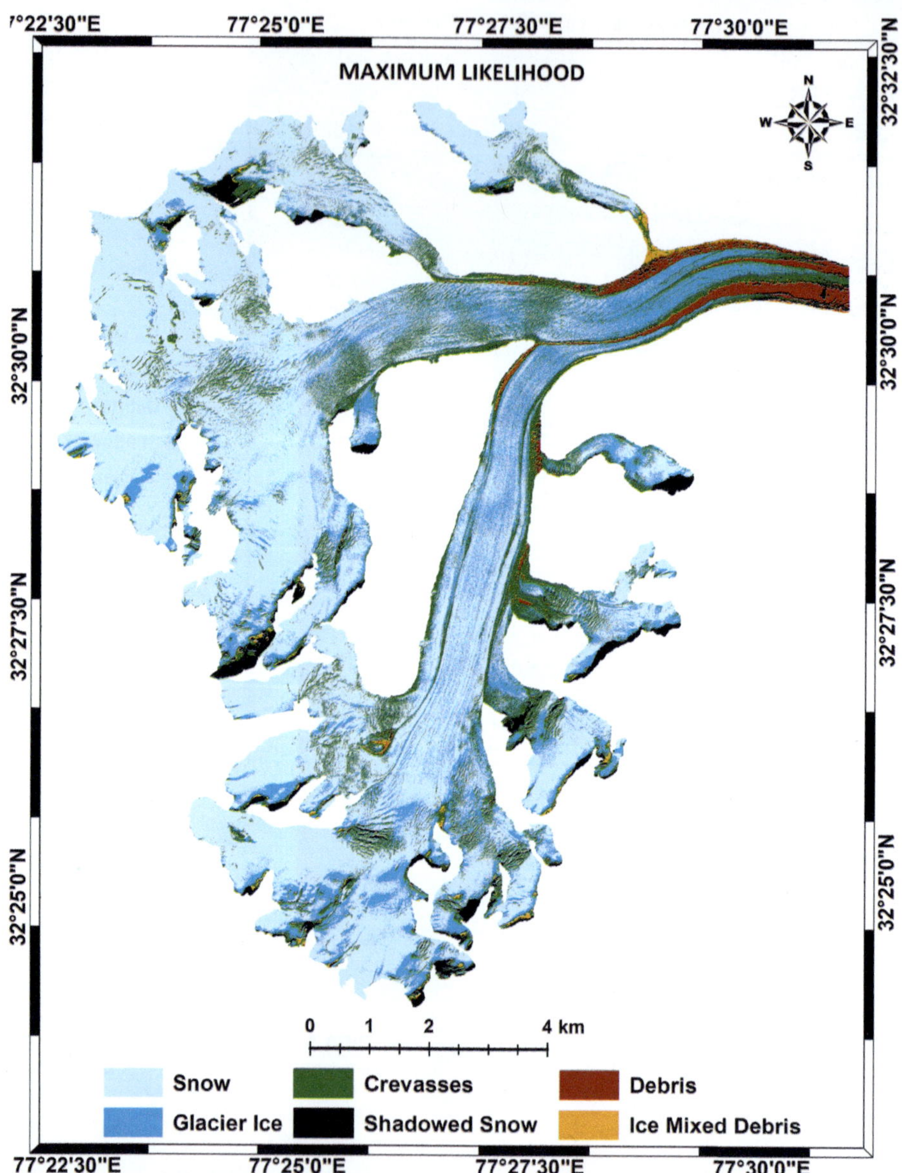

FIGURE 10.10   PBIA MXL classification thematic map for the Samudra Tapu glacier.

specific area into the target thematic classes. We found the most effective usage of the area parameter for mapping crevasses. This may be because crevasses are linear features with smaller net areas than other objects. In Jawak *et al.* (2019a) we observed the exact shape of the crevasse being retained in the segments resulting from these indices. A reduction in threshold would imply that crevasse objects lying outside of the threshold would need to be included by either specifying the area, as is the case of OBIA 2 or by including relative border (OBIA 3 and 4). The relative border threshold can be applied to crevasses as they tend to be clustered due to the motion of the glacier. Other facies have fuzzy boundaries and as such require careful examination of the rule set logic to ensure overestimation does not occur. This therefore places impetus on the operator in preparation of the rule sets. In Jawak *et al.* (2023) we observed the importance of operator bias resulting in variability in final thematic maps of facies in Ny-Ålesund. Moreover, higher resolution allows for detailed segmentation resulting in better-quality output maps. This effect is known to occur specifically for VHR imagery (Gao and Mas, 2008). Interestingly, in moving from 2 m to 0.5 m spatial resolution, OBIA is found to produce even greater improvements in thematic accuracy for facies (Jawak *et al.*, 2022b). PBIA in the current study did not deliver accuracies on par with OBIA. This may be in part due to the detailed step-wise analysis of OBIA. Moreover, poor PBIA may arise due to the inherent salt and pepper noise in satellite imagery (Ansari and Buddhiraju, 2018), and vast inter-algorithm variability reducing overall performance. WTA is an aggregate majority classifier based on the results of the other PBIA algorithms. Thus, while it utilizes the best classification results to generate its own thematic output, it may be also influenced by the misclassifications of SAM resulting in overall inferior performance.

10.4.2.1 Challenges and Possibilities

Current strategies for mapping glacier facies using optical satellite data revolve around overcoming extraneous factors such as sudden precipitation, presence of clouds and cloud induced shadows (Zhai *et al.*, 2018). Resolving cloud cover may require masking of the cloud and its associated shadow, or cloud-detection and removal methods such as the F-mask (Qiu *et al.*, 2017) and the cloud/shadow index based on spectral indices (CSD-SI) (Zhai *et al.*, 2018). Shadows in mountain areas forming due to illumination angle add complexity to classification methods as varying intensity of shadows can be interpreted as different classes by both algorithm and operator (Jawak *et al.*, 2019a). This variation in intensity can also cause a change in the derived spectral reflectance within the shadowed region (Jawak *et al.*, 2022b). Although higher radiometric resolution is suggested for overcoming shadow area complexity (Gore *et al.*, 2017), in the current study, we are limited by the lack of a higher quantized imagery. Therefore, in the current study, shadowed snow was extracted using manual digitization. The local topography itself can influence analysis of glaciers using remote sensing (De Beer and Sharp, 2009). Confusion between glacier debris and surrounding topography due to rock falls can lead to erroneous mapping. Moreover, differentiating between supraglacial debris and periglacial debris in the ablation zone can be cumbersome. In such scenarios, SAR coupled with elevation (Kaushik *et al.* 2022; Robson *et al.* 2015), SWIR bands (Bhardwaj *et al.*, 2014), and thermal information (Bhambri *et al.*, 2011) are utilized for mapping glacier debris. A combination of brightness temperature and slope in an OBIA domain has been used to differentiate supraglacial and periglacial debris (Mitkari *et al.*, 2022). In the current study, we have not incorporated any ancillary data for differentiating between IMD and debris. This indicates that at finer spatial resolution, segmentation, and spectral indices can be sufficient for mapping facies using VNIR imagery. Therefore, while local and scene-based adjustments are necessary for adjusting the application of GEOBIA and SIRs, the indices themselves may be transferable to other test sites such as Svalbard, other Arctic regions, and Antarctica. PBIA has not been effective for mapping facies. This conclusion agrees with previous findings and suggests that for VHR mapping of facies in GEOBIA, the vast object-based features in it can be explored further for reducing the reliability on ancillary data and improving thematic accuracy.

10.5 CONCLUSIONS

In this research, we evaluated the utility of OBIA and customized spectral index ratios (SIRs) for mapping glacier facies using four rule sets. Thematic results of these rule sets were compared with PBIA algorithms MHD, MXL, MD, SAM and WTA. Summarily, all OBIA rule sets outperformed the PBIA methods. OBIA 1 achieved maximum KS of 0.94 and SAM yielded the lowest KS of 0.06. Based on assessment of all accuracy measurements, the order of performance is OBIA 1 > OBIA 3 > OBIA 2 = OBIA 4 > MHD = WTA > MXL > MD > SAM. Variations in threshold were used to differentiate between each OBIA rule set. Current findings suggest that reduction in threshold necessitates a greater emphasis on contextual parameters for mapping facies. PBIA is simplistic in its execution, whereas OBIA relies on operator skill. We find that VHR VNIR mapping of facies does not require ancillary datasets but accurate segmentation and implementation of spectral indices. Although shadowed areas are problematic, we manually extracted them to avoid misclassification. Future precision mapping of facies should focus on limiting the use of ancillary data and exploring the capacity of VHR VNIR imagery.

REFERENCES

Ali, I.; Shukla, A.; Romshoo, S.A. Assessing linkages between spatial facies changes and dimensional variations of glaciers in the upper Indus Basin, western Himalaya. *Geomorphology* 2017, 284, 115–129. https://doi.org/10.1016/j.geomorph.2017.01.005

Ansari, R.A.; Buddhiraju, K.M. Noise filtering in high-resolution satellite images using composite multiresolution transforms. *PFG – J. Photogramm. Remote Sens. Geoinf. Sci.* 2018, 86, 249–261. https://doi.org/10.1007/s41064-019-00061-4

Atmospheric Correction User Guide. Available online: www.l3harrisgeospatial.com/portals/0/pdfs/envi/Flaash_Module.pdf (Accessed on 17 February 2017).

Baatz, M.; Schape, A. Multiresolution segmentation: An optimization approach for high quality multi-scale image segmentation. In *Angewandte Geographische Informations-Verarbeitung, XII*, Strobl, J.; Blaschke, T.; Griesbner, G. Eds. Wichmann Verlag: Heidelberg/Karlsruhe, Germany, 2000; pp. 12–23.

Benn, D.; Evans, D.J. 2014. *Glaciers and Glaciation*. Routledge.

Bhambri, R.; Bolch, T.; Chaujar, R. Mapping of debris-covered glaciers in the Garhwal Himalayas using ASTER DEMs and thermal data. *Int. J. Remote Sens.* 2011, 32, 8095–8119. https://doi.org/10.1080/01431161.2010.532821

Bhardwaj, A.; Joshi, P.K.; Sam, L.; Singh, M.K.; Singh, S.; Kumar, R. Applicability of Landsat 8 data for characterizing glacier facies and supraglacial debris. *Int. J. Appl. Earth Obs. Geoinf.* 2015, 38, 51–64. https://doi.org/10.1016/j.jag.2014.12.011

Bhardwaj, A.; Joshi, P.K.; Snehmani; Singh, M.; Sam, L.; Gupta, R. Mapping debris-covered glaciers and identifying factors affecting the accuracy. *Cold Reg. Sci. Technol.* 2014, 106–107, 161–174. https://doi.org/10.1016/j.coldregions.2014.07.006

Casey, K.; Kääb, A. Estimation of supraglacial dust and debris geochemical composition via satellite reflectance and emissivity. *Remote Sens.* 2012, 4, 9, 2554–2575. https://doi.org/10.3390/rs4092554

Congalton, R.; Green, K. *Assessing the Accuracy of Remotely Sensed Data: Principles and Practices.* CRC Press: Boca Raton, 2009.

De Angelis, H.; Rau, F; Skvarca, P. Snow zonation on Hielo Patagónico Sur, southern Patagonia, derived from Landsat 5 TM data. *Glob. Planet. Change* 2007 Oct 1, 59(1–4), 149–58. https://doi.org/10.1016/j.gloplacha.2006.11.032

De Beer, C.M.; Sharp, M.J. Topographic influences on recent changes of very small glaciers in the Monashee Mountains, British Columbia, Canada. *J. Glaciol.* 2009, 55, 691–700. https://doi.org/10.3189/002214309789470851

Definiens Developer 7. *Reference Book.* Definiens AG: Munich, Germany, 2007.

Description of the Machine Learning Algorithms. Available online: www.l3harrisgeospatial.com/Learn/Whitepapers/Whitepaper-Detail/ArtMID/17811/Article,ID/17299/Workflow-Tools-in-ENVI (Accessed on 24 November 2021).

Dozier, J.; Schneider, S.R.; McGinnis, Jr, D.F. Effect of grain size and snowpack water equivalence on visible and near-infrared satellite observations of snow. *Water Resour. Res.* 1981, 17(4), 1213–1221. https://doi.org/10.1029/WR017i004p01213

Gao, Y.; Mas, J.F. A comparison of the performance of pixel-based and object-based classifications over images with various spatial resolutions. *Online J. Earth Sci.* 2008, 2, 1, 27–35. https://citeseerx.ist.psu.edu/document?repid=rep1&type=pdf&doi=a5ea97fa6421b57a4963a0beed0ebad210dcf99c

Gore, A.; Mani, S.; HariRam, R.P.; Shekhar, C.; Ganju, A. Glacier surface characteristics derivation and monitoring using Hyperspectral datasets: A case study of Gepang Gath glacier, Western Himalaya. *Geocarto Int.* 2017, 34, 23–42. https://doi.org/10.1080/10106049.2017.1357766

Gupta, R.P.; Haritashya, U.K.; Singh, P. Mapping dry/wet snow cover in the Indian Himalayas using IRS multispectral imagery. *Remote Sens. Environ.* 2005, 97(4), 458–469. https://doi.org/10.1016/j.rse.2005.05.010

Haq, M.A.; Alshehri, M.; Rahaman, G.; Ghosh, A.; Baral, P.; Shekhar, C. Snow and glacial feature identification using Hyperion dataset and machine learning algorithms. *Arab. J. Geosci.* 2021, 14, 1–21. https://doi.org/10.1007/s12517-021-07434-3

Hock, R.; Hutchings, J.K.; Lehning, M. Grand challenges in cryospheric sciences: Toward better predictability of glaciers, snow and sea ice. *Front. Earth Sci.* 2017, 5, 64. https://doi.org/10.3389/feart.2017.00064

Jawak, S.; Luis, A.J. A spectral index ratio-based Antarctic land-cover mapping using hyperspatial 8-band World, View-2 imagery. *Polar Sci.* 2013, 7, 18–38. https://doi.org/10.1016/j.polar.2012.12.002

Jawak, S.D.; Wankhede, S.F.; Luis, A.J. Exploration of glacier surface facies mapping techniques using very high resolution Worldview-2 satellite data. *Proceedings* 2018, 2, 7, 339. https://doi.org/10.3390/ecrs-2-05152

Jawak, S.D.; Wankhede, S.F.; Luis, A.J. Explorative study on mapping surface facies of selected glaciers from Chandra Basin, Himalaya using World, View-2 data. *Remote Sens.* 2019a, 11, 1207. https://doi.org/10.3390/rs11101207

Jawak, S.D.; Wankhede, S.F.; Luis, A.J.; Balakrishna, K. Impact of image-processing routines on mapping glacier surface facies from Svalbard and the Himalayas using pixel-based methods. *Remote Sens.* 2022a, 14, 1414. https://doi.org/10.3390/rs14061414

Jawak, S.D.; Wankhede, S.F.; Luis, A.J.; Balakrishna, K. Effect of Image-processing routines on geographic object based image analysis for mapping glacier surface facies from Svalbard and the Himalayas. *Remote Sens.* 2022b, 14, 4403. https://doi.org/10.3390/rs14174403

Jawak, S.D.; Wankhede, S.F.; Luis, A.J.; Balakrishna, K. Multispectral characteristics of glacier surface facies (Chandra-Bhaga Basin, Himalaya, and Ny-Ålesund, Svalbard) through investigations of pixel and object-based mapping using variable processing routines. *Remote Sens.* 2022c, 14, 6311. https://doi.org/10.3390/rs14246311

Jawak, S.D.; Wankhede, S.F.; Luis, A.J.; Balakrishna, K. High-resolution remote sensing for mapping glacier facies in the Arctic. In *Advances in Remote Sensing Technology and the Three Poles*, Pandey, M.; Pandey, P. C.; Ray, Y.; Arora, A.; Jawak, S. D.; Shukla, U. K. Eds. Wiley: Hoboken, NJ, 2023; pp. 371–380. https://doi.org/10.1002/9781119787754.ch26

Jawak, S.D.; Wankhede, S.F.; Luis, A.J.; Pandit, P.H.; Kumar, S. Implementing an object-based multi-index protocol for mapping surface glacier facies from Chandra-Bhaga Basin, Himalaya. *Czech Polar Rep.*, 2019b, 9, 2, 125–140. https://doi.org/10.5817/CPR2019-2-11

Kaufman, Y.; Wald, A.; Remer, L.; Gao, B.-C.; Li, R.-R.; Flynn, L. The MODIS 2.1-_m channel-correlation with visible reflectance for use in remote sensing of aerosol. *IEEE Trans. Geosci. Remote Sens.* 1997, 35, 1286–1298. https://doi.org/10.1109/36.628795

Kaushik, S.; Singh, T.; Bhardwaj, A.; Joshi, P.K.; Dietz, A.J. Automated delineation of supraglacial debris cover using deep learning and multisource remote sensing data. *Remote Sens.* 2022, 14, 1352. https://doi.org/10.3390/rs14061352

Keshri, A.K.; Shukla, A.; Gupta, R.P. ASTER ratio indices for supraglacial terrain mapping. *Int. J. Remote Sens.* 2009, 30, 519–524. https://doi.org/10.1080/01431160802385459

Luis, A.J.; Singh, S. High-resolution multispectral mapping facies on glacier surface in the Arctic using WorldView-3 data. *Czech Polar Rep.* 2020 Aug 7, 10(1), 23–36. https://doi.org/10.5817/CPR2020-1-3

Mahmon, N.A.; Ya'Acob, N.; Yusof, A.L. Differences of image classification techniques for land use and land cover classification. In *Proceedings of the 2015 IEEE 11th International Colloquium on Signal Processing & Its Applications (CSPA), Kuala Lumpur, Malaysia, 6–8 March 2015*. Institute of Electrical and Electronics Engineers (IEEE): Piscataway, NJ, USA, 2015; pp. 90–94.

Mitkari, K.V.; Arora, M.K.; Tiwari, R.K.; Sofat, S.; Gusain, H.S.; Tiwari, S.P. Large-scale debris cover glacier mapping using multisource object-based image analysis approach. *Remote Sens.* 2022, 14, 3202. https://doi.org/10.3390/rs14133202

Paul, F.; Winsvold, S.H.; Kääb, A.; Nagler, T.; Schwaizer, G. Glacier remote sensing using Sentinel-2. Part II: Mapping glacier extents and surface facies, and comparison to Landsat 8. *Remote Sens.* 2016, 8, 575. https://doi.org/10.3390/rs8070575

Pope, A.; Rees, W.G. Impact of spatial, spectral, and radiometric properties of multispectral imagers on glacier surface classification. *Remote Sens. Environ.* 2021, 141, 1–13.

Qiu, S.; He, B.; Zhu, Z.; Liao, Z.; Quan, X. Improving Fmask cloud and cloud shadow detection in mountainous area for Landsats 4–8 images. *Remote Sens. Environ.*, 2017, 199, 107–119. https://doi.org/10.1016/j.rse.2017.07.002

Racoviteanu, A.; Williams, M.W. Decision tree and texture analysis for mapping debris-covered glaciers in the Kangchenjunga area, Eastern Himalaya. *Remote Sens.* 2012, 4(10), 3078–3109. https://doi.org/10.3390/rs4103078

Radiative Transfer Code. Available online: www.harrisgeospatial.com/docs/backgroundflaash.html (Accessed on 17 February 2017).

Rathore, B.P.; Singh, S.K.; Jani, P.; Bahuguna, I.M.; Brahmbhatt, R.; Rajawat, A.S.; Randhawa, S.S.; Vyas, A. Monitoring of snow cover variability in Chenab Basin using IRS AWiFS sensor. *J. Indian Soc. Remote Sens.* 2018, 46, 1497–1506. https://doi.org/10.1007/s12 524-018-0797-8

Robson, B.; Nuth, C.; Dahl, S.; Hölbling, D.; Strozzi, T.; Nielsen, P. Automated classification of debris-covered glaciers combining optical, SAR and topographic data in an object-based environment. *Remote Sens. Environ.* 2015, 170, 372–387. https://doi.org/10.1016/j.rse.2015.10.001

Sahu, R.; Gupta, R.D. Snow cover area analysis and its relation with climate variability in Chandra Basin, Western Himalaya, during 2001–2017 using MODIS and ERA5 data. *Environ. Monit. Assess.* 2020, 192, 489. https://doi.org/10.1007/s10661-020-08442-8

Sakai, A. Brief communication: Updated GAMDAM glacier inventory over high-mountain Asia. *Cryosphere* 2019, 13, 2043–2049. https://doi.org/10.5194/tc-13-2043-2019

Vatsal, S.; Bhardwaj, A.; Azam, M.F.; Mandal, A.; Ramanathan, A.; Bahuguna, I.; Raju, N.J.; Tomar, S.S. A comprehensive multidecadal glacier inventory dataset for the Chandra-Bhaga Basin, Western Himalaya, India. *Earth Syst. Sci. Data Discussions* 2022, 1–38. https://doi.org/10.5194/essd-2022-311

Warren, S.G. Optical properties of snow. *Rev. Geophys.* 1982, 20, 1, 67–89. https://doi.org/10.1029/RG020i001p00067

Zhai, H.; Zhang, H.; Zhang, L.; Li, P. Cloud/shadow detection based on spectral indices for multi/hyperspectral optical remote sensing imagery. ISPRS *J. Photogramm. Remote Sens.* 2018, 144, 235–253. https://doi.org/10.1016/j.isprsjprs.2018.07.006

Zhang, J.; Jia, L.; Menenti, M.; Hu, G. Glacier facies mapping using a machine-learning algorithm: The Parlung Zangbo Basin case study. *Remote Sens.* 2019, 11, 452. https://doi.org/10.3390/rs11040452

11 Assessment of Climate Change and Its Impact on Human Inhabitation of Gujarat during the Mid-Late Holocene Period

Archana Das, Aashima Sodhi,
S.P. Prizomwala and Sumer Chopra

11.1 INTRODUCTION

As per the World Meteorological Organization (WMO) report on global climate change, the year 2019 was the second warmest year recorded in the instrumental era. The alarming trend of increase in global temperature since the pre-industrial era, has triggered the eminent threat of sea-level rise and changes in precipitation pattern. The Indian Summer Monsoon (ISM) is one of the governing climatic model systems of the Earth and is known to exhibit variabilities onto various spatial and temporal timescales. Understanding the variability of the ISM has gained enough momentum in the recent past, owing to the alarming warming trends, escalating at a rapid pace. These warming trends are known to demonstrate the episodes of the climate change, which tend to have an adverse effect over the dynamics of the Earth; resulting in the melting of the glaciers, which further causes the rise in the sea levels, ultimately leading to a disrupted equilibrium of the earth. The recent studies led by the Inter-Governmental Panel on Climate Change (IPCC) (2021), has reported the rise in the global mean temperatures by 1.5°C to 2°C, owing to rapid warming in the current century corroborated with the rise in the global sea levels.

In light of this, it is important to take lessons from such changes in the past records and assessing their impact on landscapes as well as human behavior. The fundamental rule of *"Present is the key to the past"* also holds valid, when reading past records and anticipating them for future with varying magnitudes. The climatic changes in association with sea-level changes have long been subjective on human performance in terms of settlement pattern, style and adaptation methods (Crema *et al.*, 2016; Griffiths and Robinson, 2018). Different facets of climate change such as monsoonal aridity, sea-level changes; strongly impact the available natural resources and the survival

DOI: 10.1201/9781003485995-11

of mankind (Wiess and Bradley, 2001; Staubwasser *et al.*, 2003; Turney and Brown, 2007, Ponton *et al.*, 2012; Dixit *et al.*, 2018; Galili *et al.*,2019; Das *et al.*, 2022). The wrath of climate change that we are anticipating now has also been experienced by several ancient civilizations, which was counterproductive to many and led to some of their demise.

The ancient human settlements have been found to be concentrated near water-rich sources, especially in the vicinity of riverine or marine water bodies. The fluctuation in the availability of water and food made these centers vulnerable for sustenance. The downfall of the oldest Indus valley (Harappan) human civilization that flourished between ~3300 BCE to ~1900 BCE (Early to Mature Phase) in the North Western alluvial plains of the Indian subcontinent (Kenoyer, 1998; Possehl, 2002; Weber *et al.*, 2010) has received great attention in this context from climatologists, early historians and policy makers worldwide. This civilization probably evolved from the earlier food producing communities of western Baluchistan (~7000 BCE) into village farming communities/pastoral camps that eventually got settled and transformed into large urban fortified centers (Possehl, 2002). These pastoral and agricultural communities expanded geographically into the alluvial plains of northwestern Indian subcontinent during the 4300 to 3200 BCE period (Possehl, 2002; Madella and Fuller, 2006). The Early and the Mature Harappan phase subsequently witnessed a more sophistication of agricultural practices as well as organized living in fortified cultural centre in the Indus and Ghaggar-Hakra river systems (Possehl, 2002). The Mature Harappan phase, in particular (2600–1900 BCE), the Indus Valley Civilization (IVC) showed a sophisticated material as well as architectural living. The mature phase of IVC is considered to have abruptly ended ~1900–1800 BCE (i.e., 3900 yr BP), marked by the disappearance of its distinctive Harappan script (Kenoyer, 1998, 2006; Possehl, 2002). This abrupt decline has been widely linked to a sharp fall in the monsoonal strength at 2200 BCE (~4200 yr BP), i.e., beginning of the Meghalayan era (Staubwasser *et al.*, 2003; Ponton *et al.*, 2012; Dixit *et al.*,2014, 2018; Sengupta *et al.*, 2019). Several other theories like Aryan invasion, infectious diseases and change in the river dynamics, have also been debated (Weiss and Bradley, 2001; Possehl, 2002; Wright *et al.*, 2008; Giosan *et al.*, 2012). In some of recent attempts, Pokahria *et al.* (2017) brought out evidences for continuity of human sustenance and culture in Gujarat (western India) by analyzing macro-botany and stable carbon isotopic data from an archaeological site Khirsara. Likewise, marked changes in surface soils and significant changes in agricultural crop patterns in areas used by Indus inhabitants (such as banks of palaeo-channels of vanished Saraswati River) along with degradation of human prosperity and migrations have been reported (Sarkar *et al.*, 2020; Singh *et al.*, 2017; Dixit *et al.* 2018).

The West coast of India has experienced sea-level change during the Middle Holocene period (~8.2 ka BP; ~6200 BCE), eustatically followed by a marginal rate regionally up to ~4 ka BP (~2000 BCE) (Hashimi *et al.*, 1995; Das *et al.*, 2017). The relative high sea stand during 4000 BCE to 2000 BCE (i.e. Middle Holocene period) was likely ~1–2 m above the present-day level, albeit the actual magnitude still being debated (Banerji *et al.*, 2015, 2017; Das *et al.*, 2017; Makwana *et al.*, 2018). The information regarding the extent of the Middle Holocene high stand and the subsequent relative decline during the Meghalayan period (~4.2 ka BP onwards) remains elusive (Das *et al.*, 2017; Makwana *et al.*, 2019). The northwestern Indian continent, especially the Gujarat region, has remained ideal for expansion but has experienced the ultimate deurbanization of the famous Indus Valley Civilization during a large part of the middle to late Holocene period (~8.2 ka to 4.2 ka). The period witnessed the well-known 4.2 ka arid event as well as a simultaneous fall in the relative sea stand (Dixit *et al.*, 2014, 2018, Sengupta *et al.*, 2019; Das *et al.*, 2017, 2022). In the present study, we explore and discuss the impact on this climate change and its associated environmental change, which affected one of the most advanced civilizations of that period on Earth.

11.2   SCOPE OF THE PRESENT REVIEW

The present review focuses on a palaeo-archival approach to understand the combined narrative upon the impacts and intensity of the prevailing climatic perturbations, followed by the governing sea-level indexes, in response to the human settlements. The approach also sheds light on the evolved landscape dynamics of the region; thereby making an attempt to understand the changes witnessed by the Western Indian subcontinent in a holistic and profound manner. Such scenarios also reveal the potential of highlighting the intriguing questions that includes (1) to what extent did the effect of climate change contribute in governing the isostatic sea-level rise in the past? (2) its impact over the human settlements and (3) potential rise of the isostatic sea-level rise owing to the current warming trends and its response by the human society? Thus, the present review chapter aims to focus on the addressed questions based on the review of the recent research from the Gujarat region, Western India. The review also attempts to discuss the response of the Indus Valley Civilization in accordance with the sea-level change (submergence and emergence) corroborated with the climate perturbations.

11.3   STUDY AREA

11.3.1   PRESENT CLIMATE

The landscape of Gujarat is bordered by the Great Thar dessert of western India in the North. Most of the Gujarat falls under semi-arid to extremely arid zone in the north and the northwestern regions and only the southern–southeastern region is sub-humid to humid (Juyal *et al.*, 2006). The present-day climate is predominantly controlled by the Indian Summer Monsoon (ISM) with periods of precipitation restricted during June–September. Owing to the uneven spatial distribution of precipitation, several climatic zones are marked as humid; >1000 mm, sub-humid; 1000–750 mm, semi-arid; 750–300 mm and extremely arid; < 300 mm (Figure 11.1).

11.3.2   GEOLOGY AND GEOMORPHOLOGY

The sequential fragmentation of the western margin of the Indian Plate, as it collided with the Eurasian plate during the Late Mesozoic, led to the formation of three major rift systems namely Kachchh, Cambay and Narmada (Biswas, 1987). These basins

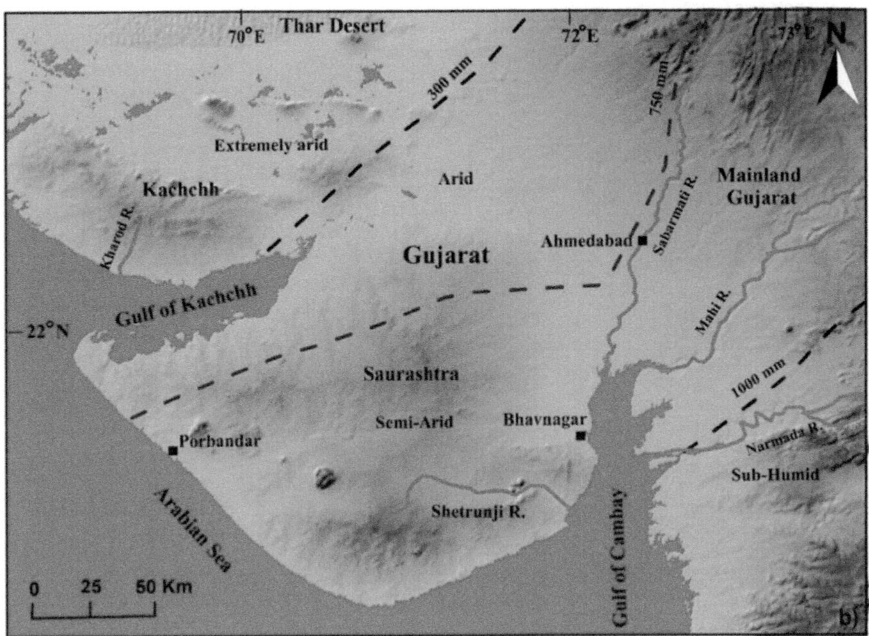

FIGURE 11.1   Physiographic map of the Gujarat region, Western India, demarcating several climatic zones and location of major rivers of the region (after Juyal *et al.*, 2006).

have thus, acted as major depocenters, accounting for a thick pile of sedimentation. The late Cretaceous period witnessed the vast Deccan Trap formation, which covered most parts of the Saurashtra and the Kachchh. This was followed by the Early Eocene transgression and subsequently the Tertiary sediments were deposited, which are mainly of marine origin (Merh, 1995). The Quaternary sediments are predominantly of fluvial origin in entire Gujarat. However, the coastal Miliolite deposits in the Saurashtra are considered to be of marine origin and inland ones are of aeoline origin (Bhatt, 2003). There is ample amount of data available that suggests that the climatic instability and the tectonic movements have shaped the present-day landscape evolution during the Quaternary period (Pant and Juyal, 1993; Srivastava *et al.*, 2001; Chamyal *et al.*, 2003; Jain and Tandon, 2003; Juyal *et al.*, 2006). Broadly, on the basis of evolution and tectonic/geological framework, the Gujarat state can be divided into three main domains namely, (1) Kachchh, (2) Saurashtra and (3) Mainland Gujarat (Figure 11.2).

The Gujarat region along the western coast of India hosts the longest coastline with 1600 km of length. The coastal landscapes of Gujarat consist of Holocene deposits in Kachchh and South Gujarat shorelines, along with the late Quaternary Miliolite deposits along the Saurashtra shoreline (Merh, 1995). Some patches of Deccan Trap basalts are also part of Saurashtra and south Gujarat shorelines. The hinterland region consists of Tertiary and Mesozoic age sandstones, shales and limestones (Merh, 1995).

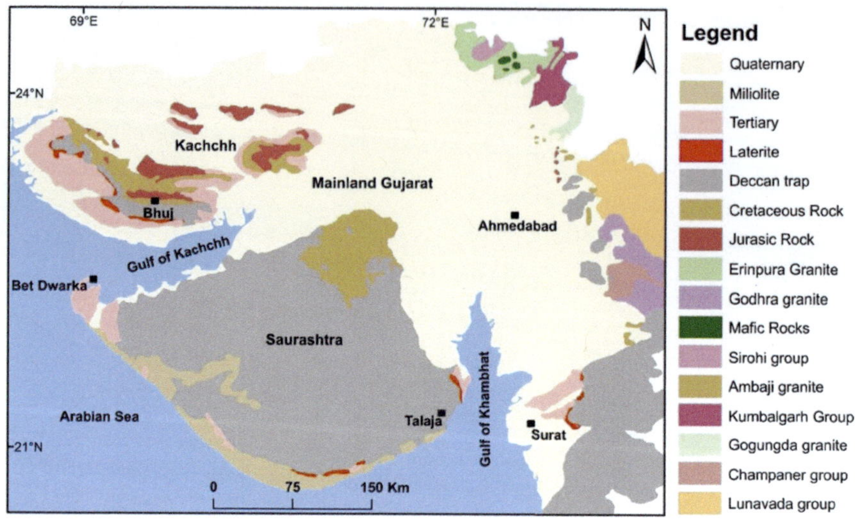

FIGURE 11.2 Detailed geological map of the Gujarat region (after Merh, 1995).

The Kachchh coastline shows the presence of wave-dominated geomorphic assemblage, viz., beach, beach-ridge, dune complex in the western end that faces the Arabian Sea (Maurya *et al.*, 2008; Prizomwala *et al.*, 2010). The coastal assemblage changes from wave dominated to tide dominated in the inner Gulf of Kachchh in the eastern segment with wide mudflats and palaeomudflats (Maurya *et al.*, 2008; Prizomwala *et al.*, 2010). While, the Saurashtra coastline shows the most varied geomorphic assemblage with rocky shoreline with narrow sandflats and mudflats in its northern coastline, along with beach, beach-ridge configuration along its western coastline and Rocky cliffs along its southern shoreline that merges into wide mudflats in inner Gulf of Khambhat in the east. The Southern Gujarat shoreline consists of sandflats and mudflats in the north to narrower beaches in the south.

11.4 RESULTS

11.4.1 HOLOCENE CLIMATE CONUNDRUM: INSIGHTS FROM THE PALAEOARCHIVES OF WESTERN INDIA

11.4.1.1 Palaeoarchives of the Mainland Gujarat Region

Since, the sedimentary records owing to their nature denotes high preservation potential of the sediments, they are considered ideal to infer the ISM variations from the region. For instance, the climatic strength inferred from the lacustrine records of the Sabarmati River basin denoted a more or less fluctuating climatic strength around ~8000 cal yr BP-1500 yr BP (Prasad *et al.*, 2014; Raj *et al.*, 2015; Sridhar *et al.*, 2020) punctuated with the episodes of reduced monsoonal strength around 7900 cal yr BP, and between 4600 yr BP and 4200 yr BP (Figure 11.3). Similarly, the sediments of the Nal Sarovar lake, denoted enhanced monsoonal strength during 4800 yr BP–3000 yr BP followed by reduction in the monsoonal strength, which persisted until 2000 yr BP (Prasad *et al.*, 1997). Similar coherency was observed from the sediments of the Mahi and Narmada river basins, representing a stronger monsoonal strength around ~3400 to 3000 cal yr BP; followed by the prevalence of weaker monsoonal conditions at around 2100 yr BP, 2850 cal yr BP and 1300 yr BP (Sridhar, 2007; Laskar *et al.*, 2013). Moreover, the terrestrial records from the region also show climatic perturbations around 4200 yr BP. The multiproxy approach investigated from the palaeochannel of the Sabarmati river basin; 3 km nearer to the Lothal archaeological site, sheds light on the hitherto impact of the coupled climate and sea-level response over the existing human resilience of its time. Based on the sediment geochemical and isotopic studies, the Lothal archaeological site had witnessed five phases of climatic perturbations with major changes occurring around 4200 yr BP with the advent of the Meghalayan Era. The region witnessed pronounced climatic and sea-level changes between 4200 and 3625 yr BP; followed by the gradual enforcement of the terrestrial-like conditions with the prevalence of the present-day-like climatic conditions during ~2070 yr BP (Das *et al.*, 2022). The attainment of the estuarine followed by the terrestrial-like conditions within the region is corroborated

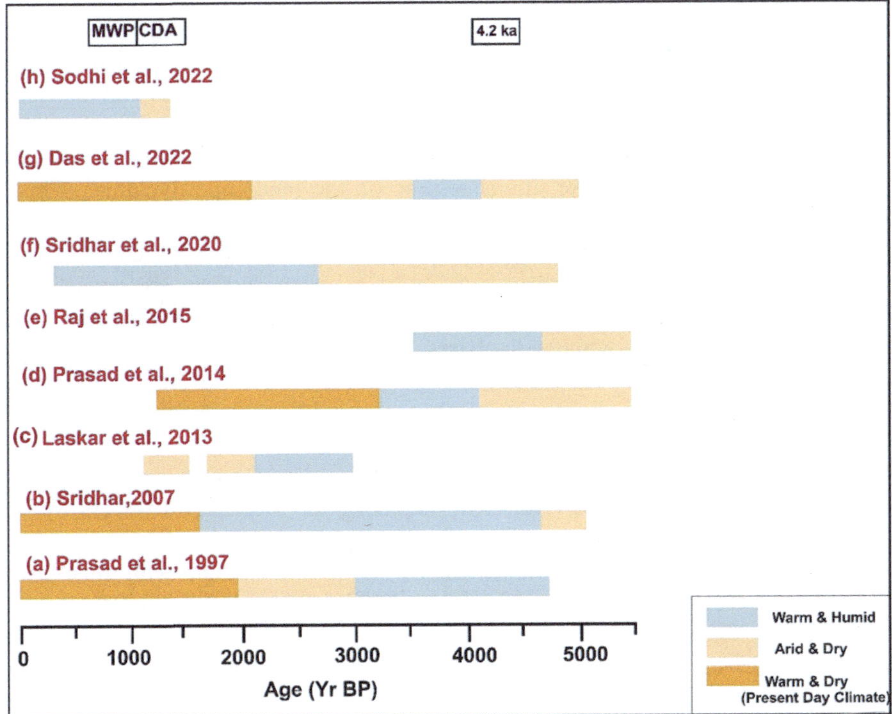

FIGURE 11.3 A regional palaeoclimatic narrative of the Mainland Gujarat region, Western India.

with the sea water retreat of 1.2±0.2 m. This retreat is said to be in accordance with major climatic and environmental changes witnessed by the region around 4200 yr BP; resulting in the gradual deurbanization of Harappan's existing at the site and ultimately affecting their sustainability (Das *et al.*, 2022). In association with the same, the climate of the last two millennia has been studied extensively owing to the short and abrupt climatic variabilities witnessed during this timeframe. Based on the multiproxy approach, the sediments of the KamaTalav region studied from the palaeomudflats of the Gulf of Khambhat, revealed signals of the Dark Ages Cold Period (DACP) and Medieval Warm Period (MWP) advocated by solar forcing as the driving mechanism (Sodhi *et al.*, 2022) (Figure 11.3).

Besides this, the fluvial dynamics of the region witnessed a completely evolved climatic and palaeoenvironmental conditions spanning between 3000 yr BP and ~2000 yr BP owing to the fluctuating discharge supply since, the fluvial archives, tend to exert strong control on the relative base-level changes, slope variations and feedback mechanisms (isostatic rebound of the erosional origin) (Wolf and Faust, 2015). Consequently, amidst the various parameters of the fluvial archives, the floodplain and terraces sequences reveal significant signals of the palaeoclimatic variations in the recent past, owing to its sensitivity towards the catchment behavior (Blum and Tornqvist, 2000; Vandenberghe *et al.*, 2010). In the testimony of the same, various palaeoclimatic and palaeoenvironmental changes have been recorded from the fluvial archives of the region, which in turn has proved complexity within the strength of the Indian Summer Monsoon. This is evident with the variabilities observed in the aggradation-incision phases and changes in the sedimentary facies characteristics (Sinha *et al.*, 2020); thereby hinting towards the evolved depositional and landscape environment of the region. The west-flowing rivers, viz., the Sabarmati, the Mahi and the Narmada have been studied in detail from the Mainland Gujarat region to understand the response of the fluvial regime on the landscape dynamics. This is further very well exhibited by the sediments of the Mahi River demonstrating a deposition of the terrace sequence, in coherence with the transiting environmental conditions from fluviatile to marine conditions between 3600 yr BP and 1700 yr BP (Kusumgar *et al.*, 1998). Such transiting conditions coevals with the fluctuating monsoonal strength that witnessed phases of predominant terrestrial input alternating with the phases of the estuarine conditions were observed in this timeframe. This also accounts for the change in the channel conditions from wider to narrower with smaller meandering channels along with reduction in the discharge conditions (Sridhar, 2007b). Similar reduction in the discharge conditions are also observed in the sediments of the Narmada river basin depicting the presence of the clayey-rich terrace sequences between 1900 yr BP and 1200 yr BP. The deposition of the terrace sequences further sheds light on the enhanced monsoonal strength with three prominent flood events occurring at 1900 yr BP, 1700 yr BP and 1200 yr BP (Raj and Yadav, 2009). Additionally, the late Holocene terrace sequences of the Sabarmati basin, are inferred from the prevalence of the scroll plains along the incised river valley within the meandering channels of the Sabarmati river basin (Srivastava *et al.*, 2001). Hence, the fluvial archives of the Mainland Gujarat region reveal a significant decrease in the discharge conditions along with the enhanced intensity of the Indian Summer Monsoon spanning from ~3600 yr BP to the present; punctuated with the episodes of the reduced monsoonal strength. Interestingly, it is to be noted that the episodes of the enhanced and reduced monsoonal strength observed from the terrestrial archives of the region are in accordance with the stream discharge conditions and stream power, thereby signifying the fact that the episodes of the reduced fluvial discharge are coherent with the episodes of the enhanced monsoonal strength and vice versa. Hence, itis feasible to infer that the archives of the mainland Gujarat region preserves signal of the Holocene timeframe climatic perturbations, which have played an essential role in shaping the landscape dynamics of the region along with its impact over the cultural establishments, which have been discussed in the later section.

11.5 PALAEOARCHIVES OF THE KACHCHH REGION

Similarly, the reconstructed and compiled palaeoclimatic archives, signifies the composite climatic evolution of the Kachchh basin, of the Western Indian subcontinent, over the Holocene timespan. The climatic evolution from the Rann archives reveals a period of transgressive marine environment, in response to the post-glacial sea-level rise at the onset of the Holocene ~10,600 yr BP as witnessed by the presence of the fining upward sequence of the sediments of the Great Rann of Kachchh (GRK) (Kumar *et al.*, 2021). The prevalence of such fine-grained sediments is plausible owing to the long-distance transport of the sediments from the proximal Himalayan source within the region (Prizomwala *et al.*, 2014a; Kumar *et al.*, 2021). Furthermore, the transportation of such fine-grained sediments is also corroborated with the strengthening of the Indian Summer Monsoon witnessed by the region during 9,000 yr BP until 6,000 yr BP (Kumar *et al.*, 2020). Similar monsoonal advances were experienced in the other parts of the Rann (Khonde *et al.*, 2017) punctuated with the drier spell of the monsoonal conditions during 8,400–7,200 yr BP, respectively (Kumar *et al.*, 2020; Khonde *et al.*, 2017; Raj *et al.*, 2021) that coincide with the globally reported 8.2 ka event. Based on the sediment geochemical studies, the fluvial archives of the region have been excellently studied, which revealed phases of enhanced monsoonal strength ranging between ~9000 yr BP and 5000 yr BP (*Bhattacharya et al.*, 2013; 2014; Prizomwala *et al.*, 2016; Das *et al.*, 2017). Instances of the same includes, the Layza Nana section from the coastline of the Kachchh (Das *et al.*, 2017) followed by the analogous stream power from the sediments of the Lotia river (Prizomwala *et al.*, 2016). Similar studies from the fluvial sequences of the Khari and the Gunwari rivers also indicated an advancement in the monsoonal strength leading to enormous sediment influx at the depocenters (Bhattacharya *et al.*, 2013; 2014). These depocenters are usually the depressed terrains that tend to preserve the excellent sediment records of the region. The Rann archives serves as the classic example of the same, which illustrates a fluctuating monsoonal strength during 6000–3000 yr BP (Pillai *et al.*, 2017;2018; Basu *et al.*, 2019; Makwana *et al.*, 2021). However, post 4200 yr BP, a climatic reversal is observed with an abrupt declinal shift, suggestive of an onset of the arid-like climatic conditions prevailing within the Rann region. This is further corroborated by the presence of relatively enhanced physical weathering, along with the presence of the formation of the aeolian miliolites, thereby signifying the signatures of the weakened monsoonal strength. The signatures of such prevailing aeolian activity is further confirmed by the formations of the dunes along the coastline of the Kachchh region, ranging somewhere between ~4,200 yr BP and 2,900 yr BP (Dabhi *et al.*, 2021; Das *et al.*, 2017). The presence of the aeolian sand within the sediments of the Layza nana of the Kharod river section at 2,900 yr BP can be affirmed to the coastal dune accretion similar to the present-day-like coastal configuration (Das *et al.*, 2017) (Figure 11.4). Apparently, with the increased concentrations of the aridity, as evident from the increased flux of Ti/Al along with the increase in the Ca/[Ti+Al+Fe] content, further infers the prolonged aridity prevailing within the sediments of the Panjor Pir, Kori Creek. The substantial amount of upsurge in the aeolian input, might have led to the enhanced wind intensity within the region, thereby leading to the greater sedimentation, ultimately resulting in the deposition of the aeolian dunes during 4,200 yr BP (Dabhi *et al.*, 2021).

The persistent aridity can also be attributed to the drying up of the rivers, which were earlier prevailing within the area thereby ceasing the coarse-grained sediment flux into the basin (Kumar *et al.*, 2020; Khonde *et al.*, 2017). Similar climatic perturbations have been divulged from the parts of the Banni Plains denoting the episodes of the reformed palaeoenvironmental and palaeoclimatic conditions. The region witnessed an overall enhanced monsoonal strength during >4800 yr BP and post ~4100 to 3000 yr BP (Makwana *et al.*, 2021). However, this period of enhanced monsoonal strength was punctuated with the peak aridity observed around 4200–4400 yr BP within the Rann (Makwana *et al.*, 2021). Owing to the abrupt aridity, changes in the vegetation pattern; with a transition from C_3 to C_4 grasses was also dominant in the region during this timeframe (Pillai *et al.*, 2017; 2018; Basu *et al.*, 2019). The period between 2,500 to the present broadly signifies the prevalence of the mesic ecosystem characterized by the coexistence of C_4 grasses and C_3 trees and shrubs with the dominance of the latter one. Whereas, a shift in the paleoecology is observed between 2,500 and 1,000 cal yr BP, wherein the vegetational pattern showed a transition towards the water-limited with dominance of the C_4 grasses along with the evidences of the fire and herbivory rangeland.

Apart from climate evolution, the Kachchh region has been subjected to marine sedimentation changes too, owing to the prevalence of the long coastline as illustrated by various studies. Various studies have reported that mere presence of the beaches and mudflats situated at an elevation of 2–3 m; higher than the present-day shoreline; demonstrates the instances of the higher sea stand than present in the historical past (Maurya *et al.*, 2008; Prizomwala *et al.*, 2010). Recent studies from the estuarine sequence of the Kharod river mouth, also suggested a higher sea stand of 2m compared to the present-day mean sea level during 6000 to 3000 yr BP (Das *et al.*, 2017). Sharma *et al.* (2020) also reported a relatively higher sea stand of 1.45±0.33 m from the similar sequence by incorporating the tectonic component of the region during 7000–5000 yr BP. Similarly, Tyagi *et al.* (2012) studied the bet and Rann sequence in the western Great Rann of Kachchh and reported dominance in marine processes during 5.5 to 2 ka period. Ngangom *et al.* (2012) also inferred the presence of the marine sedimentation in the vicinity of the Nara river channel, in the Great Rann of Kachchh, somewhere during 2200–1000 yr BP. In addition to the same, the Rann region was also said to be influenced by the marine activity until ~2000 yr

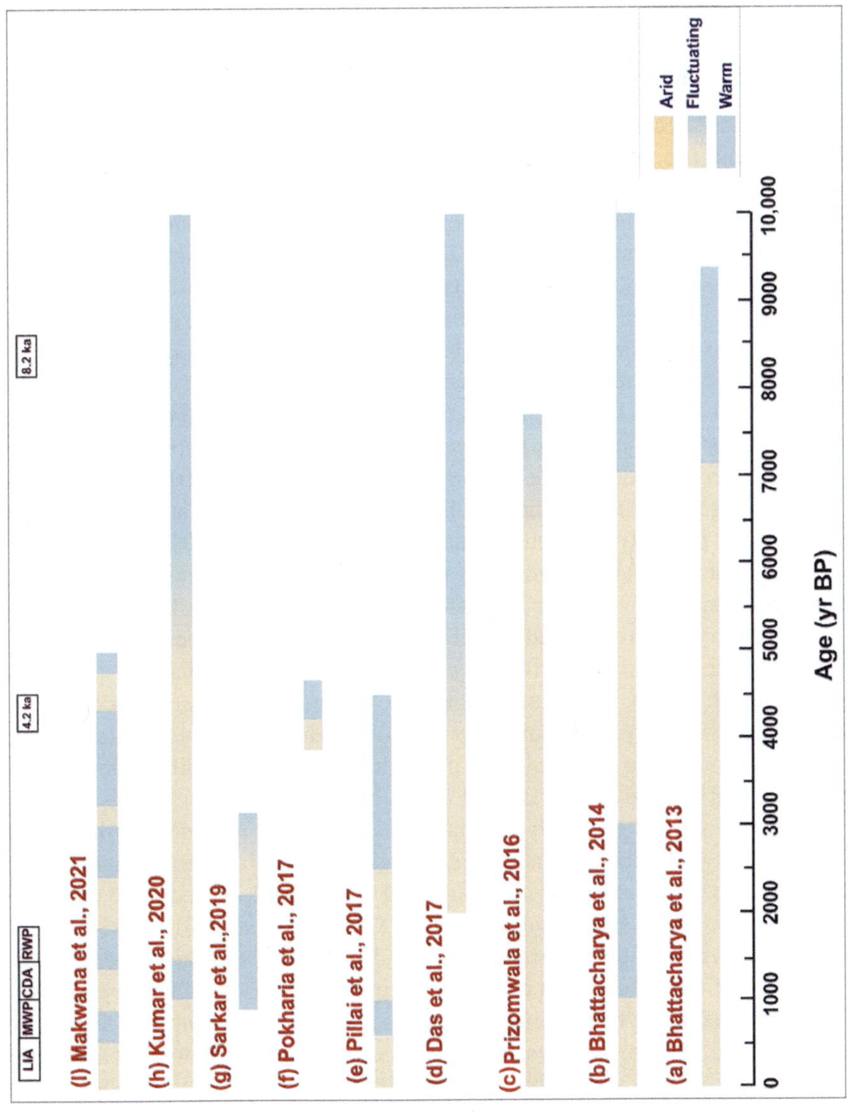

FIGURE 11.4 A regional palaeoclimatic narrative of the Kachchh region, Western India.

BP with approximate high sea stand of ~2m than present (Makwana *et al.*, 2019). From the Little Rann, Gupta (1975) suggested for the first time ~4m deep water at around 2 ka, whereas, Khonde *et al.* (2011), reported that the marine sedimentation has continued up to ~500 a. The marine indentation from the Little Rann of Kachchh, was also reported by Raj *et al.* (2021), suggestive of the predominant marine environment during ~2650 yr BP until 1500 yr BP.

11.6 PALAEOARCHIVES OF THE SAURASHTRA REGION

The landscape of the Saurashtra is an outcome of two processes, viz., the eustatic sea level changes and the processes associated with the neotectonism (Pant and Juyal, 1993). However, the fluvial archives of the region are limited and require a substantial approach to understand the landscape evolution of the region. Owing to the limited availability, it has been inferred that the landscapes of the Saurashtra region reveal the significant presence of the valley fill sequences, and terraces corresponding to the prevalence of the high sea stand and monsoonal strength (Solanki *et al.*, 2021). The Shetrunji river basin, which happens to be the largest river basin of the Saurashtra region revealed the evidences of the intensified monsoonal strength; corroborating the higher sea stand followed by the episodes of the intense erosion and incisional phases somewhere during 5500 yr BP. Such intensified episodes of the monsoonal strength is also recorded from other archives of the region (Banerji *et al.*, 2015; 2017; Thakur *et al.*, 2019), suggestive of the coherent climatic pattern approximately around 5000 yr BP. Based on the palynological studies, the sediments of the Harshad estuary indicated pronounced ISM strength prior to 5100 yr BP, which later transitioned to arid-like conditions until 1400 yr BP (Thakur *et al.*, 2019). Similar palynological studies inferred from the relict mudflats of Diu inferred the presence of the mangrove forest within the region during 4710 cal yr BP, congruent to the enhanced monsoonal strength within the region. The termination of the warmer and moister conditions was inferred during 2880 cal yr BP (Banerji *et al.*, 2015). However, this aridification was not prominent for a longer timescale and the region attained humidification spanning between 1930 and 355 cal yr BP (Banerji *et al.*, 2017).

Besides, evidences of the high sea stand has also been recorded in the coastal archives of the region. The coastal archives of western India, even though limited, have been scrutinized recently for reconstructing the palaeo high sea stand evidences (Kusumgar *et al.*, 1998; Bhatt and Bhonde, 2006; Das *et al.*, 2017; 2022). Gupta (1972) hypothesized a high sea stand of 2–3m based on dated raised beaches and dead coral reefs. Consequently, an evidence of higher sea stands of 2m around ~3400 yr BP, has been recorded from the coastline of Saurashtra by dating the dead oyster shell (Juyal *et al.*, 1995). Besides, the presence of the marine notches and shore platforms situated 2–3m above the mean sea level (msl) and at >7m above mean sea level (msl) respectively, are depicted as representatives of the palaeo-sea stand markers (Bhatt and Bhonde, 2006; Prizomwala, 2018). Additionally, the first level of notch is said to be linked to the Middle Holocene high sea stand and the second level with MIS 5e stand, inferring the presence of the relative high sea stand compared to the present-day shoreline (Pant and Juyal, 1993; Bhatt and Bhonde, 2006). Similarly, a high sea stand of 1m was also decoupled from the relict palaeomudflats of the region; using neotectonic uplift during 4780–2825 yr BP. However, these estimates are on the higher side, in light of the non-availability of realistic uplift rates from the region (Banerji *et al.*, 2015). Nevertheless, the fluvial archives of the region too, are limited and require a substantial approach to understand the landscape evolution of the region. The fluvial landscapes of the Saurashtra region reveal the prominent presence of the valley fill sequences, and terraces corresponding to the prevalence of the high sea stand and monsoonal strength (Solanki *et al.*, 2020). The Shetrunji river basin, which happens to be the largest river basin of the Saurashtra region revealed the evidences of the intensified monsoonal strength; corroborating the higher sea stand followed by the episodes of the intense erosion and incisional phases somewhere during 5500 yr BP. The episode of the intense erosion corresponds to the change in the palaeoenvironmental conditions transiting from fluviatile- to estuarine-like conditions in the region followed by a drop in the sea level along the coastline of Saurashtra and gradual attainment of the present-day-like conditions (Solanki *et al.*, 2020).

11.7 DISCUSSION

11.7.1 IMPACTS OF GEOLOGICAL IMPRINTS OVER THE CULTURAL ESTABLISHMENTS

The landscape of the western India has been a hotbed for flourishing and subsequent de-urbanization of the mighty civilizations since past 7500 BCE (Gupta, 2003). The Harappan civilization or the Indus Valley Civilization (IVC) originated from the mountainous reigns of the Harappa (present-day Pakistan) that migrated gradually into the northwestern and western parts of the Indian subcontinent. The Gujarat region has been a host to the many such centers of cultural establishments, viz., Lothal, Dholavira, Karim Shahi, Vigakot, Khirsara etc., (Figure 11.5) denoting the treasure towards the past human resilience.

Although, these centers of cultural establishments have been studied briefly with respect to archaeological context; studies pertaining to the geological imprints (palaeoclimate and palaeoenvironmental impacts) are now being the focus of the research. In the lacuna of the same, various studies have hinted towards the role of these geological footprints in the sustenance and decentralization of these cultural establishments (Pokharia *et al.*, 2017; Sengupta *et al.*, 2020; Sarkar *et al.*, 2019; Das *et al.*, 2022).

The abrupt and prolonged climatic aridity observed during 4,200 - 2,000 yr BP played a major role in changing the landscape dynamics of the Gujarat region. The ultimate drying up of the rivers corroborated with the changes in the vegetational pattern (transition from C_3 to C_4) signifies the prevailing mega drought like conditions. Such climatic settings had a coherent impact over the cultural establishments thriving within the region. Besides, the mega-scale drought activity archived during the 4,200 yr BP (4.2 ka event) is said to be the marker for the onset of the deurbanization of such cultural establishments (Pokharia *et al.*, 2017; Sengupta *et al.*, 2020). Many important cultural establishments, such as Dholavira, Kanmer, Khirsara, Lothal, etc., have been subjected to such catastrophic drought that ultimately led to their collapse or deurbanization. Lothal, situated 23km from the present-day shoreline

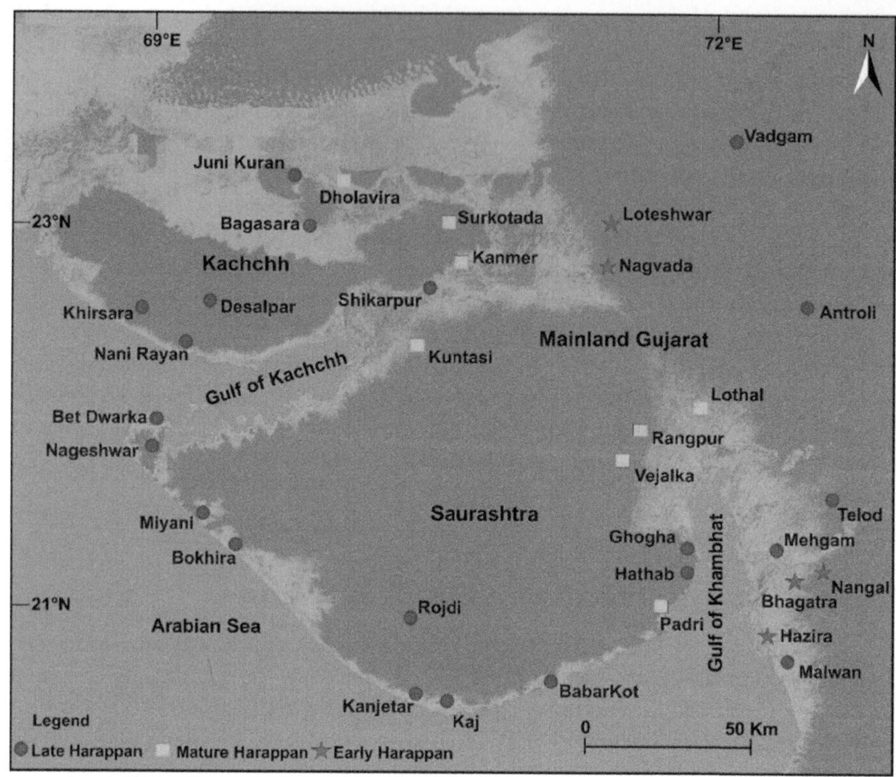

FIGURE 11.5 Map of the Western Indian subcontinent, showcasing the location of the various archaeological sites present within the region.

is believed to be an ancient dockyard (Rao, 1979; Gaur *et al.*, 2006). The site is believed to have been occupied somewhere during 5600 yr BP to 3800 yr BP. Recent investigation from the site inferred the combined role of palaeoclimate and palaeoenvironmental change as the key causative factor behind the abandonment of the site (Das *et al.*, 2022). The site is believed to have undergone multiple (five) phases of climate change based on the sediment geochemistry and stable isotopic contents; with the major change occurring with the advent of the Meghalayan Era, i.e., 4200 yr BP. The site is believed to have been under the non-terrestrial conditions somewhere during 5030 yr BP, which accords well with the Middle Holocene high sea stand BP; which gradually shifted to a transitional environment with the prevalence of the estuarine-like conditions that persisted until 2070 yr BP. The transitional environment occurred during 4200–3625 yr BP with the gradual attainment of the aridification within the region at 4200 yr BP; followed by an improvised monsoonal strength during 3625 yr BP. Such transitivity within the region also witnessed a gradual fall in the sealevel, with a sea water retreat of 1.2±0.2m ultimately affecting their sustenance and leading to a dysfunctional dockyard at Lothal. The sea water retreat in corroboration with the peak aridity might have resulted into the gradual abandonment of the settlement (Das *et al.*, 2022). Similarly, the Dholavira, which was an important culturally rich Harappan civilization, had a humid fluvial landscape somewhere around 5,500 yr BP that persisted until 4,500 yr BP. The civilization expanded until 4,500 yr BP, as evident with the continuous additions in the architectural elements. However, the period between 4,300 and 4,000 yr BP, the cultural expansion resisted owing to the widespread drought-like conditions prevailing within the region. During this climatic transition, the civilization ceased its expansion leading to the near collapse of the settlement as evident with the degeneration of the architecture, craftsmanship and material cultural typical of that period. The deterioting hydrological strength was illustrated by the concentrations of the $\delta^{18}O$ values; which eventually led to the hypothesis of the defunct river channel; thereby marking the onset of their collapse. The ultimate collapse of the civilization occurred somewhere around 3,800 yr BP (Sengupta *et al.*, 2019). However, even after the complete collapse of the Harappan settlements in the Kachchh region, some settlements tried to adapt the withstanding changing climatic trends prevailing at that time. The recent investigation made at the Karim Shahi and the Vigakot at the fringe of the Thar desert at the Great Rann of Kachchh, reveals the presence of such cultural establishments ranging from Early Iron Age to Early Historic Age (~3,100 to 2,300 yr BP) (Sarkar *et al.*, 2019). The shift in the cultural adaptation at the Karim Shahi; was evident based on the presence of the archaeological artefacts and changes in the values of the $\delta^{13}C$ and micro-botanical remains. While, the settlement at the Vigakot that occurred during the Historic to Medieval periods (~1500–900 yr BP) suggested that the climate at Vigakot was semi-arid rather than hyper-arid climate of today. The relatively enhanced climatic condition is illustrated based on the concentrations of the mollusc $\delta^{18}O$ that are considered as a true representation of the ambient precipitation. Thus, itis feasible to say that the monsoonal conditions were good enough during the Historic to Medieval periods (~1500–900 yr BP), which eventually led to the sustainment of the fluvial system that led to the thriving of the settlements at the Vigakot. However, the ultimate collapse of the same is due to the decline in the monsoonal strength and prevalence of the C_4 vegetation within the region. This decline in the monsoonal strength is owed to the southward migration of the inter-tropical convergence zone (ITCZ), which also affected the affluent fluvial system (Sarkar *et al.*, 2019). Alternatively, evidences to assess the palaeomonsoonal–human relationships have also been attempted from the Khirsara site, based on the macro-botanical remains and carbon isotopes of the soil organic matter between 4,600 and 3,900 yr BP. The study inferred a significant change in the crop pattern at 4,100 yr BP from barley-wheat-based agriculture to "drought-resistant millet-based crops" as an adaptation measure in response to the defunct fluvial system owing to the deteriorated monsoonal conditions. While, similar inferences were made from the Greater Rann of Kachchh, by delineating the sediment routing system, which further strengthened the inference that the abandonment of a fluvial system, between 5,000 and 1,000 yr BP did not have a hitherto effect over the abandonment of the cultural civilizations (Chatterjee and Ray, 2017). Besides, the near absence/abandonment of the fluvial system at the Khadir Bet further hints towards the prevalence of the prolonged aridity witnessed by the region somewhere during 5,000 yr BP, ultimately shaping the landscape dynamics of the region (Ngangom *et al.*, 2017).

Whilst, the relative rise of the sea level, owing to the climate-induced variabilities and its impact upon the cultural establishments cannot be ignored. This is relevant to the submergence of the Dwarka city owing to the rise in the sea levels somewhere back during the Early to Late Medieval period (Gaur *et al.*, 2006). This is further confirmed based on the traces of the ancient river channel of the River Gomti situated 12km seaward and at 8–10m water depth (Vora *et al.*, 1991). Further excavations inferred that the site was the part of the Gujarat coastline until the Protohistoric period; followed by their submergence (Tripathi *et al.*, 2004). This agrees well with the Middle Holocene high sea stands, which experienced an overall high sea stand somewhere between 6000 yr BP and 3000 yr BP. Besides, the recent studies at the Diu Island of the southern Saurashtra, the presence of the fish tank was also reported hinting towards the prevalence of the coastal activity during the Late Holocene period. Owing to its dysfunctional condition, its position of 0.5m above the high water line and the absence of any other signature of sea level change during last 1000 BCE, it was postulated that the Diu Island has been uplifted by about 0.5m during the last 500 years (Kazmer *et al.*, 2016). Thus, itis feasible to account that the palaeoclimatic and palaeoenvironmental conditions played a major role in the sustenance and decentralization of the cultural establishments within the region, which is very well evident with the aforementioned instances.

11.8 DELINEATING THE LANDSCAPE EVOLUTION OF THE WESTERN INDIA

The recognition and illustration of the landscape evolution is crucial and essential since it reflects changes witnessed by the region during the stipulated timespan. The present study attempts to highlight the landscape evolution of the Gujarat region in corroboration with its archival response. The fluvial archives of the region flourished somewhere during 9000 yr BP–~7000 yr BP (Bhattacharya *et al.*, 2013; 2014; Prizomwala *et al.*, 2016; Solanki *et al.*, 2020), often punctuated with an episode of the weaker monsoonal strength recorded around 8000 yr BP. This episode of the weaker monsoonal strength coincides well with the global arid event; namely 8.2 ka event (Figure 11.6). Similar instances of the climate strength are also recorded from other parts of the Indian subcontinent. The sediments from the Chandra peat, Northwest Himalaya demonstrates an intensified monsoonal strength during 11,640 yr BP–8810 yr BP followed by an episode of the colder and weaker monsoonal strength from 8810 to 8117 cal yr BP coinciding well with the 8.2 ka event using palynological and isotopic studies (Rawat *et al.*, 2015). Based on the multiproxy approach, the sediments of the Benital lake, Uttarakhand also suggested an intensified monsoonal strength during 8600 yr BP–5800 yr BP punctuated with a reduction at 8100 yr BP (Bhushan *et al.*, 2018). The stalagmite oxygen isotopic ratio from the Kotumsar cave, Central India also reveals the reduction in the monsoonal strength at 8200 yr BP and at 5900 yr BP (Band *et al.*, 2018). While, from the northwestern India, similar climatic perturbations have been recorded in the sediments of the Rewasa palaeolake, using oxygen isotopic ratio (Dixit *et al.*, 2014b). This was followed by strength of the fluctuating climatic regime throughout the western India, along with the formation of

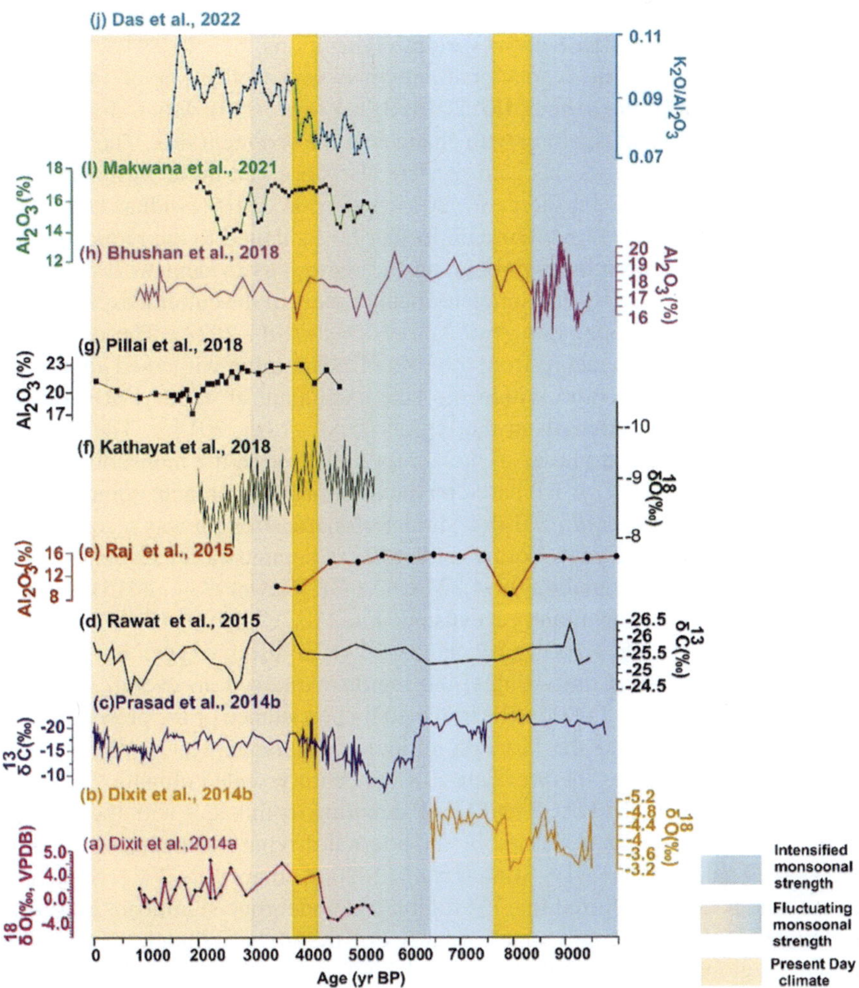

FIGURE 11.6 Palaeoclimate records (a, b, c, d, f, and h) compared has been compared with the monsoonal records of the Gujarat region (e, g, I, and j) for the Middle-Late Holocene timeframe. The vertical yellow bars are manifestations of the globally known climatic events, viz., 8.2 ka event and 4.2 ka event for the Holocene timeframe.

the coastal dune accretion, which substantiates the persistence of the aeolian activity (Das *et al.*, 2017; *Dabhi et al.*, 2021) within the region. Such prevalence of the aeolian strength accords well with the global climatic reduction at 4200 yr BP. The timeframe ~ 4,200 yr BP a.k.a. 4.2 ka event is a globally known mega drought event (Dixit *et al.*, 2014a; Kathayat *et al.*, 2018), which has been lately accepted as the formal boundary of the Late and Middle Holocene; primarily known as the Meghalayan Era (Walker *et al.*, 2018). This timeframe is known to have a significant climatic effect over the region leading to the complete drying up of the rivers, abrupt shift in the vegetational pattern along with the collapse of the various civilizations.

Pronounced shifts in the vegetational pattern is very well observed in the proxy records of the Rann (Pillai *et al.*, 2017; 2018; Basu *et al.*, 2019), which showcases the transit from C_3 to C_4 plants; along with changes in the sediment flux. The changes in the sediment flux and feeder mechanism were also observed in the sediments of the Mainland Gujarat region (Prasad *et al.*, 2014; Raj *et al.*, 2015; Sridhar *et al.*, 2020; Das *et al.*, 2022) (Figure 11.6). This can further be attributed to the reduction in the fluvial stream power prominent throughout the region; as evident with the gradual decentralization followed by the complete abandonment of the cultural establishments (Pokharia *et al.*, 2017; Sengupta *et al.*, 2019; Das *et al.*, 2022). The oxygen isotopic gastropod aragonite record from the Kotla Dahar region witnessed an increased evaporation/precipitation ratio within the lake catchment at 4100 yr BP; inferring the weaker monsoonal strength in the region (Dixit *et al.*, 2014a). The sediments of the Lonar lake revealed phases of the abrupt reduction in the monsoonal strength during 4600 yr BP to 3900 yr BP based on the multiproxy approach; coinciding with the 4.2 ka event (Prasad *et al.*, 2014b). Similar climatic strength was recorded in the sediments of the Chandra peat, Northwest Himalaya demonstrating an episode of the drier climatic strength around 4,808–4,327 cal yr BP (Rawat *et al.*, 2015). A speleothem record from the Mawmulah cave also denoted the evidences of abrupt climatic aridity spanning between 4,000 yr BP and 3,900 yr BP using oxygen isotope record (Kathayat *et al.*, 2018). Contrastingly, the region witnessed an overall improvised monsoonal strength until 3,000 yr BP followed by persistence of the present-day-like climatic conditions during the last two millennia. Interestingly, the climate of the last two millennia denotes the prevalence of the century scaled climatic oscillations within the region, viz., the DACP and MWP denoting its linkages with the climate of the North Atlantic Region governed by an external driving mechanisms (Ngangom *et al.*, 2017; Thakur *et al.*, 2019; Sridhar *et al.*, 2020; Kumar *et al.*, 2021; Sodhi *et al.*, 2022). Thus, it can be inferred that the region has undergone significant episodes of the climate amelioration with equal instances of the global climatic perturbations; denoting its global linkages with the North Atlantic oscillations.

11.9 CONCLUSIONS

The varied and versatile palaeo-archives from the Gujarat region shed light on the evolved landscape dynamics of the region, which impacted the existing cultural centers of that period. Based on the regional comparative study, it can be inferred that the palaeoclimatic and palaeoenvironmental shifts, played a significant role in the sustenance of these cultural centers. This is very well observed from the Lothal, Dholavira and Khirsara sites, which witnessed a surge in the site abandonment owing to the widespread aridification witnessed by the region. This widespread aridity is also associated with the drying up of the rivers and sea water retreat ultimately leading to their collapse. However, pertaining to the emergence and submergence of the sea-level stands, the region requires a more systematic and robust approach, with a special focus on the coastal ancient settlers of the region. Furthermore, based on the archival response, the continental archives serve the potential parameter to elucidate the response of the monsoonal strength within the region. Although, the fluvial archives too, gives a prominent evidence to study the landscape impacts but are not considered significant owing to its preservation in the discrete patches. It is also worth noting that the fluvial response in terms of aggradation or degradation is primarily determined by the sediment to water ratio, and thus aggradation can occur in both wet and dry climates, explaining, in some cases, the differential responses of river systems across climatic regimes. Furthermore, while tectonic processes control sediment production in the hinterland, its mobilization into the alluvial plains is determined by water fluxes, which are affected by climate and is reflected in the frequent switching between fluvial and aeolian deposition modes. The landscape of Gujarat henceforth illustrates an excellent palaeoecological domain that has cradled human civilizations, which have been impacted and affected by pertinent climate changes during the Holocene period.

ACKNOWLEDGEMENTS

The authors thank the guest editors for the invitation to contribute to this book. The study was carried out under part of GoG funded programme HR20K. A part of the study also forms part of doctoral thesis of AS.

REFERENCES

Band, S., Yadava, M. G., Lone, M. A., Shen, C. C., Sree, K., and Ramesh, R. (2018). High-resolution mid-Holocene Indian summer monsoon recorded in a stalagmite from the Kotumsar Cave, central India. *Quaternary International*, 479, 19–24. https://doi.org/10.1016/j.quaint.2018.01.026

Banerji, U. S., Bhushan, R., and Jull, A. J. T. (2017). Mid-late Holocene monsoonal records from the partially active mudflat of Diu Island, southern Saurashtra, Gujarat, western India. *Quaternary International*, 443, 200–210. http://dx.doi.org/10.1016/j.quaint.2016.09.060

Banerji, U. S., Pandey, S., Bhushan, R., and Juyal, N. (2015). Mid-Holocene climate and land–sea interaction along the southern coast of Saurashtra, western India. *Journal of Asian Earth Sciences*, 111, 428–439. http://dx.doi.org/10.1016/j.jseaes.2015.06.021

Basu, S., et al. (2019). Response of grassland ecosystem to monsoonal precipitation variability during the mid-late Holocene: Inferences based on molecular isotopic records from Banni grassland, western India. *PloS One*, 14(4), e0212743. https://doi.org/10.1371/journal.pone.0212743

Biswas, S. K. (1987). Regional tectonic framework, structure and evolution of the western marginal basins of India. *Tectonophysics*, 135(4), 307–327.

Bhatt, N., and Bhonde, U. (2006). Geomorphic expression of late Quaternary sea level changes along the southern Saurashtra coast, western India. *Journal of Earth System Science*, 115(4), 395–402.

Bhatt, N., and Bhonde, U. A. (2003). Quaternary fluvial sequences of south Saurashtra, western India. *Current Science*, 84(8), 1065–1071.

Bhattacharya, F., Rastogi, B. K., Ngangom, M., Thakkar, M. G., and Patel, R. C. (2013). Late Quaternary climate and seismicity in the Katrol Hill Range, Kachchh, western India. *Journal of Asian Earth Sciences*, 73, 114–120. http://dx.doi.org10.1016/j.jseaes.2013.04.030

Bhattacharya, F., Rastogi, B. K., Thakkar, M. G., Patel, R. C., and Juyal, N. (2014). Fluvial landforms and their implication towards understanding the past climate and seismicity in the northern Katrol Hill Range, western India. *Quaternary International*, 333, 49–61. http://dx.doi.org/10.1016/j.quaint.2014.03.002

Blum, M. D., and Törnqvist, T. E. (2000). Fluvial responses to climate and sea-level change: A review and look forward. *Sedimentology*, 47, 2–48. https://doi.org/10.1046/j.1365-3091.2000.00008.x

Chamyal, L. S., Maurya, D. M., and Raj, R. (2003). Fluvial systems of the drylands of western India: A synthesis of late Quaternary environmental and tectonic changes. *Quaternary International*, 104(1), 69–86. http://dx.doi.org/10.1016%2FS1040-6182(02)00136-2

Chatterjee, A., and Ray, J. S. (2017). Sources and depositional pathways of mid-Holocene sediments in the Great Rann of Kachchh, India: Implications for fluvial scenario during the Harappan culture. *Quaternary International*, 443, 177–187. http://dx.doi.org/10.1016/j.quaint.2017.06.008

Crema, E. R., Habu, J., Kobayashi, K., and Madella, M. (2016). Summed probability distribution of ^{14}C dates suggests regional divergences in the population dynamics of the Jomon period in eastern Japan. *PloS One*, 11(4), e0154809. https://doi.org/10.1371/journal.pone.0154809

Cullen, H. M., Demenocal, P. B., Hemming, S., Hemming, G., Brown, F. H., Guilderson, T., and Sirocko, F. (2000). Climate change and the collapse of the Akkadian empire: Evidence from the deep sea. *Geology*, 28(4), 379–382. https://doi.org/10.1130/0091-7613(2000)28%3C379:CCATCO%3E2.0.CO;2

Dabhi, M., Thakkar, A., Chavan, A., Chauhan, G., Bhagora, R., Chauhan, N., ... and Thakkar, M. G. (2021). Mid-late Holocene climatic reconstruction from coastal dunes of the western Kachchh, India. *Quaternary International*, 642, 29–40. https://doi.org/10.1016/j.quaint.2021.09.011

Das, A., Prizomwala, S. P., Makwana, N., and Thakkar, M. G. (2017). Late Pleistocene-Holocene climate and sea level changes inferred based on the tidal terrace sequence, Kachchh, western India. *Palaeogeography, Palaeoclimatology, Palaeoecology*, 473, 82–93. http://dx.doi.org/10.1016/j.palaeo.2017.02.026

Das, A., Sodhi, A., Vedpathak, C. D., Prizomwala, S. P., Agnihotri, R., Makwana, N., Joseph, J., Patel, N., Chopra, S., and Kumar, M. R. (2022). Evidence for seawater retreat with advent of Meghalayan era (~ 4200 a BP) in a coastal Harappan settlement. *Geochemistry, Geophysics, Geosystems*, 23, e2021GC010264. https://doi.org/10.1029/2021GC010264

Dixit, Y., Hodell, D. A., Giesche, A., Tandon, S. K., Gázquez, F., Saini, H. S., Skinner, L. C., Mujtaba, S. A., Pawar, V., Singh, R. N., and Petrie, C. A. (2018). Intensified summer monsoon and the urbanization of Indus Civilization in northwest India. *Scientific Reports*, 8(1), 1–8.

Dixit, Y., Hodell, D. A., and Petrie, C. A. (2014). Abrupt weakening of the summer monsoon in northwest India ~4100 yr ago. *Geology*, 42(4), 339–342. https://doi.org/10.1130/G35236.1

Dixit, Y., Hodell, D. A., Sinha, R., and Petrie, C. A. (2014). Abrupt weakening of the Indian summer monsoon at 8.2 kyr BP. *Earth and Planetary Science Letters*, 391, 16–23. https://doi.org/10.1016/j.epsl.2014.01.026

Galili, N., Shemesh, A., Yam, R., Brailovsky, I., Sela-Adler, M., Schuster, E. M., ... and Halevy, I. (2019). The geologic history of seawater oxygen isotopes from marine iron oxides. *Science*, 365(6452), 469–473. https://doi.org/10.1126/science.aaw9247

Gaur, A. S., Sundaresh, S., and Tripati, S. (2006). Evidence for Indo-Roman trade from Bet Dwarka waters, west coast of India. *International Journal of Nautical Archaeology*, 35(1), 117–127. https://doi.org/10.1111/j.1095-9270.2005.00080.x

Griffiths, S., and Robinson, E. (2018). The 8.2 ka BP Holocene climate change event and human population resilience in northwest Atlantic Europe. *Quaternary International*, 465, 251–257. https://doi.org/10.1016/j.quaint.2017.10.017

Gupta, A. K., Anderson, D. M., and Overpeck, J. T. (2003). Abrupt changes in the Asian southwest monsoon during the Holocene and their links to the North Atlantic Ocean. *Nature*, 421(6921), 354–357. https://doi.org/10.1038/nature01340

Gupta, S. K. (1972). Chronology of the raised beaches and inland coral reefs of the Saurashtra coast. *The Journal of Geology*, 80(3), 357–361.

Hashimi, N. H., Nigam, R., Nair, R. R., and Rajagopalan, G. (1995). Holocene sea level fluctuations on western Indian continental margin: An update. *Journal of the Geological Society of India*, 46, 157–162.

Hernández, A., Martin-Puertas, C., Moffa-Sánchez, P., Moreno-Chamarro, E., Ortega, P., Blockley, S., Cobb, K. M., Comas-Bru, L., Giralt, S., Goosse, H., Luterbacher, J., and Xu, G. (2020). Modes of climate variability: Synthesis and review of proxy-based reconstructions through the Holocene. *Earth-Science Reviews*, 209, 103286. https://doi.org/10.1016/j.earscirev.2020.103286

Jain, M., and Tandon, S. K. (2003). Fluvial response to late Quaternary climate changes, western India. *Quaternary Science Reviews*, 22(20), 2223–2235. https://doi.org/10.1016/S0277-3791(03)00137-9

Juyal, N., Chamyal, L. S., Bhandari, S., Bhushan, R., and Singhvi, A. K. (2006). Continental record of the southwest monsoon during the last 130 ka: Evidence from the southern margin of the Thar Desert, India. *Quaternary Science Reviews*, 25(19–20), 2632–2650. https://doi.org/10.1016/j.quascirev.2005.07.020

Kathayat, G., Cheng, H., Sinha, A., Berkelhammer, M., Zhang, H., Duan, P., ... and Edwards, R. L. (2018). Evaluating the timing and structure of the 4.2 ka event in the Indian summer monsoon domain from an annually resolved speleothem record from northeast India. *Climate of the Past*, 14(12), 1869–1879. https://doi.org/10.5194/cp-14-1869-2018

Kázmér, M., Bhatt, N., Ukey, V., Prizomwala, S., Taboroši, D., and Székely, B. (2016). Archaeological evidence for modern coastal uplift at Diu, Saurashtra Peninsula, India. *Geoarchaeology*, 31(5), 376–387.

Kenoyer, J. M. (1998). *Ancient Cities of the Indus Valley Civilization*. Oxford: Oxford University Press.

Khadkikar, A. S., Basavaiah, N., Gundurao, T. K., and Rajshekhar, C. (2004). Palaeoenvironments around the Harappan Port of Lothal, Gujarat, western India. *Journal of Indian Geophysical Union*, 8(1), 49–53.

Khonde, N., Maurya, D. M., Singh, A. D., Chowksey, V., and Chamyal, L. S. (2011). Environmental significance of raised rann sediments along the margins of Khadir, Bhanjada and Kuar Bet islands in Great Rann of Kachchh, western India. *Current Science*, 101(11), 1429–1434.

Khonde, N. N., Maurya, D. M., and Chamyal, L. S. (2017). Late Pleistocene–Holocene clay mineral record from the Great Rann of Kachchh basin, western India: implications for palaeoenvironments and sediment sources. *Quaternary International*, 443, 86–98. http://dx.doi.org/10.1016/j.quaint.2016.07.024

Kusumgar, S., Raj, R., Chamyal, L. S., and Yadav, M. G. (1998). Holocene paleoenvironmental changes in the lower Mahi basin, western India. *Radiocarbon*, 40(2), 819–823.

Lamb, H. H. (1965). The early medieval warm epoch and its sequel. *Palaeogeography, Palaeoclimatology, Palaeoecology*, 1, 13–37.

Laskar, A. H., Yadava, M. G., Sharma, N., and Ramesh, R. (2013). Late-Holocene climate in the Lower Narmada valley, Gujarat, western India, inferred using sedimentary carbon and oxygen isotope ratios. *The Holocene*, 23(8), 1115–1122.

Madella, M., and Fuller, D. Q. (2006). Palaeoecology and the Harappan civilization of South Asia: Reconsideration. *Quaternary Science Reviews*, 25 (11–12), 1283–1301. https://doi.org/10.1016/j.quascirev.2005.10.012

Makwana, N., Prizomwala, S. P., Chauhan, G., Phartiyal, B., and Thakkar, M. G. (2019). Late Holocene palaeo-environmental change in the Banni plains, Kachchh, western India. *Quaternary International*, 507, 197–205. https://doi.org/10.1016/j.quaint.2018.11.02

Makwana, N., Prizomwala, S. P., Das, A., Phartiyal, B., Sodhi, A., and Vedpathak, C. (2021). Reconstructing the climate variability during the last 5000 years from the Banni Plains, Kachchh, western India. *Frontiers in Earth Science*, 9, 581. https://doi:10.3389/feart.2021.679689

Mann, M. E. (2007). Climate over the past two millennia. *Annual Review of Earth and Planetary Sciences*, 35(1), 111–136. https://doi:10.1146/annurev.earth.35.031306.140042

Maurya, D. M., Thakkar, M. G., Patidar, A. K., Bhandari, S., Goyal, B., and Chamyal, L. S. (2008). Late Quaternary geomorphic evolution of the coastal zone of Kachchh, western India. *Journal of Coastal Research*, 24(3), 746–758. https://doi:10.2112/05-0500

Mayewski, P. A., Rohling, E. E., Stager, J. C., Karlén, W., Maasch, K. A., Meeker, L. D., and Steig, E. J. (2004). Holocene climate variability. *Quaternary Research*, 62(3), 243–255. https://doi:10.1016/j.yqres.2004.07.001

Merh, S. S. (1995). *Geology of Gujarat*. Geological Society of India, GSI Publications, Vol. 2, no. 1.

Ngangom, M., Bhandari, S., Thakkar, M. G., Shukla, A. D., and Juyal, N. (2017). Mid-Holocene extreme hydrological events in the eastern Great Rann of Kachchh, western India. *Quaternary International*, 443, 188–199. http://dx.doi.org/10.1016/j.quaint.2016.10.017

Ngangom, M., Thakkar, M. G., Bhushan, R., and Juyal, N. (2012). Continental–marine interaction in the vicinity of the Nara River during the last 1400 years, Great Rann of Kachchh, western India. *Current Science*, 103(11), 1339–1342.

Pant, R. K., and Juyal, N. (1993). Late Quaternary coastal instability and sea level changes: New evidences from Saurashtra coast, western India. *Zeitschrift für Geomorphologie*, 37(1), 29–40. https://doi.org/10.1127/zfg/37/1993/29

Pillai, A. A., Anoop, A., Prasad, V., Manoj, M. C., Varghese, S., Sankaran, M., and Ratnam, J. (2018). Multi-proxy evidence for an arid shift in the climate and vegetation of the Banni grasslands of western India during the mid-to late-Holocene. *The Holocene*, 28(7), 1057–1070. https://doi.org/10.1177/0959683618761540

Pillai, A. A., Anoop, A., Sankaran, M., Sanyal, P., Jha, D. K., and Ratnam, J. (2017). Mid-late Holocene vegetation response to climatic drivers and biotic disturbances in the Banni grasslands of western India. *Palaeogeography, Palaeoclimatology, Palaeoecology*, 485, 869–878. https://doi.org/10.1016/j.palaeo.2017.07.036

Pokharia, A. K., Agnihotri, R., Sharma, S., Bajpai, S., Nath, J., Kumaran, R. N., and Negi, B. C. (2017). Altered cropping pattern and cultural continuation with declined prosperity following abrupt and extreme arid event at ~4,200 yrs BP: Evidence from an Indus archaeological site Khirsara, Gujarat, western India. *PloS One*, 12(10), e0185684. https://doi.org/10.1371/journal.pone.0185684

Ponton, C., Giosan, L., Eglinton, T. I., Fuller, D. Q., Johnson, J. E., Kumar, P., and Collett, T. S. (2012). Holocene aridification of India. *Geophysical Research Letters*, 39(3), L03704. https://doi.org/10.1029/2011GL050722

Possehl, G. (2002). *The Indus Civilization: A Contemporary Perspective*. Lanham, Maryland: Alta Mira Press, p. 100.

Prasad, S., Kusumgar, S., and Gupta, S. K. (1997). A mid to late Holocene record of palaeo-climatic changes from Nal Sarovar: A palaeodesert margin lake in western India. *Journal of Quaternary Science: Published for the Quaternary Research Association*, 12(2), 153–159.

Prasad, V., Farooqui, A., Sharma, A., Phartiyal, B., Chakraborty, S., Bhandari, S., and Singh, A. (2014a). Mid-late Holocene monsoonal variations from mainland Gujarat, India: A multi-proxy study for evaluating climate culture relationship. *Palaeogeography, Palaeoclimatology, Palaeoecology*, 397, 38–51. http://dx.doi.org/10.1016/j.pal aeo.2013.05.025

Prasad, S., Anoop, A., Riedel, N., Sarkar, S., Menzel, P., Basavaiah, N., and Stebich, M. (2014b). Prolonged monsoon droughts and links to Indo-Pacific warm pool: A Holocene record from Lonar Lake, central India. *Earth and Planetary Science Letters*, 391, 171–182. http://dx.doi.org/10.1016/j.epsl.2014.01.043

Prizomwala, S. P., Bhatt, N., Basavaiah, N. (2014). Understanding the sediment routing system along the Gulf of Kachchh coast, western India: Significance of small ephemeral rivers. *Journal of Earth System Science,* 123(1), 121–133.

Prizomwala, S. P., Das, A., Chauhan, G., Solanki, T., Basavaiah, N., Bhatt, N., ... and Rastogi, B. K. (2016). Late Pleistocene–Holocene uplift driven terrace formation and climate-tectonic interplay from a seismically active intraplate setting: An example from Kachchh, western India. *Journal of Asian Earth Sciences*, 124, 55–67. http://dx.doi.org/10.1016/ j.jseaes.2016.04.013

Prizomwala, S. P., Shukla, S. B., and Bhatt, N. (2010). Geomorphic assemblage of the Gulf of Kachchh coast, western India: Implications in understanding the pathways of coastal sediments. *Zeitschrift für Geomorphologie*, 54, 31–46. https://doi.org/10.1127/0372-8854/2010/0054-0003

Prizomwala, S. P., Yadav, G., Bhatt, N., and Sharma, K. (2018). Late Pleistocene relative sea-level changes from Saurashtra, west coast of India. *Current Science*, 115(12), 2297–2300. https://doi.org/10.1002/gea.21575

Raj, R., Chamyal, L. S., Prasad, V., Sharma, A., Tripathi, J. K., and Verma, P. (2015). Holocene climatic fluctuations in the Gujarat Alluvial Plains based on a multiproxy study of the Pariyaj Lake archive, western India. *Palaeogeography, Palaeoclimatology, Palaeoecology*, 421, 60–74. https://doi.org/10.1016/j.palaeo.2015.01.004

Raj, R., Tripathi, J. K., Kumar, P., Singh, S. K., Phartiyal, B., Sharma, A., et al. (2021). Palaeoclimatic and sea-level fluctuations from the last deglaciation to late Holocene from western India: Evidence from multiproxy studies. *Journal of Asian Earth Sciences*, 214, 104777. https:// doi.org/10.1016/j.jseaes.2021.104777

Raj, R., and Yadava, M. G. (2009). Late Holocene uplift in the lower Narmada basin, western India. *Current Science,* 96(7), 00113891.

Rao, S. R. (1979). *Lothal: A Port Town (1955–62)*. New Delhi: Archaeological Survey of India.

Rawat, S., Gupta, A. K., Sangode, S. J., Srivastava, P., and Nainwal, H. C. (2015). Late Pleistocene–Holocene vegetation and Indian summer monsoon record from the Lahaul, northwest Himalaya, India. *Quaternary Science Reviews*, 114, 167–181. https://doi.org/ 10.1016/j.quascirev.2015.01.032

Sarkar, A., Mukherjee, A. D., Sharma, S., Sengupta, T., Ram, F., Bera, M. K., ... and Juyal, N. (2020). New evidence of early Iron Age to Medieval settlements from the southern fringe of Thar Desert (western Great Rann of Kachchh), India: Implications to climate-culture co-evolution. *Archaeological Research in Asia*, 21, 100163. https://doi.org/10.1016/j.ara.2019.100163

Sengupta, T., Deshpande Mukherjee, A., Bhushan, R., Ram, F., Bera, M. K., Raj, H., Dabhi, A. J., Bisht, R. S., Rawat, Y. S., Bhattacharya, S. K., Juyal, N., and Sarkar, A. (2020). Did the Harappan settlement of Dholavira (India) collapse during the onset of Meghalayan stage drought? *Journal of Quaternary Science*, 35(3), 382–395. https://doi.org/10.1002/ jqs.3178

Sharma, S., Chauhan, G., Shukla, A. D., Nambiar, R., Bhushan, R., Desai, B. G., Pandey, U., Dabhi, M., Bhandari, S., Bhosale, S., Lakhote, A., and Juyal, N. (2021). Causes and implications of mid- to late Holocene relative sea-level change in the Gulf of Kachchh, western India. *Quaternary Research*, 100, 98–121. https://doi.org/10.1017/qua.2020.86

Singh, A., Thomsen, K. J., Sinha, R., Buylaert, J. P., Carter, A., Mark, D. F., Mason, P. J., Densmore, A. L., Murray, A. S., Jain, M., Paul, D., and Gupta, S. (2017). Counter-intuitive influence of Himalayan river morphodynamics on Indus Civilization urban settlements. *Nature Communications*, 8(1), 1–14. https://doi.org/10.1038/s41467-017-01643-9

Sinha, R., Singh, A., and Tandon, S. K. (2020). Fluvial archives of north and northwestern India as recorders of climatic signatures in the late Quaternary: Review and assessment. *Current Science*, 119(2), 232–243.

Sodhi, A., Das, A., Prizomwala, S. P., Vedpathak, C., and Makwana, N. (2022). Centennial-scale linkages between the Indian Summer Monsoon and the solar irradiation from the Gulf of Khambhat (western India). *Quaternary International*, 631, 82–92. https://doi.org/10.1016/j.quaint.2022.05.011

Solanki, T., Prizomwala, S. P., Makwana, N., and Solanki, P. M. (2021). Assessing the climatically triggered aggradation-incision processes in a dryland environment during the late Quaternary period from Shetrunji River basin (Saurashtra), India. *Quaternary International*, 585, 70–84. https://doi.org/10.1016/j.quaint.2020.07.007

Sridhar, A. (2007). A mid-late Holocene flood record from the alluvial reach of the Mahi River, western India. *Catena*, 70(3), 330–339.

Sridhar, A., Thakur, B., Basavaiah, N., Seth, P., Tiwari, P., and Chamyal, L. S. (2020). Lacustrine record of high magnitude flood events and climate variability during mid to late Holocene in the semiarid alluvial plains, western India. *Palaeogeography, Palaeoclimatology, Palaeoecology*, 542, 109581. https://doi.org/10.1016/j.palaeo.2019.109581

Srivastava, P., Juyal, N., Singhvi, A. K., Wasson, R. J., and Bateman, M. D. (2001). Luminescence chronology of river adjustment and incision of Quaternary sediments in the alluvial plain of the Sabarmati River, north Gujarat, India. *Geomorphology*, 36(3–4), 217–229.

Staubwasser, M., Sirocko, F., Grootes, P. M., and Segl, M. (2003). Climate change at the 4.2 ka BP termination of the Indus valley civilization and Holocene south Asian monsoon variability. *Geophysical Research Letters*, 30(8), 1425. https://doi.org/10.1029/2002gl016822

Thakur, B., Seth, P., Sharma, A., Pokharia, A. K., Spate, M., and Farooqui, S. (2019). Linking past cultural developments to palaeoenvironmental changes from 5000 BP to present: A climate-culture reconstruction from Harshad estuary, Saurashtra, Gujarat, India. *Quaternary International*, 507, 188–196.

Tripati, S., Gaur, A. S., and Sundaresh (2004). Marine archaeology in India. *Man and Environment*, 29(1), 28–41.

Turney, C. S., and Brown, H. (2007). Catastrophic early Holocene sea level rise, human migration and the Neolithic transition in Europe. *Quaternary Science Reviews*, 26(17–18), 2036–2041. https://doi.org/10.1016/j.quascirev.2007.07.003

Tyagi, A. K., Shukla, A. D., Bhushan, R., Thakker, P. S., Thakkar, M. G., and Juyal, N. (2012). Mid-Holocene sedimentation and landscape evolution in the western Great Rann of Kachchh, India. *Geomorphology*, 151, 89–98. https://doi.org/10.1016/j.geomorph.2012.01.018

Vandenberghe, J., Cordier, S., and Bridgland, D. R. (2010). Extrinsic and intrinsic forcing of fluvial development: Understanding natural and anthropogenic influences. *Proceedings of the Geologists' Association*, 121(2), 107–112. https://doi.org/10.1016/j.pgeola.2010.05.002

Vora, K.H., Gaur, A.S., Sundaresh., and Tripathi, S. (2006). Archaeological Sites as Indicators of Ancient Shorelines. *Society for Marine Archaeology*, 82–86.

Vora, K. H., Naik, D. K., Ganesan, P., and Moraes, C. (1991). Offshore extension of Gomati River, Dwarka. *Journal of Marine Archaeology*, 2, 32–38.

Walker, M., Head, M. J., Berkelhammer, M., Björck, S., Cheng, H., Cwynar, L., Fisher, D., Gkinis, V., Long, A., Lowe, J., Newnham, R., and Weiss, H. (2018). Formal ratification of the subdivision of the Holocene Series/Epoch (Quaternary System/Period): Two new Global Boundary Stratotype Sections and Points (GSSPs) and three new stages/subseries. *Episodes Journal of International Geoscience*, 41(4), 213–223. https://doi.org/10.18814/epiiugs/2018/018016

Weber, S., Kashyap, A., and Harriman, D. (2010). Does size matter: The role and significance of cereal grains in the Indus civilization. *Archaeological and Anthropological Science*, 2, 35–43.

Weiss, H., and Bradley, R. S. (2001). What drives societal collapse? *Science*, 291(5504), 609–610.

Wolf, D., and Faust, D. (2015). Western Mediterranean environmental changes: Evidences from fluvial archives. *Quaternary Science Reviews*, 122, 30–50. http://dx.doi.org/10.1016/j.quascirev.2015.04.016

Wright, R. P., Bryson, R., and Schuldenrein, J. (2008). Water supply and history: Harappa and the Beas regional survey. *Antiquity* 8, 37–48.

Glossary

Term	Glossary
8.2 ka event	The 8.2 ka event was a sudden decrease in global temperatures that occurred approximately 8,200 years before the present, lasting for the next two to four centuries. It was the largest and most abrupt climatic event of the Holocene epoch. The cooling was likely caused by a large meltwater pulse from the final collapse of the Laurentide Ice Sheet in north-eastern North America, which disrupted the North Atlantic thermohaline circulation and reduced northward heat transport
Allergic responses	Allergic responses, also known as hypersensitivity reactions, are inappropriate reactions of the immune system to harmless substances, triggering various symptoms. These reactions can range from mild, such as sneezing, watery eyes, and a runny nose, to severe and life-threatening, known as anaphylactic reactions
Andaman Sea	The Andaman Sea is a marginal sea of the north-eastern Indian Ocean bounded by the coastlines of Myanmar and Thailand along the Gulf of Martaban and the west side of the Malay Peninsula. It is separated from the Bay of Bengal to its west by the Andaman Islands and the Nicobar Islands.
Annual land use change rate (ALUCR)	The annual land use change rate refers to the percentage or amount of land area that undergoes a change in its use or cover over one year. It is an important metric for understanding the dynamics and impacts of human activities on the Earth's surface.
Aquifer	An aquifer is a geological formation that can store and provide water. It consists of one or more layers of permeable rocks where groundwater accumulates. The water contained in aquifers is of good quality. Aquifers form when rainwater seeps through various layers of permeable rocks until it reaches an impermeable layer. The impermeable layer can be made of clay, granite, quartzite, etc., and the aquifer's water accumulates above it.
Arctic sea ice	The Arctic Ocean is the smallest and shallowest of the world's five oceanic divisions which is known as one of the coldest of oceans. The Arctic Ocean is surrounded by Eurasia and North America and is mostly covered by sea ice throughout the year, especially in winter. In the Arctic, the annual minimum sea ice extent occurs in September and the annual sea ice maximum extent occurs in March

Term	Glossary
Atmosphere	An atmosphere is a layer of gases that envelop an astronomical object, held in place by the gravity of the object. It can consist of various gases and is essential for sustaining life on planets like Earth. The composition of an atmosphere can vary depending on the astronomical object, with planets like Earth having a mix of nitrogen, oxygen, carbon dioxide, and trace gases. The atmosphere plays a crucial role in protecting organisms from the harmful effects of sunlight, ultraviolet radiation, solar wind, and cosmic rays, thereby safeguarding against genetic damage
Average annual precipitation	Average annual precipitation refers to the total amount of precipitation, including both rainfall and the water equivalent of snowfall, that falls on average over a one-year period.
Base-flow	Base-flow, also known as drought flow, groundwater recession flow, low flow, low-water flow, low-water discharge, or fair-weather runoff, is the portion of streamflow that is sustained between precipitation events, fed to streams by delayed pathways. It represents the flow that reaches a stream essentially as groundwater, distinct from direct runoff from rainfall. Base-flow is crucial for sustaining human populations and ecosystems, especially in watersheds that do not rely on snowmelt.
Bay of Bengal	The Bay of Bengal is a large body of water that is part of the Indian Ocean, located to the south of India, Bangladesh, and Myanmar. It is characterized by its strategic location between these countries and is known for its significance in regional trade, climate patterns, and marine biodiversity. The Bay of Bengal serves as a vital water body for the surrounding nations, influencing weather systems, supporting fisheries, and playing a crucial role in the livelihoods of millions of people living along its coastlines.
Biodiversity	Biodiversity, also known as biological diversity or biotic diversity, encompasses the variety of life on Earth at all levels, from genes to ecosystems. It includes not only species considered rare, threatened, or endangered but also every living organism, from humans to microbes, fungi, and invertebrates.
Carbon dioxide	Carbon dioxide (CO_2) is a colourless, odourless gas that is present in the atmosphere. It is a chemical compound consisting of one carbon atom covalently bonded to two oxygen atoms

(continued)

Term	Glossary
Cardiovascular disorders (CVDs)	Cardiovascular disorders, also known as heart and blood vessel diseases, are a group of conditions that affect the heart and blood vessels
Chhotanagpur Gneissic Complex (CGC)	A geological formation located in the eastern part of the Satpura Mountain Belt of India. It is characterized by polyphase deformation and metamorphism dating back to the Proterozoic Eon. The structural and tectonic analyses of the CGC have revealed a complex history of regional evolution, including multiple phases of deformation, metamorphism, and magmatism.
Climate change	Climate change refers to long-term shifts in temperatures and weather patterns. These changes can be natural, resulting from factors like volcanic eruptions or variations in solar radiation, but they can also be influenced by human activities, particularly the release of greenhouse gases into the atmosphere. Human-induced climate change, driven by activities such as burning fossil fuels and deforestation, has led to an increase in global temperatures, changes in precipitation patterns, rising sea levels, and more frequent extreme weather events.
Coldwater fish	Those species of fish that thrive in relatively cold water.
Condensation	The process by which a substance changes from a gaseous state to a liquid state. As a gas cools, it loses heat, or thermal energy. The particles that make up the gas move more slowly; that allows the attractive forces among them to increase, causing droplets of liquid to form. The steps in this process are the opposite of what occurs in evaporation, the formation of a gas from a liquid.
Cryosphere	The portions of Earth's surface where water is in solid form, encompassing sea ice, lake ice, river ice, snow cover, glaciers, ice caps, ice sheets, and frozen ground like permafrost. It plays a crucial role in regulating climate and sea levels, with significant impacts on surface energy and moisture fluxes, clouds, the water cycle, and atmospheric and oceanic circulation.
Cyprinids	The largest and most diverse fish family, and the largest vertebrate animal family overall, with about 3,000 species.
Deccan Trap	The Deccan Traps, located in west-central India, is one of the largest volcanic features on Earth, characterized by a large shield volcano formation. It consists of numerous layers of solidified flood basalt that are over 2,000 meters thick, covering an area of about 500,000 sq. km and having a volume of approximately 1,000,000 cubic km.

Term	Glossary
El Nino	A reoccurring cycle that develops in the Pacific Ocean along the west coast of North and South America, bringing warm waters to replace typically cold waters. El Nino is Spanish for the boy child but refers to the Christ Child because the cycle occurs in the winter around Christmas.
El Nino Southern Oscillation (ENSO)	A global climate phenomenon characterized by variations in winds and sea surface temperatures over the tropical Pacific Ocean. ENSO consists of three phases: Neutral, La Niña, and El Niño. La Niña and El Niño are opposite phases that occur when specific ocean and atmospheric conditions are met.
Electrical conductivity (EC)	Electrical conductivity, also known as specific conductance, is a fundamental property that measures a material's ability to conduct electric current. It is a reciprocal of electrical resistivity and represents how easily electricity can flow through a substance. Conductivity is typically measured in siemens per meter (S/m) and is influenced by factors such as temperature, material composition, impurities, and pressure.
Electrolyte imbalance	An electrolyte imbalance, also known as an electrolytic imbalance, refers to a condition where the levels of one or more electrolytes in the body are either too high or too low. Electrolytes are minerals that carry an electric charge when dissolved in body fluids and play essential roles in various bodily functions, including regulating heart and neurological function, fluid balance, and acid-base balance.
Equilibrium line altitude (ELA)	A critical concept in glaciology, representing the average elevation on a glacier where accumulation (snowfall) equals ablation (melting and evaporation) over a one-year period. It marks the boundary between the accumulation zone, where snowfall exceeds melting, and the ablation zone, where melting exceeds snowfall.
Evaporation	Evaporation is the process by which a liquid, such as water, transitions from its liquid state to a gaseous state at temperatures below its boiling point. This occurs when the molecules in the liquid gain enough kinetic energy to escape the liquid-gas interface and enter the surrounding gas phase.
Evapotranspiration	Evapotranspiration is the combined process of water evaporation from land and water surfaces, and transpiration from plants, which transfers water from the Earth's surface into the atmosphere
Fast-Line-of-Sight Atmospheric Analysis of Spectral Hypercubes (FLAASH)	A first-principles atmospheric correction tool that corrects visible through shortwave infrared (VSWIR) spectral imagery for the effects of the atmosphere

(continued)

Term	Glossary
Follicle stimulating hormone	A hormone produced by the anterior pituitary gland that plays a crucial role in sexual development and reproduction in both males and females.
Food-borne diseases	Illnesses caused by consuming contaminated food or beverages. They are usually infectious or toxic in nature and can be caused by bacteria, viruses, parasites, or chemical substances
Glacial mass loss	The net decrease in the total mass of a glacier over time, which occurs when the rate of ablation (loss of ice through melting, calving, sublimation, etc.) exceeds the rate of accumulation (snowfall, avalanching, etc.)
Global burden of disease (GBD)	The most comprehensive worldwide observational epidemiological study to date, led by the Institute for Health Metrics and Evaluation (IHME).
Global water availability	The total amount of freshwater resources that are accessible and can be utilized worldwide for various purposes, such as human consumption, agriculture, industry, and ecosystem maintenance. It encompasses both surface water (rivers, lakes, reservoirs) and groundwater resources
Greenhouse gas (GHG)	Gases in the atmosphere that absorb and trap infrared radiation, causing the greenhouse effect that warms the Earth's surface.
Groundwater flow	The movement of water through the saturated subsurface
Groundwater resources	The total amount of freshwater stored underground in aquifers and other geological formations
Groundwater withdrawal	The extraction of water from underground aquifers for various purposes such as irrigation, industrial use, and municipal water supply
Heat loss	The transfer of thermal energy from a warmer object or system to a cooler surrounding environment. It occurs through the mechanisms of conduction, convection, and radiation
Hydrological cycle	Also known as the water cycle, is the continuous movement of water on, above, and below the Earth's surface. It is a fundamental process that sustains life on our planet
Indian Ocean Dipole	A coupled ocean-atmosphere phenomenon in the tropical Indian Ocean characterized by an irregular oscillation of sea surface temperatures between the western and eastern parts of the ocean.
Indian Plate	It is a major tectonic plate that includes the Indian subcontinent, the surrounding Indian Ocean, and parts of the Southern Ocean.

Term	Glossary
Indian Summer Monsoon (ISM)	A seasonal wind pattern that brings heavy rainfall to the Indian subcontinent during the summer months, typically from June to September.
Indus Valley Civilization	Also known as the Harappan Civilization, was an ancient urban culture that flourished in the Indus River basin on the Indian subcontinent from around 3300 to 1300 BCE. Key facts about the Indus Valley Civilization
Infiltration	The process by which water on the surface enters the soil
Intergovernmental Panel on Climate Change (IPCC)	The United Nations body for assessing the science related to climate change
International Geosphere-Biosphere Program (IGBP)	A major international scientific research program that ran from 1987 to 2015. Its primary focus was to study global environmental change and the interactions between the earth's biological, chemical, and physical processes.
International Human Dimensions Program (IHDP)	The International Human Dimensions Programme on Global Environmental Change (IHDP) was an international science program that existed from 1996 to 2014. Its main goals were to promote and coordinate research on the human aspects of global environmental change
Irrigation return flow	The portion of applied irrigation water that is not consumed by crops or lost to evaporation and instead flows back into surface water bodies or groundwater aquifers
La Nina	A climate pattern characterized by unusually cold ocean temperatures in the Equatorial Pacific, the opposite of El Niño which features unusually warm ocean temperatures
Land surface temperature	The temperature of the Earth's surface as measured from space using satellite sensors
Land use land cover change	The alteration of the Earth's terrestrial surface over time due to both human activities and natural processes
Landslide	The downward and outward movement of slope-forming materials such as rock, debris, or soil under the force of gravity.
Late Mesozoic	Refers to the latter part of the Mesozoic Era, which lasted from approximately 252 to 66 million years ago. It encompasses the Jurassic and Cretaceous Periods.
Late Quaternary	To the latter part of the Quaternary Period, which spans from approximately 2.58 million years ago to the present. It encompasses the Pleistocene Epoch (2.58 million to 11,700 years ago) and the Holocene Epoch (11,700 years ago to the present).
Leutinising hormone	A hormone produced by the anterior pituitary gland that plays a crucial role in the reproductive system of both males and females

(continued)

Term	Glossary
Medieval Warm Period (MWP)	The Medieval Climate Optimum or Medieval Climatic Anomaly, was a time of warm climate in parts of the world, particularly in the North Atlantic region, that lasted from around 950 to 1250 AD
Mid-Late Holocene	The period spanning approximately 7,000 to 1,000 years ago, during the current Holocene geological epoch.
Miliolite deposits	Sedimentary rock formations composed primarily of the fossilized shells of small marine organisms called miliolids
MODerate resolution atmospheric TRANsmission (MODTRAN)	A computer program designed to model atmospheric propagation of electromagnetic radiation for the spectral range from the middle ultraviolet to the far infrared (0.2 to 100 μm or 100-50,000 cm−1
Non-vector borne diseases	In contrast to vector-borne diseases that are transmitted by arthropod vectors, non-vector borne diseases are spread through other means, such as contaminated water, food, or direct contact, rather than by a living organism vector
Object-Based Image Analysis (OBIA)	A technique that segments an image into meaningful objects, rather than analyzing it at the individual pixel level
Ocean acidification	The ongoing reduction in the pH levels of the Earth's oceans, primarily due to the absorption of atmospheric carbon dioxide
Ozone	Ozone is a triatomic molecule composed of three oxygen atoms (O_3).
pH	A scale used to specify the acidity or basicity of an aqueous solution
Pixel-based image analysis (PBIA)	A traditional and widely used approach for classifying and interpreting remote sensing imagery. It involves assigning a class or category to each individual pixel in the image based solely on its spectral properties, which represent the reflectance of electromagnetic radiation at different wavelengths
Pollutants	A substance or energy introduced into the environment causing undesired effects or adversely affecting a resource's usefulness.
Precipitation	Any form of water that falls from the sky and reaches the Earth's surface
Quality rating	A systematic assessment and classification of the overall quality or validity of a study, review, or other type of research.
Remote sensing (RS)	Acquisition of information about an object or phenomenon without physical contact
respiratory tract infections	Infectious diseases that affect the upper or lower respiratory tract

Term	Glossary
Sea level	Mean sea level is an average surface level of Earth's coastal bodies of water used as a standard reference point for various measurements
Sea surface salinity	The concentration of dissolved salts in the upper layer of the ocean, typically measured in practical salinity units (PSU) or parts per thousand (ppt)
Soil evaporation	The process by which water is lost from the soil surface to the atmosphere in the form of water vapor. It is a key component of the water cycle and plays a crucial role in determining the water balance of a soil-plant-atmosphere system
Submarine groundwater discharge	The flow of water from the seabed to the coastal ocean, regardless of the fluid composition (fresh groundwater, recirculated seawater, or a mixture). It is a ubiquitous coastal process driven by a combination of climatic, hydrogeologic, and oceanographic factors
Surface and groundwater interaction	Surface water (including rivers, lakes, reservoirs, wetlands, and estuaries) interacts with groundwater in a variety of ways. This interaction occurs through the exchange of water between the surface water body and the underlying aquifer.
Surface runoff	The unconfined flow of water over the ground surface, typically occurring when excess rainwater, stormwater, meltwater, or other sources cannot sufficiently rapidly infiltrate in the soil
Thermal stress	The mechanical stress created by any change in temperature of a material.
Total dissolved solid	A measure of the combined content of all inorganic and organic substances dissolved in a liquid, such as water. It is typically expressed in milligrams per liter (mg/L) or parts per million (ppm).
Transpiration	Transpiration is the process in which water is lost as water vapor from the aerial parts of plants through stomata.
United Nations Framework Convention on Climate Change (UNFCCC)	An international treaty adopted in 1992 with the goal of preventing "dangerous anthropogenic interference with the climate system".
Universal Transverse Mercator projection	A map projection system is used for assigning coordinates to locations on the Earth's surface.
urbanization	The process by which towns and cities are formed and becomes larger as more people begin living and working in central areas.
Vector-borne diseases	Illnesses caused by parasites, viruses, and bacteria that are transmitted by vectors, which are living organisms that can transmit infectious pathogens between humans or from animals to humans

(continued)

Term	Glossary
Water quality index (WQI)	A tool used to provide a single number that expresses overall water quality at a certain location and time based on several water quality parameters
Waterborne diseases	Illnesses caused by pathogenic microorganisms that are transmitted through contaminated water. They are a major global health issue, causing an estimated 1.5 million deaths annually, with a disproportionate impact on developing countries
World Economic Forum	An international non-governmental organization, think tank, and lobbying organization based in Cologny, Switzerland. It was founded in 1971 by German engineer Klaus Schwab.
World Health Organization (WHO)	A specialized agency of the United Nations responsible for international public health. It is headquartered in Geneva, Switzerland and has regional and field offices worldwide.

Index